Sustainable Engineering

Drawing on multidisciplinary perspectives from engineering, economics, business, science, and human behavior, this text presents an unrivaled introduction to how engineering practice can contribute to sustainable development. Varied approaches for assessing the sustainability of engineering and other human activities are presented in detail, and potential solutions to meet key challenges are proposed, with an emphasis on those that require engineering skills. Each concept and approach is supported by mathematical representation, solved problems, real-world examples, and self-study exercises. The topics covered range from introductory material on the nature of sustainability, to more advanced approaches for assessment and design. The prerequisites for each chapter are clearly explained so the text can be adapted to meet the needs of students from a range of backgrounds. Software tutorials, project statements and solutions, lecture slides, and a solutions manual accompany the book online, making this an invaluable resource for courses in sustainable engineering, as well as a useful reference for industry practitioners.

Bhavik R. Bakshi is the Morrow Professor of Chemical and Biomolecular Engineering at The Ohio State University (OSU). He also holds appointments in civil, environmental, and geodetic engineering at OSU and as a Visiting Professor at the Indian Institute of Technology in Mumbai, India. He has developed and taught a course on sustainable engineering for 20 years at OSU, and shorter versions at institutions such as MIT, IIT Bombay, McGill University, and South China University of Technology.

Sustainable Engineering

Principles and Practice

Bhavik R. Bakshi

The Ohio State University

CAMBRIDGE
UNIVERSITY PRESS

CAMBRIDGE
UNIVERSITY PRESS

University Printing House, Cambridge CB2 8BS, United Kingdom

One Liberty Plaza, 20th Floor, New York, NY 10006, USA

477 Williamstown Road, Port Melbourne, VIC 3207, Australia

314-321, 3rd Floor, Plot 3, Splendor Forum, Jasola District Centre, New Delhi - 110025, India

79 Anson Road, #06-04/06, Singapore 079906

Cambridge University Press is part of the University of Cambridge.

It furthers the University's mission by disseminating knowledge in the pursuit of education, learning and research at the highest international levels of excellence.

www.cambridge.org
Information on this title: www.cambridge.org/9781108420457
DOI:10.1017/9781108333726

First published 2019

A catalogue record for this publication is available from the British Library

Library of Congress Cataloging in Publication data
Names: Bakshi, Bhavik R., author.
Title: Sustainable engineering principles and practice /
Bhavik R. Bakshi, Ohio State University.
Description: Cambridge, United Kingdom; New York, NY,
USA: Cambridge University Press, 2019.|
Includes bibliographical references and index.
Identifiers: LCCN 2018048004 | ISBN 9781108420457 (hardback)
Subjects: LCSH: Sustainable engineering. | Environmental protection. |
Conservation of natural resources. | Energy conservation.
Classification: LCC TA163.B35 2019 | DDC 620.0028/6–dc23
LC record available at https://lccn.loc.gov/2018048004

ISBN 978-1-108-42045-7 Hardback

Additional resources for this publication at www.cambridge.org/bakshi

To my parents

Brief Contents

Contents

PART IV Solutions for Sustainability   341

Preface

Engineering aims to enhance human well-being, but also contributes to the degradation of ecosystems and depletion of resources. These negative side-effects often appear as unexpected surprises and unintended harm. Given the essential role played by technology in the modern world, it is important to ensure that engineering decisions and activities contribute positively to the economy, society, and environment. This book aims to help engineers in meeting this goal. It has grown out of an elective course that I have developed and taught for 20 years, and shorter versions have been co-taught at various institutions across the world.

The challenge of sustainable development is of a transdisciplinary nature, and no single discipline is in a position to address it by itself. Therefore, this book adopts an approach that cuts across disciplinary boundaries, mainly to understand the challenges and also to learn about possible transdisciplinary solutions that include engineering. Such a broad understanding motivates various engineering methods that help in assessing sustainability and in designing systems that contribute positively to the three values of ecology, society, and economy. Knowing the quantitative bias of most engineers, this book relies on equations and quantitative examples to convey various methods in a rigorous but understandable manner for students with a range of mathematical skills and backgrounds.

Key Features

The key features of this book include:

- a multidisciplinary introduction to the challenges of sustainable development and potential solutions;
- systematic and quantitative coverage of methods for assessing the sustainability of technological alternatives and for devising solutions;
- nearly 100 solved problems throughout the book;
- boxes describing practical examples relevant to each topic;
- a clear introduction to the content of each chapter in the chapter introduction, and a listing of key ideas and concepts at the end;
- exercises at the end of each chapter; and
- an introduction to relevant software and its use for solving practical problems available on the companion website.

Content of this Book

This book is organized in four major parts.

- *Introduction and Motivation.* Part I introduces the subject and motivates the need for sustainable engineering. It considers the basis of human well-being and the status of the goods and services that are essential for our well-being. Then the meaning of sustainable development is discussed along with the challenges in achieving sustainability.

- *Reasons for Unsustainability.* Part II explores reasons for the unsustainability of human activities. It describes how activities in multiple disciplines, such as economics, business, science, engineering, ethics, and behavior, can contribute to unsustainable decisions. Such cross-disciplinary understanding of the problem is essential for finding effective solutions.

- *Sustainability Assessment.* Part III describes methods for assessing the sustainability of human activities and technologies. It introduces various popular techniques and ways to solve practical problems. This material is written in such a way that its mathematical rigor may be adjusted as desired.

- *Solutions for Sustainability.* Part IV focuses on solutions for achieving sustainability. It describes approaches related to engineering, such as: techno-economic analysis and process design; methods inspired from ecology such as biomimicry, industrial symbiosis, and techno-ecological synergy; and solutions from economics, policy, and societal transformation.

- *Companion Website.* This contains information about software for applying various methods covered in the text, statements for individual or group projects, and solutions to selected projects. Updated chapters and errata are also available.

Ways of Using this Book

The content of this book has been used primarily for junior and senior engineering undergraduate students and for graduate students. The chapters in Parts I and II provide the motivation and understanding of the issues facing sustainable development, and should be included. Chapters from Parts III and IV may be chosen depending on the students' backgrounds and interests. Chapters that require a more advanced background include the following.

- Chapter 10 is mathematically more challenging as it covers the rigorous framework of sustainability assessment methods. A proper understanding of this framework requires a background in basic linear algebra. However, the chapter has been written in a manner such that the framework may also be understood without a knowledge of linear algebra. If this chapter is excluded from the syllabus, it is still possible to use the other chapters since the concepts relevant to Chapter 10 are covered in a simpler manner (without linear algebra) in Chapter 9. Also, most of the examples throughout the book are solved with and without the use of linear algebra.

- Chapters 13 and 14 rely on the concepts of entropy and the second law of thermodynamics. If students do not have the necessary background, these chapters may be excluded without any loss of continuity.

This book may be used for various types of courses:

- A semester-long course could include the chapters in Parts I and II, and selected chapters from Parts III and IV. Homework exercises and a larger group project are also recommended.
- This material may also be covered over two semesters. The first semester could focus on Parts I and II and a few chapters from Part III. This course would be on Sustainability Assessment. The second semester would focus on the remaining chapters of Part III and Part IV, along with a substantial course project. This would be on Solutions for Sustainability.
- Many universities have a freshman-level course on sustainable engineering. For such a course, the syllabus could include chapters from Parts I and II, and selected chapters from Part III, such as Chapter 11 on Footprint Assessment and Chapter 12 on Energy Analysis. Chapters 19 and 20 from Part IV may also be included.

Acknowledgments

Such a book is impossible without direct and indirect contributions from a large number of individuals. It would be difficult to name them all!

I am grateful to the students who have taken this course. Their role has been indispensable. This includes more than 500 students at OSU. It also includes students at the Institute of Chemical Technology, Mumbai and in short courses at the Massachusetts Institute of Technology, the Indian Institute of Technology, Mumbai, TERI University, New Delhi, the South China University of Technology, Guangzhou, and McGill University, Montreal.

The teaching associates for this course and various graduate students and postdoctoral researchers have helped in many ways, such as by refining the content, suggesting and developing teaching material and homework problems, preparing solutions to the exercise problems, and providing constructive suggestions and feedback. They include Jorge Hau, Nandan Ukidwe, Daniel Arthur, Yi Zhang, Jun-Ki Choi, Anil Baral, Vikas Khanna, Geoffrey Grubb, Shweta Singh, Robert Urban, Nathan Cruze, Berrin Kursun, Laura Woods, Erin Gibbemeyer, Rebecca Hanes, Sachin Jadhao, Prasad Mandade, Deepika Singh, Varsha Gopalakrishnan, Shelly Bogra, Xinyu Liu, Tapajyoti Ghosh, Kyuha Lee, Michael Charles, and Utkarsh Shah. Several graduate and undergraduate students have contributed problems to various chapters. They have been acknowledged near each problem. Class projects were provided by several corporations, organizations, and individuals, which are too numerous to list individually. I am thankful for their guidance, support, and interest. The chapters were proofread by Xinyu Liu, Tapajyoti Ghosh, Kyuha Lee, Michael Charles, Utkarsh Shah, and Harshal Bakshi. Their diligence has helped to improve clarity and to correct many errors.

The material in this book has benefited from conversations and discussions with several colleagues and friends. Joseph Fiksel has been a collaborator for many years from whom I have learned about many things, including corporate sustainability and resilience. He has contributed a section to the chapter on business and sustainability. Tim Gutowski and Dusan Sekulic have been co-instructors of the short courses at MIT. The many lively discussions that I have had with them on sustainable engineering and other topics have been entertaining and educational, and have influenced the content of many chapters. Brian Fath has helped me learn some basics of ecology and has provided valuable feedback on relevant chapters. Yogendra Shastri has provided opportunities to co-teach short courses in India. Mary Evelyn Tucker and John Grimm were instrumental in introducing me to the fascinating intersection of religion and sustainability. They have contributed material to those chapters. Several anonymous reviewers have also provided valuable feedback at various stages of development of the book.

My teachers have taught me a lot and got me started in the direction of sustainable engineering. George Stephanopoulos introduced me to the versatility and rigor of process systems engineering. John Ehrenfeld introduced me to the field of industrial ecology.

Steve Elliott, Nicola Chapman, and others at Cambridge University Press have played an important role in making sure that this book reaches you.

My parents have contributed in more ways than I know; their love and understanding, and curiosity about the natural world, to name a few. My wife, Mamta, and son, Harshal have tolerated many "lectures" on sustainability and have been immensely patient over the hundreds of weekends it took to develop this book. My extended family, including nieces and nephews. After all, this is for their generation and beyond.

Bhavik R. Bakshi

PART I
Introduction and Motivation

In the first part of this book, we will focus on the motivation for learning about sustainable engineering. We will address some basic questions, such as:

- What needs to be sustained and why?
- What is the state of human development and what are its side-effects?
- What do human activities and well-being depend on, and what is the status of these resources?
- What is sustainable development? How do we make decisions that are sustainable?

We will learn that sustaining our well-being requires goods and services from nature. Many human activities, including engineering, are causing the degradation of these goods and services, which cannot be sustained for long. Sustainability is not just about the environment, but needs to consider the economy and society as well as the environment for the sake of current and future generations.

1 The Basis of Human Well-Being

> Well-being denotes a state of the world that is intrinsically, and not
> merely instrumentally, valuable to human beings.
>
> Anna Alexandrova [1]

What does our well-being depend on? This question is fundamental to any effort for improving human well-being and for sustaining it now and in the future. We start by focusing on answering this question, since once we understand the factors that form the basis of human well-being, we can determine whether our activities can continue to sustain our well-being, and then devise approaches and solutions for ensuring sustainability. This insight forms the basis of sustainable engineering and the rest of this book. In this chapter, we will learn about historical trends in human development based on different ways of defining this concept, and then explore the basis of our well-being.

1.1 Trends in Human Development

The modern human, *Homo sapiens sapiens*, is among the most successful species on this planet today. One indication of our success is the growth in our population, as depicted in Figure 1.1. Current estimates are that global population will continue to grow during this century and will increase from the current 7.2 billion to between 9.6 billion and 12.3 billion by 2100 [2, 3]. Of course, success as a species cannot be measured by population numbers alone, as quality of life also matters. It turns out that we have done very well on that front as well.

A popular measure of societal well-being commonly used by economists is gross domestic product (GDP). It is the monetary value of all goods and services produced. On the basis of this measure of wealth, our overall standard of living has increased dramatically over the centuries all over the world, as shown in Figure 1.2. GDP only includes monetary aspects and is often criticized because it fails to capture factors such as health and education, which are essential components of our well-being.

Indicators of human well-being more comprehensive than GDP have also been developed. Among these, the human development index (HDI) combines factors such as health and education with income, as described in Box 1.1. Figure 1.3

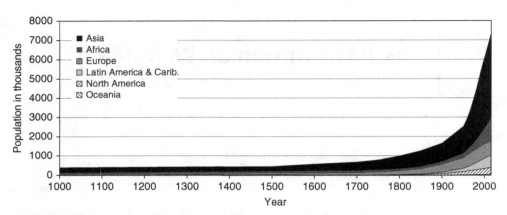

Figure 1.1 Human population over the centuries.

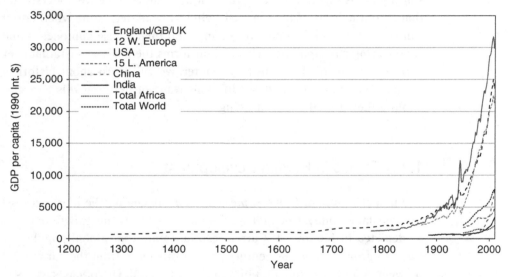

Figure 1.2 Trend in global gross domestic product (GDP) in terms of international dollars adjusted for purchasing parity [4].

shows the trend in the HDI for several countries. Using this more comprehensive measure of well-being, we can again see that humanity has done well, since the HDI has been increasing in almost all countries of the world. Of course, there are localized challenges such as war, drought, and poverty that do get reflected in these indicators. Thus, the drop in the HDI for Botswana, Cameroon, and Tajikistan reflects political troubles. However, such countries are few, and the drop is usually temporary. On the whole, as the GDP and HDI trends indicate, things are going well for us, and seem to have continually improved over the centuries, particularly in the last 50 years.

BOX 1.1 Human Development Index (HDI) [5].

The HDI was created to emphasize that people and their capabilities should be the ultimate criteria for assessing the development of a country, not economic growth alone. The HDI can also be used to question national policy choices, asking how two countries with the same level of GNI [gross national income] per capita can end up with different human development outcomes. These contrasts can stimulate debate about government policy priorities.

The Human Development Index (HDI) is a summary measure of average achievement in key dimensions of human development: a long and healthy life, being knowledgeable, and have a decent standard of living. The HDI is the geometric mean of normalized indices for each of the three dimensions.

The health dimension is assessed by life expectancy at birth, the education dimension is measured by mean of years of schooling for adults aged 25 years and more and expected years of schooling for children of school entering age. The standard of living dimension is measured by gross national income per capita. The HDI uses the logarithm of income, to reflect the diminishing importance of income with increasing GNI. The scores for the three HDI dimension indices are then aggregated into a composite index using geometric mean. The HDI simplifies and captures only part of what human development entails. It does not reflect on inequalities, poverty, human security, empowerment, etc.

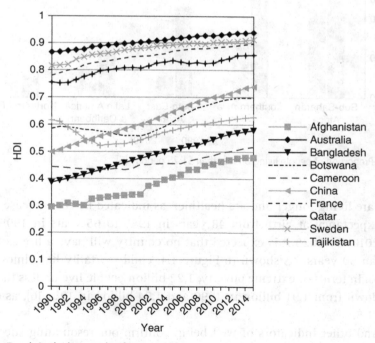

Figure 1.3 Trends in the human development index.

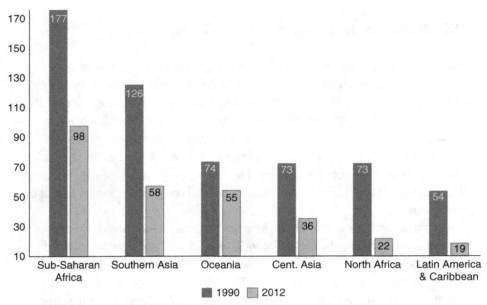

Figure 1.4 Under-five-years child mortality rate. Deaths per 1000 live births [6].

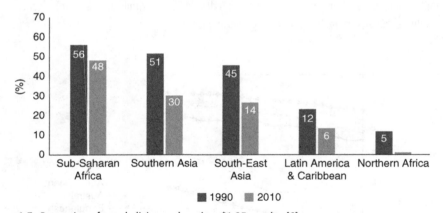

Figure 1.5 Proportion of people living on less than $1.25 per day [6].

People are living longer and are healthier, as indicated by the increase in average life expectancy at birth from 48 years in 1955 to 65 years in 1995 and 71 years in 2010. By 2025 it is expected that no country will have a life expectancy of less than 50 years. As shown in Figure 1.4, child mortality has almost halved since 1990. In terms of extreme poverty, 1.22 billion people live on less than $1.25 per day, down from 1.91 billion in 1990 and 1.94 billion in 1981 [6], as depicted in Figure 1.5.

These and other indicators of well-being confirm our resounding success as a species. However, as we will learn in Chapter 2, our success has also resulted in an increasingly large environmental impact. This raises questions about whether we will be able to sustain this level of well-being and enhance it, particularly for those of us who are not doing too well and for future generations. This includes the 1.22 billion still living in extreme poverty, the 1.6 billion who live without access to electricity, the 1.1 billion with inadequate access to water, and the 2.6 billion lacking basic sanitation. If we are to sustain this standard of living and enhance it in parts of the world where it is needed, it would help to know the underlying foundation of our well-being. Sustaining the gains in our well-being and extending them to those who have not yet benefited requires that this foundation is preserved.

1.2   What Does Human Well-Being Depend On?

The constituents of human well-being include material needs, health and education, opportunity, community, and security. Understanding what provides basic and critical support to ensure availability of these constituents to individuals and societies is essential to determine whether the trends discussed in the previous section can be sustained, and whether the benefits of development can become more widespread. Broadly speaking, human activities and well-being depend on the following three categories of goods and services:

- *Economic* goods and services include things like equipment, energy supply, market for products, industrial waste treatment, food, and transportation. These have monetary value and are traded in markets, which means that people pay money to obtain them.
- *Societal* goods and services include labor, educational institutions, intellectual capital, legal system, government, and culture. These involve individuals or groups of people. Some societal goods and services, such as labor, have monetary value, while others, like culture, do not.
- *Ecological* goods and services come directly from nature and include minerals, water, air, sunlight, biomass, ocean and river currents, wind, pollination, soil formation, carbon sequestration, and disease regulation. These are usually considered to have no monetary value since they are not traded in markets. There is no monetary transaction with nature since we do not pay trees for the oxygen they provide or the Earth for the crude oil we take from it.

Owing to the importance of these three categories of goods and services, they have been called the *triple bottom line* or triple values. But are the categories equally important? Or are some categories dependent on the availability of others?

Let us consider the source of economic goods and services. As depicted in Figure 1.6, economic activities transform ecosystem goods and services into economic

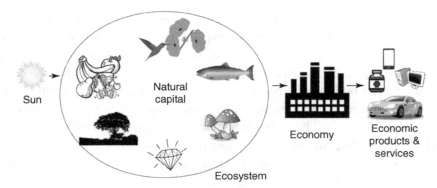

Figure 1.6 Ecosystems form the basis of all economic activities.

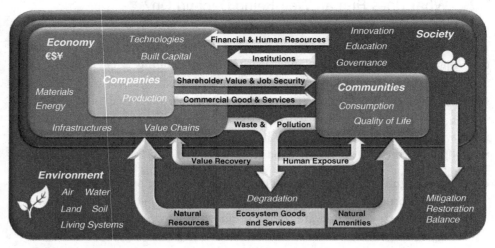

Figure 1.7 Interaction between ecosystems, society, and economy. Figure courtesy of J. Fiksel. Adapted from [7].

goods and services. In addition to inputs from nature, economic activities also require societal goods and services such as institutions and governance, as shown in Figure 1.7. Without goods and services from society and nature, economic activities are not possible. Societal well-being requires goods and services from nature because the people that constitute society need natural amenities. Thus, both societal and economic activities rely on ecosystems. Ecosystems can thrive even without societal or economic goods and services, but neither societal nor economic activities are possible without ecosystem goods and services. Thus, ecosystem goods and services form the foundation for economic and societal goods and services.

However, couldn't one argue that human innovation, which has been able to overcome all kinds of constraints imposed by nature by finding substitutes for many ecosystem goods and services, will be able to develop substitutes for ecosystems and all their goods and services? Examples of such success include climate-controlled buildings to maintain comfortable surroundings regardless of the weather outside, vaccinations to overcome the scourge of disease, artificial lighting to overcome darkness, genetically modified crops to fight pests, water treatment processes to replace natural water purification by wetlands, plastics and metals to replace wood, the International Space Station for human activities in space, and many others. We have developed artificial organs, synthesized molecules that never existed before, and may even be on the verge of synthesizing life itself and creating autonomous intelligent machines. Advances in science and technology have allowed us to find and extract more resources, and when faced with scarcities we have been able to find substitutes. For example, for lighting, electricity has replaced kerosene, which replaced whale oil; artificial fertilizers have replaced guano (accumulated bird droppings), which was being mined in South America and shipped to farmers in Europe, and so on.

Given such advances due to human ingenuity, are ecosystems truly essential for economic and societal activities? Can we not artificially synthesize ecosystems and their goods and services, or find substitutes? Because, if we can, then we could overcome any limit imposed by their deterioration. We could even build self-sustaining artificial biospheres to inhabit outer space and other planets. These questions have intrigued many, and an attempt to develop a self-contained and self-sustaining system was made in the Biosphere 2 project in the early 1990s. As described in Box 1.2, this experiment showed that with the technology available at that time, there was no substitute for Biosphere 1, which is planet Earth. In addition, history abounds with examples where entire societies and civilizations have collapsed owing to the loss of goods and services from nature [8, 9]. One such example is described in Box 1.3.

BOX 1.2 Biosphere 2 Project

A self-contained site, shown in Figure 1.8, was built in the Sonoran desert near Tucson, Arizona in the early 1990s. It contained various types of ecosystems, including a rain forest, ocean, and desert, and areas for agriculture and human habitation. The buildings were designed to have close to zero exchange with the surroundings of everything except sunlight. Eight "biospherians" spent two years in this complex. It was called Biosphere 2, since planet Earth is Biosphere 1.

While many new scientific insights were obtained from this work, the goal of building an artificial biosphere was never realized. Among the many problems encountered during this experiment was the inability to maintain the desired atmospheric composition: the carbon dioxide concentration kept on going up even though fossil fuels were not being used. What this experiment showed is that even with the benefits of modern science and our ability to artificially synthesize many things, we are still not able to produce the goods and services that we get from nature. In other words, ecosystem goods and services are truly the foundation of all human (and planetary) activities. Without them, humanity cannot be sustained! Thus, sustainability is ultimately about ensuring the availability of ecosystem goods and services for present and future generations.

BOX 1.3 The Lessons of Easter Island [10]

The story of Easter Island is about how a thriving civilization can collapse owing to the loss of ecosystem goods and services. This island in the Pacific Ocean is among the most remote inhabited islands on Earth. Polynesians arrived there in the fifth century and found a volcanic island with limited resources. The fresh water was in dormant volcanoes in the center of the island, and the soil was of poor quality, making it difficult to grow much more than sweet potatoes which, with the chickens that they brought with them, formed their diet. From the 20–30 settlers, the population gradually expanded and societies developed along with their culture and rituals. Even though food was limited, acquiring it was easy, and the plenty of spare time helped create a complex and elaborate society. One of the Easter Island rituals involved the construction of huge stone statues along the coast. Each statue was carved into a human head and torso, was about 20 feet tall, and weighed several tons. The stone was obtained from a quarry in the center of the island and transported on logs to the coast. It is estimated that over 600 such statues were constructed. The reason for the collapse of this civilization seems to have been the massive environmental destruction due to the felling of trees to provide logs to transport the rocks to the coast. This was in addition to the trees that must have been cleared for agriculture, fuel for cooking, and boats for fishing. Scientific studies have shown that the island was completely deforested by about 1600, and owing to the scarcity of trees, the islanders replaced wood with stones and then reeds for constructing their homes. Deforestation also meant the loss of soil due to erosion, which reduced their agricultural yields. The inability to build boats from logs also limited their ability to catch fish or escape from the island. By 1600, their civilization had started to collapse, and the population fell from a peak of about 7000 inhabitants. When the first Europeans arrived in 1722, they found a primitive society of about 3000 living in reed huts, with perpetual warfare between the clans, which were resorting to cannibalism.

Figure 1.8 Site of the Biosphere 2 project. Photograph by John de Dios (Wikimedia Commons CC-BY-3.0).

Thus, there is little doubt about the essential role of ecological systems in sustaining human well-being. Human ingenuity can certainly find substitutes for some individual goods and services, but replacing entire ecological systems and the biosphere involves complex interactions that are beyond our current understanding and engineering ability. Even substitutes for individual goods result in unexpected surprises and unintended harm, as we will see in subsequent chapters such as Chapter 3. Therefore, at this point, our best and only source of ecosystem goods and services to support our well-being is the natural biosphere itself, shown in Figure 1.9. Even if future advances in science and technology are able to develop a Biosphere 2, it still makes sense to protect and preserve our current biosphere, for if the advances toward Biosphere 2 do not happen or are not quick enough, the consequences could be dire indeed.

1.3   Ecosystem Goods and Services

Ecosystem goods and services are the benefits that humans derive from nature. They are the flows derived from natural capital, and provide the foundation for all human activities. Natural capital includes natural resources such as minerals, fertile soil, forests, and wetlands. We cannot survive for more than a few minutes without oxygen – a molecule provided by green plants as a byproduct of

Figure 1.9 Earth rise as captured by *Apollo 8* astronauts on December 24, 1968. One of the astronauts, Jim Lovell said, "The vast loneliness is awe-inspiring and it makes you realize just what you have back there on Earth" [11].

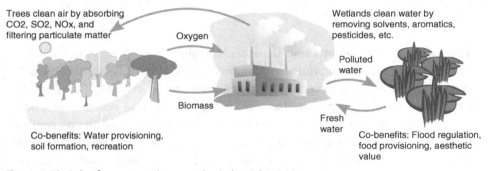

Figure 1.10 Role of ecosystems in supporting industrial activities.

photosynthesis, which in turn depends on carbon dioxide, a byproduct of respiration. While such dependence on nature is quite obvious, ecosystems play a much bigger role which is usually ignored in our decisions.

Consider a typical company or industrial process and its interaction with the economy, society, and environment, as shown in Figure 1.7. The dependence of this process on raw materials is well-understood, as is the market for the products and byproducts, and the technologies for dealing with wastes. Societal systems that train workers, protect property rights, and regulate markets also exist. In addition to these economic and societal goods and services, the manufacturing process also relies on a variety of ecosystem goods and services, with some specific interactions shown in Figure 1.10. The availability of water relies on the hydrological cycle, while the mitigation of emissions to air and water relies on ecosystem services of air and water quality regulation. This includes the role of air and water currents in dissipating and diluting pollutants, the ability of trees to remove not just carbon dioxide but also particulate matter and the oxides of sulfur and nitrogen, and the ability of water bodies such as wetlands to degrade and remove solvents, oils, greases, pesticides, phosphates, heavy metals, pharmaceuticals, etc. In addition, ecosystems like trees and wetlands also provide many co-benefits to society, such as flood regulation, water provisioning, and opportunities for recreation. If emissions exceed the ability of nature to absorb and mitigate them, then human and ecological systems may suffer damage and lose their ability to provide various services. Like industrial processes, agricultural activities also rely on ecosystem services such as primary production, regulation of soil fertility, biogeochemical cycles of carbon and nitrogen, availability of fresh water, and the water cycle. Fossil resources are the product of ancient goods and services from nature, but even their current availability relies on the geological cycle that concentrates them in the Earth's crust and makes them available for easier extraction. Industry also benefits from the aesthetic and cultural aspects of ecosystems; one example is the use of nature-inspired images in corporate logos, such as the apple in Apple's logo, the shell in Shell's logo, the peacock in the logo of NBC, birds in the logos of many airlines such as Singapore Airlines and Lufthansa, the butterfly in Microsoft's logo, and the mouse in Disney's logo.

The large variety of goods and services provided by nature may be categorized as shown in Table 1.1 and described below:

- *Provisioning services* are products from ecosystems such as food, fresh water, fuel, ornamental resources, genetic resources, biochemicals, and pharmaceuticals. These are most familiar owing to their direct role in human activities.
- *Regulation and maintenance services* are benefits from the maintenance of ecological processes by the regulation of air quality, water quality, climate, soil fertility, pests, and diseases.
- *Cultural services* are the non-material benefits that people get from nature in the form of spiritual and religious values, cultural diversity, educational values, aesthetic values, sense of place, social relations, and recreation and ecotourism.

These services play a critical role in enabling and enhancing human well-being. As shown in Figure 1.11, ecosystem goods and services provide security, basic materials for a good life, health, and social relations, which in turn enable freedom of choice and action. Most of these connections are quite obvious, such as the effect of access to nutritious foods and clean water on our well-being. Ecosystem services are also known to have a direct impact on health and the feeling of well-being, since living close to nature reduces overall mortality, cardiovascular disease,

Table 1.1 Classification of ecosystem services.

Section	Division	Group
Provisioning (biotic)	Biomass	Cultivated terrestrial plants for nutrition, materials, or energy
		Cultivated aquatic plants for nutrition, materials, or energy
		Reared animals for nutrition, materials, or energy
		Reared aquatic animals for nutrition, materials, or energy
		Wild plants (terrestrial and aquatic) for nutrition, materials, or energy
		Wild animals (terrestrial and aquatic) for nutrition, materials, or energy
	Genetic material from all biota	Genetic material from plants, algae, fungi, or animals
Regulation and maintenance (biotic)	Transformation of biochemical or physical inputs to ecosystems	Mediation of wastes or toxic substances of anthropogenic origin by living processes
		Mediation of nuisances of anthropogenic origin
	Regulation of physical, chemical, and biological conditions	Regulation of baseline flows and extreme events
		Life cycle maintenance, habitat, and gene pool protection
		Pest and disease control
		Regulation of soil quality
		Water conditions
		Atmospheric composition and conditions
Cultural (biotic)	Direct, in-situ, and outdoor interactions with living systems that depend on presence in the environmental setting	Physical and experiential interactions with natural environments
		Intellectual and representative interactions with natural environments
	Indirect, remote, often indoor interactions with living systems that do not require presence in the environment	Spiritual, symbolic, and other interactions with natural environments
		Other biotic characteristics that have a non-use value

Provisioning (abiotic)	Water	Surface water used for nutrition, materials, or energy
		Ground water for use for nutrition, materials, or energy
		Other aqueous ecosystem outputs
	Non-aqueous natural abiotic ecosystem outputs	Mineral substances used for nutrition, materials, or energy
		Non-mineral substances or ecosystem properties use for nutrition, materials, or energy
Regulation and maintenance (abiotic)	Transformation of biochemical or physical inputs to ecosystems	Mediation of waste, toxics, and other nuisances by non-living processes
		Mediation of nuisances of anthropogenic origin
	Regulation of physical, chemical, and biological conditions	Regulation of baseline flows and extreme events
		Maintenance of physical, chemical, and abiotic conditions
Cultural (abiotic)	Direct, in-situ, and outdoor interactions with natural physical systems that depend on presence in the environmental setting	Physical and experiential interactions with natural abiotic components of the environment
		Intellectual and representative interactions with abiotic components of the natural environment
	Indirect, remote, often indoor interactions with physical systems that do not require presence in the environment	Spiritual, symbolic, and other interactions with the abiotic components of the natural environment

Source: Adapted from [13].

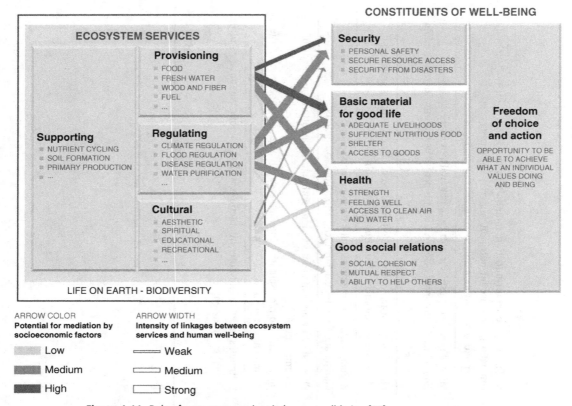

Figure 1.11 Role of ecosystem services in human well-being [14].

and depressive symptoms. In addition, the incidence of many modern diseases and allergies, particularly in high-income countries, is due to lack of exposure to microorganisms that occur in nature [12]. Thus, exposure to green spaces and biodiversity can reduce the occurrence of such diseases and enhance well-being. Ecosystems can also enhance the resilience of communities to extreme events. For example, coastal wetlands and mangroves can protect communities from flooding during storms and hurricanes.

1.4   What about Saving the Planet?

So far, our primary focus in this chapter has been on human well-being. This may come as a surprise to you if you were thinking that sustainability is about "saving the planet." This is a common rhetoric that is often heard. However, "saving the planet" is not the goal of sustainable engineering and other efforts toward sustainable development. Even those who talk about saving the planet usually focus on human well-being. They say that the planet does not need to be saved. It has sustained itself for billions of years, despite dramatic changes, and will most likely continue to do so for many more millennia. Sustainability is ultimately about human beings and and our well-being, not just in the short run, but in the long run as well. As we have learned in this chapter, human well-being is strongly dependent on healthy ecosystems, so sustaining ourselves does require saving the planet.

1.5   Summary

Human well-being has improved over the last several centuries, as indicated by enhanced indicators such as population, GDP, and the HDI. This improvement is due to our ability to utilize goods and services from economic, societal and ecological systems. Among these three, ecosystems are most important since they are needed to support the economy and society. Goods and services from nature are categorized as provisioning, regulating, and cultural. Despite various efforts, it has not been possible to generate them by artificial systems. Thus, the availability of healthy ecosystems is essential for sustaining the well-being of current and future generations.

Key Ideas and Concepts

- Human well-being
- Human Development Index
- Ecosystem goods and services

- Gross domestic product
- Triple values
- Biosphere 2

1.6 Review Questions

1. What are some components of human well-being?
2. Define economic, societal, and ecosystem goods and services, with examples of each.
3. How has the poverty rate changed since 1990?
4. What are Biospheres 1 and 2?
5. What roles do ecosystems play in our feeling of well-being?

Problems

1.1 Describe some similarities and differences between gross domestic product and the human development index. Why is the latter index considered to be a better indicator of human well-being?

1.2 Identify some economic, societal, and ecosystem services that the following systems depend on, and describe the dependence: (a) a coal-burning power plant; (b) driving a car; (c) intensive agriculture.

1.3 Consider a typical single-family dwelling that consists of a house surrounded by a yard. Identify some of the ecosystem services that the yard could provide to the residents of the house and to the larger community.

1.4 The triple value concept conveys that it is important for corporations to consider economic, social, and environmental goals in their decisions. Do companies consider goals in these three categories to be equally important? Explain your response with examples.

1.5 Give examples of how the following activities rely on cultural ecosystem services: (a) the use of logos by corporations; (b) religious and spiritual practices; (c) recreational activities.

1.6 What are the Sustainable Development Goals (SDGs) of the United Nations? Do they account for the critical role of ecosystems in human well-being?

1.7 Which ecosystem goods and services did the residents of Easter Island depend on for building their statues? How did the loss of these goods and services contribute to their demise?

1.8 Identify some ecosystem services for which technological substitutes are not yet available. Will we be able to develop technological substitutes for these services?

1.9 Search the health literature to determine how ecosystems help us in maintaining and improving our health. Focus on aspects such as mental and emotional well-being, resistance to allergens, and the development of new medicines such as antibiotics.

1.10 Ethanol is a required additive in gasoline in many countries. The main steps in producing ethanol from corn include the agricultural step, where corn is grown, and the manufacturing step, where corn sugar is converted to ethanol by fermentation. For these two steps, discuss the ecosystem services that each activity depends on and how each activity impacts ecosystem services.

References

[1] A. Alexandrova. Well-being as an object of science. *Philosophy of Science*, 79(5):678–689, 2012.

[2] P. Gerland, A. E. Raftery, H. Ševčíková, et al. World population stabilization unlikely this century. *Science*, 346(6206):234–237, 2014.

[3] United Nations World population prospects: the 2017 revision. `https://esa.un
.org/unpd/wpp/Graphs/Probabilistic/POP/TOT/`, 2017, accessed
November 22, 2018.

[4] The Maddison Project. `www.ggdc.net/maddison/maddison-project/
home.htm`, 2013 version, accessed November 22, 2018.

[5] United Nations Development Program. Human Development Index. `http://hdr
.undp.org/en/content/human-development-index-hdi`, accessed
March 25, 2018.

[6] World Bank. World Development Indicators 2013. `http://data.worldbank
.org`, 2013, accessed March 25, 2018.

[7] J. Fiksel. *Resilient by Design: Creating Businesses that Adapt and Flourish in a
Changing World.* Island Press, 2015.

[8] J. Diamond. *Collapse: How Societies Choose to Fail or Succeed.* Penguin Books,
2006.

[9] J. A. Tainter. *The Collapse of Complex Societies.* Cambridge University Press, 1988.

[10] C. Ponting. *A New Green History of the World: The Environment and the Collapse of
Great Civilizations.* Random House, 2011.

[11] NASA. Earth rise. `www.nasa.gov/multimedia/imagegallery/image_
feature_1249.html`, accessed March 25, 2018.

[12] G. A. Rook. Regulation of the immune system by biodiversity from the natural
environment: an ecosystem service essential to health. *Proceedings of the National
Academy of Sciences*, 110(46):18360–18367, 2013.

[13] R. Haines-Young and M. Potschin. Common international classification of ecosystem
services (CICES) v5.1 and guidance on the application of the revised structure.
Technical report, European Environmental Agency, 2018.

[14] Millennium Ecosystem Assessment. *Ecosystems and Human Well-being: A
Framework for Assessment.* Island Press, 2003.

2 Status of Ecosystem Goods and Services

Every country can be said to have three forms of wealth: material, cultural and biological. The first two we understand very well, because they are the substance of our everyday lives. Biological wealth is taken much less seriously. This is a serious strategic error, one that will be increasingly regretted as time passes.

E. O. Wilson [1]

As we learned in Chapter 1, ecosystems provide the goods and services that are essential for sustaining human activities. Without inputs from nature, neither societal nor economic activities can be sustained. Therefore, the well-being of ecosystems is of critical importance for the well-being of human beings. In this chapter, we will explore the status of ecosystem goods and services. We will consider how human development has relied on various natural resources, and the impact of this reliance on their supply and on our environment. Such insight is needed to assess whether the enhancement in human well-being over the previous decades can be sustained. It will also help in identifying underlying reasons for trends in ecosystem services and potential solutions to ensure their sustained availability. We will first consider the status of essential ecosystem goods such as fuels, materials, water, and food, and then selected ecosystem services such as the regulation of climate, air and water quality, primary productivity, and pollination. We will end this chapter with a look at the overall global status of ecosystem goods and services.

2.1 Fuels

Until the Industrial Revolution, the main fuel used by humanity was biomass. This changed dramatically in the last 200 years owing to the dominance of fossil fuels. These fuels are important ecosystem goods, produced from ancient biomass that was buried and transformed by planetary processes in an oxygen-starved reducing environment. The resulting products of coal, natural gas, and crude oil are highly concentrated hydrocarbons and carbon that have a high fuel value and can be transformed quite easily into many other products.

Recent trends in fuel consumption, shown in Figure 2.1, depict this dominant role of fossil fuels. These fuels are nonrenewable because their rate of extraction is

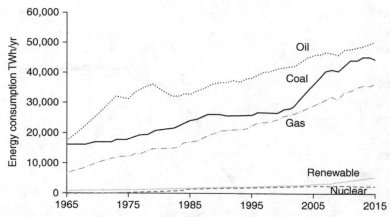

Figure 2.1 Global use of fuel resources [2]. Wikimedia Commons CC-BY-SA-4.0.

much greater than their rate of production. Thus, the consumption of nonrenewable resources must result in their depletion over time.

While the overall trend for fossil fuel use in Figure 2.1 shows a continuous increase, attention to specific geographic locations shows the depletion of some sources and the discovery of others. In general, the total quantity produced from individual oil wells follows an S-curve or logistic curve, which means that the rate of production from a well tends to peak followed by a decline. On the basis of such observations for US wells, Hubbert surmised that the overall production of oil will also peak [5], and he correctly predicted a peak for US production. His prediction and the actual US oil production are shown in Figure 2.2. Beyond the peak, production cannot keep up with demand, indicating a resource scarcity, and likely price increase. Many researchers predicted a global peak in the early part of this century, but production data in Figure 2.2 for the USA and in Figure 2.3 for the world seem to indicate that the peak may not have been reached yet. Some reasons for the delay in the peak include new discoveries, such as those of shale oil and gas, new technologies for extracting them such as horizontal drilling and hydraulic fracturing (fracking), and reduced consumption due to economic recessions or the adoption of alternate technologies. However, ultimately, for any nonrenewable resource, not just fossil fuels, if extraction continues, a peak is inevitable. To date, the primary ecological degradation associated with fossil fuels has not been their depletion, but the impacts of their use at such a large scale on services such as air quality and climate regulation, as we will see in Sections 2.6 and 2.7.

2.2 Materials

Figure 2.4 summarizes global resource use trends for several materials. Total use is shown by gray lines and use per capita by black lines. As can be seen, the total use

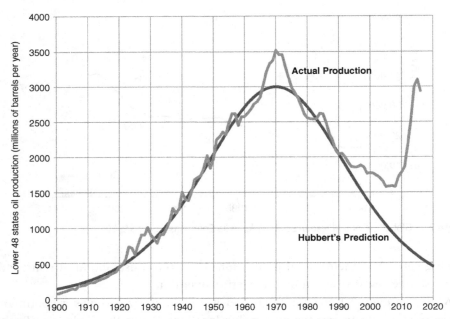

Figure 2.2 Hubbert's prediction in 1956 about the peak of US oil production versus actual production [3].

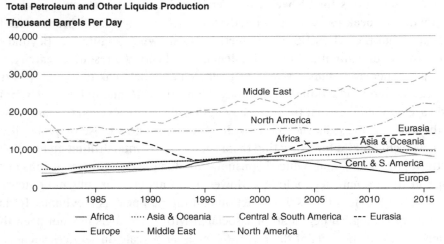

Figure 2.3 World oil production by regions [4].

for most resources has been increasing, while the per capita use of many resources has been declining. This decline could be due to resource constraints such as the resource being past the peak. Like fossil fuels, any resource that is used in a nonrenewable manner, with a rate of use larger than the rate of replenishment,

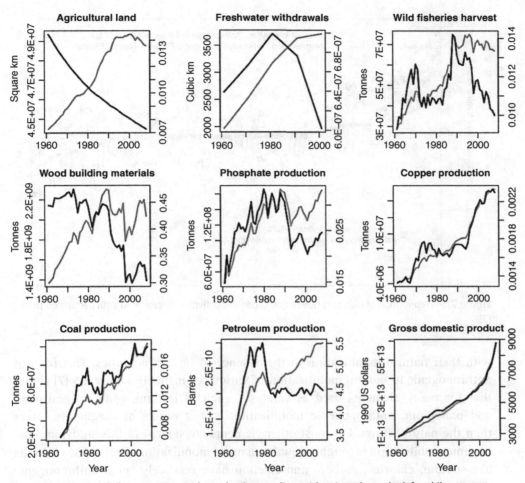

Figure 2.4 Total global resource use is shown by the gray line, with axis scale on the left, while resource use per capita is shown by the black line with axis scale on the right [6].

will show a peak. In this figure, the resources of fresh water, phosphate, wood building materials, and wild fisheries harvest seem to show a peak. As discussed in Section 2.1, after the peak, a resource is expected to have difficulty meeting demand and its consumption is likely to decline, particularly for those resources that lack substitutes. Phosphates and fresh water are examples of such resources. As has happened with crude oil in the USA, new technologies and discoveries can postpone the ultimate peak, but these approaches are usually more expensive owing to the depleting quality of the resource and greater difficulty in extraction.

Another way of understanding the extent of human dependence on materials is by comparing the mobilization of various elements by human activities

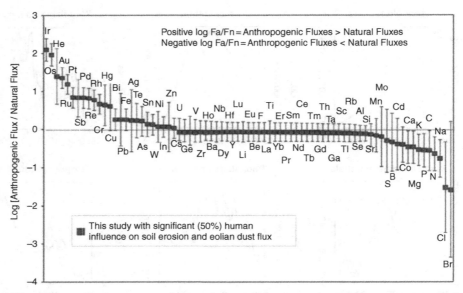

Figure 2.5 Perturbation of elements due to anthropogenic activities. Reproduced with permission from [7].

with their natural mobilization in the absence of human activities. This ratio of anthropogenic to natural mobilization of various elements in Figure 2.5 [7] shows that for many elements used as catalysts, such as iridium, osmium, platinum, and palladium, anthropogenic mobilization is 1–2 orders of magnitude larger than the natural flows. For most other elements considered in this study, anthropogenic mobilization is roughly equal to natural mobilization. Only a few elements like sodium, chlorine, carbon, and calcium have relatively small anthropogenic mobilization. It is the introduction of mobilized elements such as lead, arsenic, mercury, etc. into the environment that causes damage to ecosystem services and human health. If they circulated within the anthropogenic system without escaping into the environment, then their mobilization would not be such an important environmental issue.

In addition to our reliance on naturally occurring materials, we have synthesized over 50,000 new molecules in the last few decades. Their introduction into ecosystems, even in small quantities, can cause significant harm because ecosystems would have never encountered them before, and may be unable to benefit from or neutralize their presence. Examples include pesticides, heavy metals, and refrigerants. Box 2.1 summarizes the history of one class of compounds that was banned due to its effect on the ecosystem service of protecting the Earth from ultraviolet radiation. The effect of these chlorofluorocarbon (CFC) compounds on stratospheric ozone is shown in Figure 2.6.

BOX 2.1 History of Chlorofluorocarbons (CFCs)

When first developed in the 1930s by Thomas Midgeley at General Motors, CFCs were touted as miracle compounds that were non-toxic and stable, with many desirable properties. The ozone hole was predicted by Molina and Rowland [8] and then observed and monitored by satellites [9], resulting in images like those in Figure 2.6. The dark region shows the ozone hole over the Antarctic, which typically appears during the Antarctic winter.

The ozone layer in the stratosphere protects the Earth by blocking the ultraviolet part of sunlight. This regulating ecosystem service has deteriorated because of the use of ozone-depleting agents such as CFCs and other substances that release a chlorine or bromine atom into the Antarctic stratosphere. These compounds had, and some still have, a variety of uses, including as refrigerants, cleaning agents, fungicides, and flame retardants.

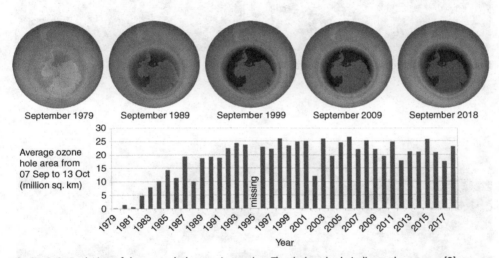

Figure 2.6 Evolution of the ozone hole over Antarctica. The darker shade indicates less ozone [9].

Another category of materials that is ubiquitous in modern life is plastics. Figure 2.7 shows regions of the world's oceans where waste plastics have accumulated. These are expansive floating islands of large and small plastic materials that disrupt marine life. Figure 2.8 shows such items on an otherwise pristine Pacific island. Due to the effect of the ocean and weather, most plastic is much smaller than the light-colored pieces visible in this figure. While these waste materials may be aesthetically undesirable to humans, they also hurt marine life by entering the foodchain and accumulating in species at the top of the foodchain, such as birds and other marine species.

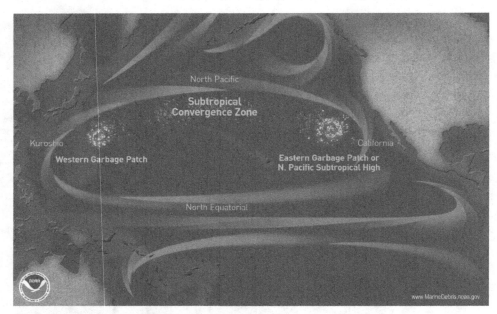

Figure 2.7 Oceanic gyres where plastic trash has accumulated.

Figure 2.8 Plastic debris and trash is light-colored on an otherwise pristine-looking seashore on the Big Island of Hawaii. Photo courtesy of Harshal Bakshi.

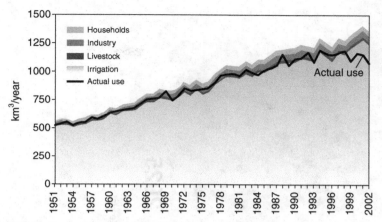

Figure 2.9 Global consumptive use of water [10].

2.3   Water

With increasing population, consumption, and development, the human with-drawal of fresh water has also increased, as shown in Figure 2.9. Most of the water is used for growing food, a consequence of agricultural intensification to meet global demand.

One of the consequences of this high water use is increasing water stress in many parts of the world. As shown in Figure 2.10, water stress, defined as the ratio of water consumed to the available renewable water, is expected to increase across the world in the near future. Figure 2.11 shows the fraction of global agriculture production under high or extremely high water stress. In high-stress areas, this fraction is 40 percent, while in areas under extremely high stress it is 80 percent or higher. Such stress levels indicate high or extremely high vulnerability to disruptions in water availability, and the potential for human conflict.

Satisfying the human demand for freshwater has also resulted in severe ecological disruption in most major rivers of the world. This is depicted in Figure 2.12, where dark to light shades indicate strongly impacted, moderately impacted, and unimpacted large river systems [13]. Dams hold back over 6500 km^3 of water, which is about 15 percent of the annual river runoff in the world. Generating electricity, enabling irrigation, and controlling floods are positive impacts of dams. However, they also cause ecological deterioration due to habitat destruction and the disruption of animal migration patterns. The resettlement of human populations outside the catchment area also has many negative societal side-effects.

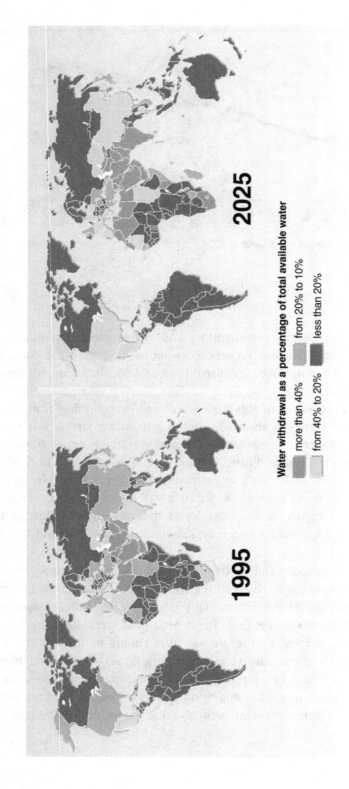

Figure 2.10 Global water stress - past and future [11].

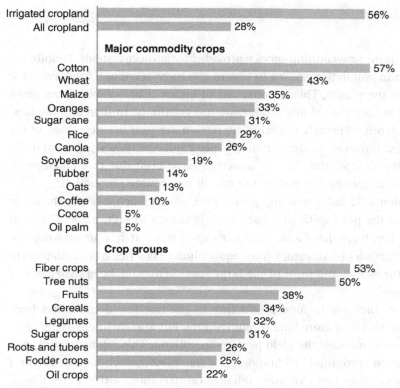

Irrigated cropland — 56%
All cropland — 28%

Major commodity crops

Cotton — 57%
Wheat — 43%
Maize — 35%
Oranges — 33%
Sugar cane — 31%
Rice — 29%
Canola — 26%
Soybeans — 19%
Rubber — 14%
Oats — 13%
Coffee — 10%
Cocoa — 5%
Oil palm — 5%

Crop groups

Fiber crops — 53%
Tree nuts — 50%
Fruits — 38%
Cereals — 34%
Legumes — 32%
Sugar crops — 31%
Roots and tubers — 26%
Fodder crops — 25%
Oil crops — 22%

Figure 2.11 Fraction of crops under water stress [12].

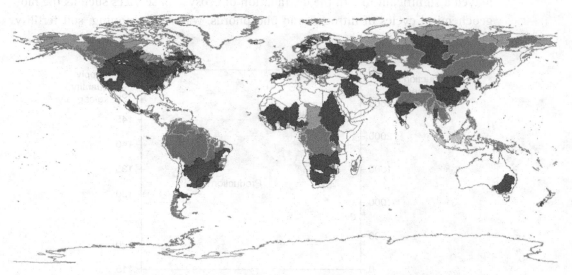

Figure 2.12 Impact of dams in large river systems across the world. From dark to light gray, strongly affected, moderately affected, and unaffected rivers, and regions with not enough data. The white regions have no large river systems. Reproduced with permission from [13].

2.4   Food

Food is an important provisioning service provided by agroecosystems. Despite the increase in human population shown in Figure 1.1, there is enough food available for everyone on the planet. This is conveyed in Figure 2.13, which shows grain production and supply [14]. While production has continued to increase, the supply quantity in terms of cereals available per person per year has decreased in the last two decades, however. A smaller supply per capita increases price volatility and vulnerability to scarcities. So far, scarcities have been due to challenges in food distribution or purchasing power, not due to inadequate production.

The total global wild fish catch and production from aquaculture are shown in Figure 2.14, and the per capita wild fish catch is shown in Figure 2.4. The total wild fish catch has leveled off and, on a per capita basis, it has already peaked. An increasing fraction of fish comes from aquaculture. Fisheries have collapsed in many parts of the world, and most of the large specimens have been fished out of the world's oceans.

The impressive increase in food production in the last few decades has been enabled not only by using more land for agriculture, but also by intensifying agricultural activities to increase the yield per hectare of land. This intensification was the result of "green revolution" technologies of the 1970s, which included hybrid high-yield crop varieties and extensive reliance on irrigation, artificial fertilizers, and pesticides. This industrialized farming has had many side-effects, which have played a significant role in the degradation of ecosystem services such as the biogeochemical cycles of nitrogen and phosphorus, water provisioning, soil fertility

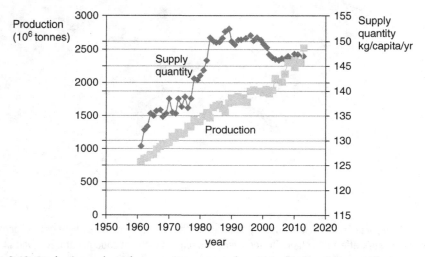

Figure 2.13 Production and supply per capita per year of cereals, excluding those used for beer.

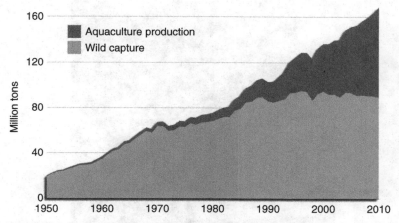

Figure 2.14 Wild fish catch and aquaculture production (Wikimedia Commons CC-BY-SA-3.0).

regulation, and pollination, as described in subsequent sections. Of course, it has played a key role in feeding the over seven billion inhabitants of our planet.

2.5 Soil

Soil plays an essential role in food production and in the biogeochemical cycles of carbon, nitrogen, and phosphorus. Human activities move large quantities of earth by construction, urbanization, and agriculture. One example is shown in Figure 2.15. It is estimated that the amount of earth moved intentionally and unintentionally by human activities over the last 5000 years could build a 4000 m mountain range 100 km long and 40 km wide [16]. Most of this earth movement has happened in the last several decades. The unintended movement of soil is usually due to negative impacts of intensive farming. Erosion rates in the central USA exceed 2000 m/Myr (Myr = million years), while on the Loess plateau in China they are 10,000 m/Myr. Natural rates of soil production are estimated to be between 50 to 200 m/Myr and the rate of soil loss in the USA before European contact is estimated to have been 21 m/Myr. Across the world, soils are severely degraded in many regions, as shown in Figure 2.16.

2.6 Air Quality Regulation

Ecosystems are able to regulate air quality by transporting, absorbing, and mitigating all kinds of emissions. An emission becomes a pollutant when its concentration

Figure 2.15 Soil degradation in a village near Vadodara in Gujarat, India. Most of the soil was probably trucked away to support the local construction boom. The Banyan tree (*Ficus benghalensis*) was not cut, owing to its religious significance. Photo courtesy of Harshal Bakshi.

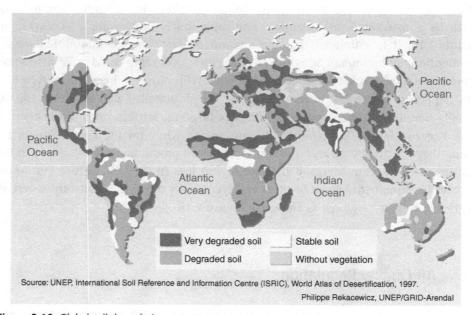

Figure 2.16 Global soil degradation. `http://grida.no/resources/5507/`

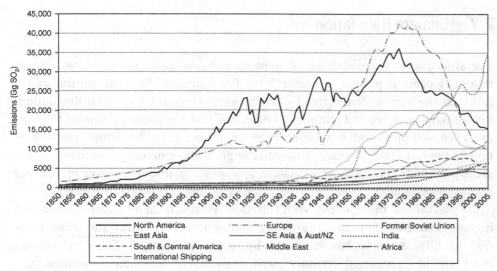

Figure 2.17 Historical trend of sulfur dioxide emissions. Reproduced with permission from [18].

exceeds nature's capacity to use or capture it. One major impact of fossil fuel use has been the emission of sulfur and nitrogen oxides in quantities that overwhelm nature's ability to regulate air quality. This is conveyed in the global trend of sulfur dioxide emission shown in Figure 2.17. From this figure we can see that emissions in Europe and North America peaked in the 1970s. This happened because of strict regulations driven by public pressure, the greater use of oil and natural gas instead of coal, and the shift of manufacturing to other countries, mainly China. However, emissions have been increasing in developing countries, particularly China and India. It is also interesting to note that emissions from international shipping have also been increasing and are larger than those from Europe.

The impact of such emissions is well documented. It includes the acidification that affects the productivity of lakes and forests and destroys ancient monuments such as St. Paul's Cathedral in London, the Acropolis in Athens, and the Taj Mahal in Agra. At its worst, these emissions have killed entire forests, and there have even been incidents of birds falling from the sky due to the pollution. It is also estimated that air pollution in North America and Europe has killed 25–40 million people, roughly the same as both World Wars combined [17], and these impacts have been shifting to the eastern part of the planet. However, history indicates that this may be one problem that can be solved by the judicious use of technology, economics, and policies, which is what is likely to happen in the developing world over time as its citizens demand better air quality.

2.7   Climate Regulation

The atmospheric concentration of some greenhouse gases over the last several thousand years is plotted in Figure 2.18. These graphs show the natural variation and the very different trend of these gases since the Industrial Revolution. Recent data, depicted in Figure 2.19 [20], show that the concentration of CO_2 has exceeded 400 ppm, which is about 40 percent more than the highest concentration in the last 800,000 years. This trend implies that emission of carbon dioxide greatly exceeds the capacity of nature to capture it through processes such as photosynthesis.

For methane, the current concentration is much more than double the highest concentration in almost one million years, while the concentration of nitrous oxide has increased by about 14 percent. These concentrations are in "uncharted territory" and most scientific studies conclude that anthropogenic emissions of greenhouse gases are exceeding nature's capacity to regulate Earth's climate in the way it has been regulated over many millennia.

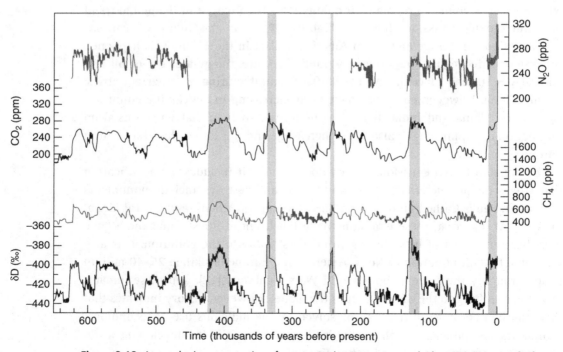

Figure 2.18 Atmospheric concentration of some greenhouse gases over the last 650,000 years [19].

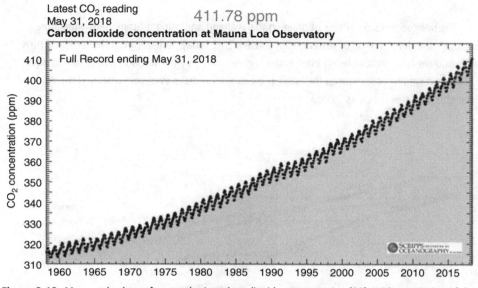

Figure 2.19 Measured values of atmospheric carbon dioxide concentration [20]. With permission of the Scripps Institution of Oceanography, UC San Diego.

2.8 Water Quality Regulation

Ecosystems such as wetlands, rivers, and lakes have a natural ability to regulate water quality. Wetlands have been called nature's kidneys owing to their ability to remove a diverse array of pollutants from water. They also provide many other services, including flood regulation, carbon sequestration, and food provisioning. However, wetlands are being lost all over the world, as shown in Figure 2.20. In the USA, most wetland loss happened between 1950 and 1970. Some states, such as Ohio and California, have lost over 90 percent of their wetlands. Box 2.2 describes the loss of one such wetland.

BOX 2.2 The Black Swamp

The Black Swamp was a large wetland that extended from northeast Indiana through northwest Ohio to Lake Erie, as shown in Figure 2.21. It covered approximately 4000 km^2 and was gradually drained in the second half of the nineteenth century. At that time, wetlands were considered to be worthless areas that should be drained for human uses such as farming. Over the years and with much effort, the trees in the swamp were cut, roads were built, and the swamp was drained to result in fertile farmland that is among the most intensively farmed areas in Ohio.

The unexpected side-effects of draining this wetland and using it for modern agriculture include the deterioration of water quality in Lake Erie. Today, fertilizer from the farmland runs off into the lake without being intercepted and detoxified by the wetland. The resulting algal blooms have not only reduced the lake's productivity but have even contaminated the water of neighboring cities such as Toledo.

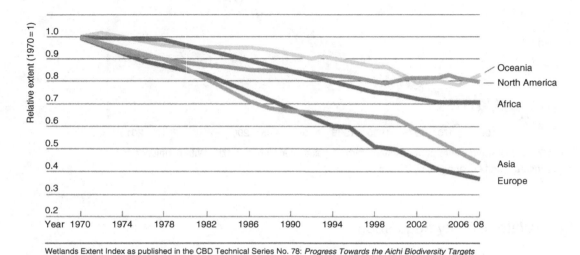

Wetlands Extent Index as published in the CBD Technical Series No. 78: *Progress Towards the Aichi Biodiversity Targets*

Figure 2.20 Extent of wetlands has declined in all continents. Reproduced with permission from [21]

In addition to such direct loss of the habitat that provides the water quality regulation ecosystem service, other activities are also contributing to the degradation of this ecosystem service. A prominent example is the use of artificial nitrogen and phosphorus fertilizers and the effect of their runoff into water bodies. Nitrogen and phosphorus are essential nutrients for most living things. Even though air is about 78 percent nitrogen, this form of nitrogen is not reactive and cannot be used by plants. The reactive form of nitrogen consists of various salts and is an essential but scarce resource. For many centuries, humanity relied on traditional methods such as crop rotation and nutrient cycling through plant residues, manure, and composting to replenish soil nutrient content. In many areas, bird droppings or guano was also used as fertilizer. It is estimated that such methods could support only about three billion people, or less than half of the current world population.

Development of the Haber–Bosch process for converting atmospheric nitrogen into reactive nitrogen in the form of ammonia by the reaction

$$N_2 + 3H_2 \rightarrow 2NH_3$$

is one of the main reasons why food production has been able to keep up with population growth. This has been possible by increasing our consumption of artificial

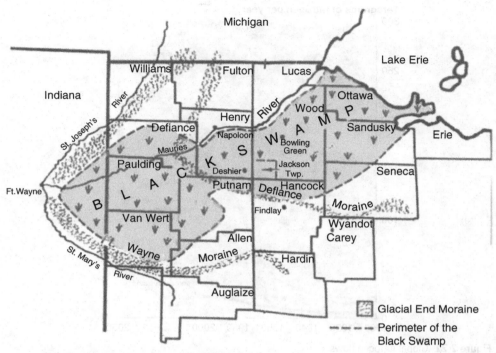

Figure 2.21 Historic location of the Black Swamp.

fertilizers more than four-fold in the last four decades, as shown in Figure 2.22. Unfortunately, all this fertilizer does not get used by plants; some is lost to the air by conversion to nitrous oxide (N_2O), and some is lost to water as dissolved nitrates. These anthropogenic flows have grown to become of the same order of magnitude as the flow of reactive nitrogen due to natural processes, as shown in Figure 2.22. Furthermore, nitrous oxide is a greenhouse gas whose contribution to climate change is about 300 times greater than carbon dioxide by mass. Its concentration in the atmosphere is included in Figure 2.18. Nitrogen runoff into water contributes to harmful algal blooms and eutrophic or dead zones. The runoff of phosphorus fertilizers also contributes to this phenomenon, which is now widespread all over the world, as shown in Figure 2.23. Most phosphorus is available from phosphate mines as a nonrenewable resource. Its trend is included in Figure 2.4.

2.9 Net Primary Productivity

Life on Earth is possible because of the ability of plants to convert sunlight into biomass. The net primary productivity (NPP) is the amount of carbon fixed by

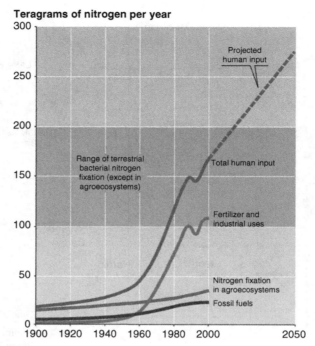

Figure 2.22 Global nitrogen flows.

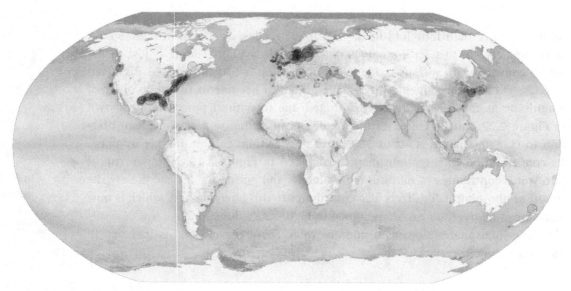

Figure 2.23 Aquatic dead zones around the world [22]. The size of a circle indicates the size of the corresponding dead zone.

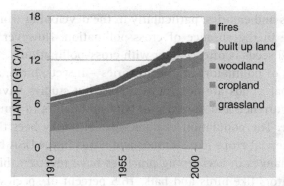

Figure 2.24 Trend in human appropriation of net primary productivity (Gt C/yr). Reproduced with permission from [23].

plants per year. With increasing population and per capita consumption, people have been appropriating a larger fraction of the planet's NPP, as shown in Figure 2.24. This is being done by transforming the Earth's surface for our use through agriculture and urbanization. It includes increasing the dependence of food on primary productivity by intensive animal farming, which involves using grains as animal feed, and by appropriating other biomass such as wood. In the previous century, it is estimated that human appropriation of net primary productivity increased from 6.9 Gt carbon per year (Gt C/yr) in 1910 to 14.8 Gt C/yr in 2005, that is from 13 percent to 25 percent of the potential NPP of the planet. This is a substantial impact of one species on the planet. With increasing population and economic growth, it can be seen how human appropriation of NPP will grow, particularly since the NPP on Earth has an upper limit due to a limit on the amount of sunlight incident on Earth, which limits the amount of primary production.

2.10 Pollination

Animal pollinators such as bees, butterflies, birds, and bats provide a service that enhances the production and nutritional content of foods such as almonds, apples, oranges, cucumbers, pumpkin, rapeseed, soybean, alfalfa, and cotton. The global economic value provided by insect pollinators is estimated to have been over $215 billion in 2005, representing 9.5 percent of the global value of agricultural production [24]. Insects are estimated to pollinate about 75 percent of crop species and up to 94 percent of wild flowering plants. Most of the staple crops like rice and wheat rely on wind pollination, but foods that provide essential micronutrients such as vitamins are insect-pollinated. Beans are one such crop that provide essential nutrients and calories, particularly in the developing world. Many plants will self-pollinate in the absence of cross-pollination. However, the nutritional value of fruits and seeds is often higher with cross-pollination.

The population of pollinators in all categories – insects, birds, and mammals – is declining. Global trends are not yet available, but studies have documented a decline in the occurrence and diversity of wild pollinators in North America and Northwest Europe. The population of domesticated honey bees that are used for pollinating commercial crops like almonds and citrus in the USA had been declining, but recent changes in beekeeping practices have reversed this trend. Among vertebrate pollinators like birds and bats, 16.5 percent of species are threatened, according to the International Union for Conservation of Nature. The volume of pollinator-dependent crops has increased by 300 percent over the last 50 years, making many societies dependent on this ecosystem service. However, their yield has a lower stability than that of pollinator-independent crops [25].

Reasons for the decline in pollinators include land-use intensification, climate change, alien species, and pests and pathogens. Land-use intensification includes large-scale factory farming which destroys or fragments natural habitats. In addition, systemic pesticides such as neonicotinoids that spread throughout the plant tissue, including nectar and pollen, have been shown to confuse bees by impairing their brain function and reducing their foraging performance. The adoption of genetically modified crops resistant to weedicides has had the side-effect of wiping out weeds that are pollinator habitat and used to grow between crop rows and on buffer strips between farms. Mortality has also increased owing to viruses spread among bees by mites and other factors [26].

2.11   Biodiversity

The diversity of life on Earth is essential for maintaining its resilience and its ability to provide goods and services. The planet has experienced five mass extinctions so far, and according to biologists we are in the middle of a sixth such event. The previous such event was the disappearance of dinosaurs 65 million years ago. The current rate of species extinctions is estimated to be up to 1000 times greater than the natural rate, and the rate of increase is summarized in Figure 2.25. Some species that have become extinct in the last 100 years or are critically endangered are shown in Figure 2.26, and the story of one extinct species, the passenger pigeon, is described in Box 2.3. In addition to species extinction, population extinction is also taking place across the planet for common and endangered species. It is estimated that half the animal individuals that once lived on the planet are gone. Biologists have also

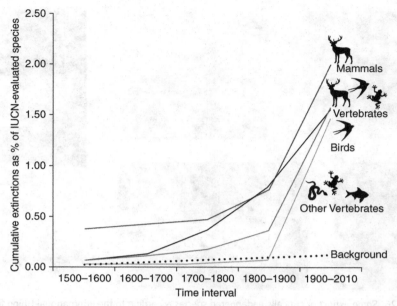

Figure 2.25 Cumulative vertebrate species considered extinct or extinct in the wild. Reproduced with permission from [27].

found that most of the large specimens in the oceans have been fished out. Such defaunation or "biological annihilation" is considered to be a precursor to species extinction and loss of the services that ecosystems can provide for human well-being.

BOX 2.3 Extinction of the Passenger Pigeon

The last passenger pigeon died in Cincinnati Zoo around September 1, 1914. Her species, shown in Figure 2.26, is estimated to have had a peak population of five billion across North America only about 60 years before her death. They lived in huge flocks east of the Rockies and were said to cover the sun when they flew. However, in about a generation there were none left. The reason for their extinction was hunting and loss of habitat. Their social behavior of flying in huge flocks, which worked well against predators, made them vulnerable to hunters. People hunted them for food and sport in huge numbers. Many also considered them to be agricultural pests.

The professionals and amateurs together outflocked their quarry with brute force. They shot the pigeons and trapped them with nets, torched their roosts, and asphyxiated them with burning sulfur. They attacked the birds with rakes, pitchforks, and potatoes. They poisoned them with whiskey-soaked corn. [28]

Figure 2.26 Some extinct or critically endangered species according to the International Union for Conservation of Nature. Clockwise from top left: (a) thylacine or Tasmanian tiger, extinct; (b) passenger pigeon, extinct (by James St. John (Wikimedia Commons CC BY 2.0)); (c) scimitar horned orynx, extinct in the wild (Wikimedia Commons CC BY-SA 3.0); (d) Florida nutmeg tree, critically endangered (Wikimedia Commons CC BY-SA 3.0); (e) thorny spikethumb frog, critically endangered (by Josiah H. Townsend (Wikimedia Commons CC BY-SA 3.0)); (f) Siamese tiger perch, critically endangered (courtesy of the USDA-NRCS PLANTS Database).

2.12   Overall Status

The Millennium Ecosystem Assessment [29] assessed the status of ecosystem goods and services and their role in human well-being. It found that among 24 ecosystem services, 15 are degraded, 5 are mixed, and only 4 have been enhanced. The key messages and conclusions from this study are listed in Box 2.4.

The results of another study [30], which focused on determining a "safe operating space" for humanity and where we are with respect to this space, focused on several anthropogenic impacts, whose results are summarized in Figure 2.27. This study indicates that several planetary boundaries are being exceeded. For example, the biodiversity loss is significantly above the safe limit, as is the disruption of the nitrogen and phosphorus cycles. These systems are in a zone of high risk. Climate change and land system change are in a zone of uncertainty or increasing risk.

The impact of human activities on the planet is already so large that many Earth scientists consider the present to belong to a new geological epoch. Epochs are periods in the Earth's history whose distinct signature can be detected from rock formations, ice cores, and tree rings. The Holocene epoch began about 12,000 years ago at the end of the ice age. Over the last few centuries, the planetary impact of human activities has become so dominant that it is leaving a unique signature that will be seen in geological records, just as the end of the ice age can be seen as a threshold change. This new epoch is being called the Anthropocene, or the "age of man" [31]. Humanity has certainly never experienced such an epoch, which means that we are most likely in "uncharted waters." Thus, the Earth system may behave in unpredictable ways that are very different from what we have come to expect and know over the last few millennia.

BOX 2.4 Statement of the Millennium Ecosystem Assessment Board [29]

The statement from the Board identifies 10 key messages and conclusions that can be drawn from the assessment:

- Everyone in the world depends on nature and ecosystem services to provide the conditions for a decent, healthy, and secure life.
- Humans have made unprecedented changes to ecosystems in recent decades to meet growing demands for food, fresh water, fiber, and energy.
- These changes have helped to improve the lives of billions, but at the same time they weakened nature's ability to deliver other key services such as purification of air and water, protection from disasters, and the provision of medicines.
- Among the outstanding problems identified by this assessment are: the dire state of many of the world's fish stocks; the intense vulnerability of the 2 billion people living in dry regions to the loss of ecosystem services, including water supply; and the growing threat to ecosystems from climate change and nutrient pollution.
- Human activities have taken the planet to the edge of a massive wave of species extinctions, further threatening our own well-being.
- The loss of services derived from ecosystems is a significant barrier to the achievement of the Millennium Development Goals to reduce poverty, hunger, and disease.
- The pressures on ecosystems will increase globally in coming decades unless human attitudes and actions change.
- Measures to conserve natural resources are more likely to succeed if local communities are given ownership of them, share the benefits, and are involved in decisions.
- Even today's technology and knowledge can reduce considerably the human impact on ecosystems. They are unlikely to be deployed fully, however, until ecosystem services cease to be perceived as free and limitless, and their full value is taken into account.

- Better protection of natural assets will require coordinated efforts across all sections of governments, businesses, and international institutions. The productivity of ecosystems depends on policy choices on investment, trade, subsidy, taxation, and regulation, among others.

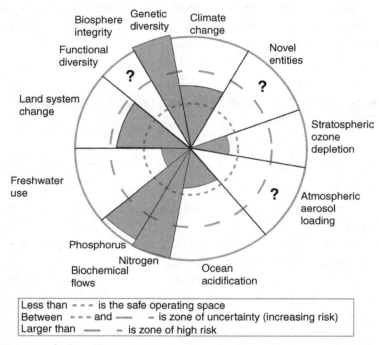

| Less than --- is the safe operating space |
| Between --- and ⸺ - is zone of uncertainty (increasing risk) |
| Larger than ⸺ - is zone of high risk |

Figure 2.27 Status of selected human impacts on the environment with respect to the likely safe operating zone. Adapted from [30]

2.13   Summary

In this chapter we learned about the status of goods and services that nature provides. We considered goods such as water, fuels, and materials, and services such as the regulation of air and water quality and climate, pollination, and net primary productivity. The overall conclusion is that a large number of essential ecosystem services are degraded or declining. The primary reason behind this trend is the pressure imposed by human activities. We learned in Chapter 1 that goods and services from nature form the foundation of human well-being. Therefore, their deterioration raises doubts about the sustainability of human activities and the well-being of current and future generations.

Key Ideas and Concepts

- Peak oil
- Ozone hole
- Eutrophication
- Planetary boundaries
- Water stress
- Atmospheric CO_2 concentration
- Primary productivity
- Anthropocene

2.14 Review Questions

1. Explain the concept of peak oil. Why is this concept relevant to all nonrenewable resources?
2. What is the ratio of anthropogenic to natural mobilization of the following elements: (a) gold, (b) titanium, (c) chlorine?
3. Discuss the current atmospheric concentration of carbon dioxide, methane, and nitrous oxide with respect to their concentration in the last 600,000 years.
4. Define net primary productivity. What fraction of NPP is being appropriated by human beings?
5. What is the Anthropocene?

Problems and Applications

2.1 Read the article available at `https://nyti.ms/1xjUWbc` and answer the following questions.
  1. Is cod a renewable or nonrenewable resource? Justify your answer.
  2. Sketch the approximate catch of cod over time. Does the resource peak hypothesis apply to the cod harvest? What has been the role of technology in the presence of multiple peaks?
  3. Do you think global oil production will follow a similar trend to that of the cod harvest? Why or why not?
2.2 Determine the status of wetlands in the world, with emphasis on the state or region where you live. What are the services that wetlands provide?
2.3 Historical information about the Great Black Swamp of Ohio is available on various websites. Using this information, answer the following questions.
  1. Why was it drained?
  2. What were the costs and benefits of draining this swamp?
  3. Could there be a connection between the loss of ecosystem services from this swamp and the algal blooms in Lake Erie?
2.4 One of the statements from the Millennium Ecosystem Assessment (see Box 2.4) is as follows: "Even today's technology and knowledge can

reduce considerably the human impact on ecosystems. They are unlikely to be deployed fully, however, until ecosystem services cease to be perceived as free and limitless, and their full value is taken into account." Why does this statement imply that ecosystem services are perceived as free and limitless? Give examples in support or against this statement.

2.5 As shown in Figure 2.18, the atmospheric concentrations of methane and nitrous oxide have increased in the last several decades. Determine and discuss some of the reasons for this increase.

2.6 Describe the reasoning behind the theory of peak oil. From consideration of Figure 2.2, has US oil production peaked? Explain the reasons behind the recent increase in US oil production. Is this increase likely to peak?

2.7 From the measured values of atmospheric CO_2 concentration in Figure 2.19, determine whether the rate of CO_2 accumulation in the atmosphere is increasing or decreasing. Why does this trend depict oscillatory behavior?

2.8 Using Figure 2.12 and [13], find a large river system close to your location. Determine the extent to which it has been impacted by the construction of dams. What have been some of the benefits and costs of such development?

2.9 Think about fruits, vegetables, and grains that you consume during a typical day. Determine which of these items are most vulnerable to (a) water stress, and (b) the decline of pollinators. If the price of these vulnerable food products increases, are there substitutes that you and your community could rely on?

2.10 Search the "Red List of Threatened Species" compiled by the International Union for Conservation of Nature (IUCN) and find species on this list that are in the region where you live. This list may be found by searching the IUCN's website.

2.11 According to the Energy Information Administration, CO_2 emissions in the USA in 2007 were 6.0218 billion metric tons (Gt). Calculate the land area that needs to be planted with trees to sequester this emission. Ohio forests sequester carbon at a rate of 130 metric tons/hectare per 100 years. This corresponds to 400 trees/hectare. How does this area compare with the area of the USA? What does this say about the sustainability of US CO_2 emissions?

2.12 The consumption of fossil fuels in 2005 was 7.5 Gt carbon. It has been estimated that 7000 g of ancient biomass was needed to form 1 g of fossil fuel [32]. The current net primary productivity of the Earth is estimated to be 105 Gt carbon per year. This is the average amount of biomass carbon that grows on the Earth per year. The current terrestrial phytomass (plant matter) stores on Earth are estimated to be 500–600 Gt carbon. Using this data, answer the following questions.

1. Calculate the amount of ancient biomass needed to produce the fuel consumed in 2005.
2. How many years would it take to produce this amount of biomass on the basis of the current net primary productivity?
3. How does the quantity calculated in item 1 compare with the current terrestrial phytomass stores?
4. What does this imply about the possibility of replacing our current rate of fuel use by biofuels?

References

[1] E. O. Wilson. *The Biodiversity of Life*. Harvard University Press, 1992.

[2] BP. BP statistical review of world energy. http://bp.com/ statisticalreview, 2012, accessed November 30, 2013.

[3] US Energy Information Administration. Petroleum and other liquids https:// www.eia.gov/petroleum/data.php, accessed January 9, 2019.

[4] US Energy Information Administration. International energy statistics. www.eia .gov/cfapps/ipdbproject/IEDIndex3.cfm, accessed January 9, 2019.

[5] M. King Hubbert. *Energy Resources*. National Research Council, 1962.

[6] J. R. Burger, C. D. Allen, J. H. Brown, W. R. Burnside, A. D. Davidson, T. S. Fristoe, M. J. Hamilton, N. Mercado-Silva, J. C. Nekola, J. G. Okie, and W. Zuo. The macroecology of sustainability. *PLoS Biology*, 10(6):e1001345, 2012.

[7] I. S. Sen and B. Peucker-Ehrenbrink. Anthropogenic disturbance of element cycles at the earth's surface. *Environmental Science & Technology*, 46(16):8601–8609, 2012.

[8] M. J. Molina and F. S. Rowland. Stratospheric sink for chlorofluoromethanes: chlorine atom-catalysed destruction of ozone. *Nature*, 249:810–812, 1974.

[9] National Aeronautics and Space Agency. Ozone hole watch. http:// ozonewatch.gsfc.nasa.gov, accessed January 9, 2019.

[10] P. Doll, K. Fiedler, and J. Zhang. Global-scale analysis of river flow alterations due to water withdrawals and reservoirs. *Hydrology and Earth Systems Science*, 13:2413–2432, 2009.

[11] UNEP/GRID-Arendal Maps and Graphics Library. Increased global water stress. http://grida.no/resources/5625, Phillippe Rekacewiscz, 2009.

[12] World Resources Institute. Portion of agricultural production under high or extremely high stress, October 2013.

[13] C. Nilsson, C. A. Reidy, M. Dynesius, and C. Revenga. Fragmentation and flow regulation of the world's large river systems. *Science*, 308(5720):405–408, 2005.

[14] Food and Agriculture Organization. FAOSTAT. `http://faostat3.fao.org`, accessed July 8, 2017.

[15] FAO. The state of world fisheries and aquaculture 2016, contributing to food security and nutrition for all. `www.fao.org/3/a-i5555e.pdf`, 2016, accessed November 22, 2018.

[16] R. LeB. Hooke. On the history of humans as geomorphic agents. *Geology*, 28(9):843–846, 2000.

[17] J. R. McNeill. *Something New Under the Sun*. Norton, 2000.

[18] S. J. Smith, J. van Aardenne, Z. Klimont, R. J. Andres, A. Volke, and S. Delgado Arias. Anthropogenic sulfur dioxide emissions: 1850-2005. *Atmospheric Chemistry and Physics*, 11:1101–1116, 2011.

[19] R. K. Pachauri and A. Reisinger, editors. *Climate Change 2007: Synthesis Report*. IPCC, 2007.

[20] Scripps Institution of Oceanography. The keeling curve. `http://keelingcurve.ucsd.edu`, accessed May 31, 2018.

[21] P. W. Leadley, C. B. Krug, R. Alkemade, et al. Progress towards the Aichi biodiversity targets: an assessment of biodiversity trends, policy scenarios and key actions. Secretariat of the Convention on Biological Diversity. `www.cbd.int/doc/publications/cbd-ts-78-en.pdf`, 2014, accessed November 22, 2018.

[22] NASA Earth Observatory. Aquatic dead zones. `https://earthobservatory.nasa.gov/images/44677/aquatic-dead-zones`, January 1, 2008 accessed January 9, 2019.

[23] F. Krausmann, K.-H. Erb, S. Gingrich, et al. Global human appropriation of net primary production doubled in the 20th century. *Proceedings of the National Academy of Sciences*, 110(25):10324–10329, 2013.

[24] S. G. Potts, J. C. Biesmeijer, C. Kremen, et al. Global pollinator declines: trends, impacts and drivers. *Trends in Ecology & Evolution*, 25(6):345–353, 2010.

[25] IPBES. The Assessment Report of the Intergovernmental Science-Policy Platform on Biodiversity and Ecosystem Services on Pollinators, Pollination and Food Production. Secretariat of the Intergovernmental Science-Policy Platform on Biodiversity and Ecosystem Services, 2016.

[26] A. J. Vanbergen and the Insect Pollinators Initiative. Threats to an ecosystem service: pressures on pollinators. *Frontiers in Ecology and the Environment*, 11(5):251–259, 2013.

[27] G. Ceballos, P. R. Ehrlich, A. D. Barnosky, et al. Accelerated modern human-induced species losses: entering the sixth mass extinction. *Science Advances*, 1(5), 2015.

[28] B. Yeoman. Why the passenger pigeon went extinct. *Audubon*, May–June 2014.

[29] Millennium Ecosystem Assessment. *Living Beyond our Means: Natural Assets and Human Well-Being*. Island Press, 2005.

[30] W. Steffen, K. Richardson, J. Rockström, et al. Planetary boundaries: guiding human development on a changing planet. *Science*, 347(6223), 2015.

[31] P. J. Crutzen and E. F. Stoermer. The "anthropocene." *Global Change Newsletter*, 41:17–18, 2000.

[32] J. S. Dukes. Burning buried sunshine: human consumption of ancient solar energy. *Climatic Change*, 61(1–2):31–44, 2003.

3 Sustainability: Definitions and Challenges

> We are coming of age on a finite planet and only just now recognizing that it is finite. So how we manage infinite aspirations of a species that's been on this explosive trajectory, not just of population growth but of consumptive appetite – how can we make a transition to a stabilized and still prosperous relationship with the Earth and each other – is the story of our time.
>
> Andrew Revkin [1]

In Chapter 1 we learned about the tremendous success of our species in enhancing our numbers and well-being, while Chapter 2 focused on the impact of this enhancement on the degradation of the ecosystem goods and services that form the foundation of our well-being. This dichotomy, conveyed in Figure 3.1, raises doubts about whether human well-being can be maintained and enhanced by the approaches we have adopted over the last few centuries. Future growth in population and consumption are likely to further increase the pressure on ecosystems through greater demand for energy, materials, water, and food. The global population is predicted to grow, as shown in Figure 3.2 [2]. According to the middle scenario, the population in 2025 will be about one billion more than now, reaching 8.1 billion. It will increase to 9.6 billion by 2050 and 10.9 billion by 2100. In addition, the effects of climate change are likely to get worse in many regions over time. These trends indicate that without significant changes in human activities, the degradation of ecosystems is also likely to continue, and may further jeopardize human well-being.

In the last few decades, as global trends in ecological deterioration have become more apparent, concerns about the sustainability of human activities have been increasing. Today, most governments, corporations, academic institutions, and disciplines are interested in sustainable development, and are attempting to understand and address its needs. In this chapter we will focus on understanding the nature of environmental problems and the meaning of sustainable development. This will yield insight into the challenges in determining the sustainability of any product or process, which will be useful for developing pragmatic approaches for ensuring that engineering and other human activities contribute positively to sustainable development. We will learn about these approaches in subsequent chapters. We will see that sustainable development is unlike most engineering problems and does not fit neatly into any existing discipline. It poses challenges that engineering and other disciplines have rarely encountered.

Figure 3.1 Cartoon in the *Washington Post* on March 29, 2005 after the Millennium Ecosystem Assessment was published. Image courtesy of Andrews McMeel Syndication.

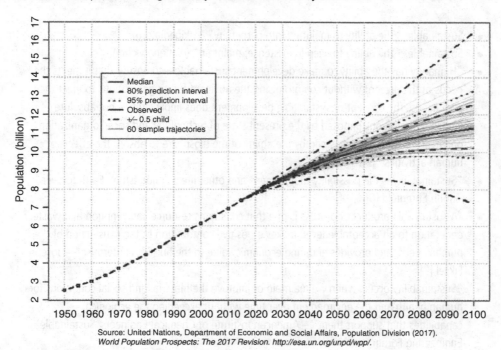

Source: United Nations, Department of Economic and Social Affairs, Population Division (2017).
World Population Prospects: The 2017 Revision. http://esa.un.org/unpd/wpp/.

Figure 3.2 Projected global population [2].

3.1   Definitions

The classic definition of sustainable development comes from a small book published in 1987 by the United Nations World Commission on Environment and Development, commonly called the Brundtland Commission [3]. According to this definition, sustainable development ensures that humanity,

> meets the needs of the present without compromising the ability of future generations to meet their own needs.

With increasing realization of the need to pursue sustainability, much has been written about this concept in diverse disciplines. Several typical definitions and statements related to sustainable development are collected in Box 3.1, including the full version of the quote given above. The quotes from Jefferson [4] and Gandhi [5] indicate the relevance of the concepts underlying sustainable development even at that time.

BOX 3.1 Definitions and Statements Relevant to Sustainable Development.

- Sustainable: "The ability to maintain into perpetuity" (wikipedia.org).
- Sustainable: "The ability to keep in existence; maintain" (dictionary.com).
- "Humanity has the ability to make development sustainable – to ensure that it meets the needs of the present without compromising the ability of future generations to meet their own needs. The concept of sustainable development does imply limits – not absolute limits but limitations imposed by the present state of technology and social organization on environmental resources and by the ability of the biosphere to absorb the effects of human activities" [3].
- "Sustainability is the possibility that humans and other life will flourish on Earth forever." – John Ehrenfeld [6].
- "A sustainable product or process is one that constrains resource consumption and waste generation to an acceptable level, makes a positive contribution to the satisfaction of human needs, and provides enduring economic value to the business enterprise" – Joseph Fiksel [7].
- "Sustainability occurs when we maintain or improve the material and social conditions for human health and the environment over time without exceeding the ecological capabilities that support them" – American Institute of Chemical Engineers, Sustainable Engineering Forum.
- "Then I say the earth belongs to each of these generations during its course, fully, and in their own right. The second generation receives it clear of the debts and incumbrances of

the first, the third of the second, and so on. For if the first could charge it with a debt, then the earth would belong to the dead and not to the living generation. Then, no generation can contract debts greater than may be paid during the course of its own existence" – Thomas Jefferson [4].

- "I suggest that we are thieves in a way. If I take anything that I do not need for my own immediate use, and keep it, I thieve it from somebody else. I venture to suggest that it is the fundamental law of Nature, without exception, that Nature produces enough for our wants from day to day, and if only everybody took enough for himself and nothing more, there would be no pauperism in this world, there would be no man dying of starvation in this world. But so long as we have got this inequality, [so long] we are thieving" – Mahatma Gandhi [5].

Careful reading of these quotes helps to identify the following common characteristics of these definitions.

- Sustainability is anthropocentric: its focus is primarily on human well-being.
- Sustainable development is about the present and the future: it emphasizes the short- and long-term aspects of satisfying human needs.
- Nature plays an essential role in sustaining human activities.
- Limits on the sustainability of human activities are imposed by the ability of ecosystems to provide resources and absorb impacts.

Despite this plethora of definitions, developing criteria by which the sustainability of any activity or product can be quantified with scientific assurance has proved to be extremely difficult. For example, the original definition given above by the Brundtland Commission [3] requires knowledge about the needs of future generations, which are difficult to determine. Similarly, other definitions require information about the acceptable levels of resource use and waste generation, and the ecological capabilities for supporting human activities, which are unknown or difficult to know. Is there something special and challenging about defining sustainability, and if so, are there practical approaches that could be devised to guide decisions toward the goal of sustainable development? These questions are considered in the rest of this chapter.

3.2 Nature of Environmental Problems

Given the important role of the environment in sustaining our well-being, we need to understand the nature of environmental problems so that we can devise appropriate approaches for solving them. In this section, we consider the history of some previous efforts for addressing environmental challenges.

Over the last few centuries, societies have become increasingly concerned about the environmental impact of human activities, and what this impact may hold for us in the future. Among the earliest such concerns was the prediction of food shortages by Rev. Thomas Malthus in 1798, based on his insight that population increases exponentially while crop production, by bringing more land into cultivation, increases linearly [8]. Similar predictions were made in the 1960s in the work of Ehrlich [9] and the Limits to Growth study [10]. These predictions even resulted in a famous bet between the ecologist Paul Ehrlich and the economist Julian Simon, which that is described in Box 3.2.

BOX 3.2 The Ehrlich–Simon Bet

Publication of the study on limits to growth [10] led to many warnings about the potential doom of civilization. Prominent among these "doomsayers" or "alarmists" was Paul Ehrlich of Stanford University and author of the best-seller *The Population Bomb* [9]. In this book and his other writings, Ehrlich worried that the increase in population would result in severe resource scarcities. Such dire predictions were countered by others, including the economist Julian Simon of the University of Maryland. Simon and other "doomslayers" or "cornucopians" argued that such scarcities will not occur, because of market forces and the development and availability of alternatives. They claimed that humanity would not face any scarcities. Thus, Ehrlich had a physical and biological perspective which indicated that consuming resources at a rate faster than their replenishment will result in scarcities. In contrast, Simon had an economic and behavioral perspective which indicated that scarcities are reflected in prices to which people adapt and respond by developing alternatives.

In 1980, Ehrlich, along with physics professors John Harte and John Holdren, wagered with Simon that the price of a basket of metals (chromium, copper, nickel, tin, tungsten) would increase between 1980 and 1990, indicating that these metals became more scarce. The bet was for $200 per metal. During this period, prices for three of the five metals – tin, copper, and tungsten – decreased, so Ehrlich, Harte, and Holdren lost the wager.

After winning the bet, Simon wanted to wager again on the basis of his claim that any trend related to human well-being will continue to improve. Ehrlich and Steven Schneider, a climatologist at Stanford, offered a wager based on environmental trends about climate change, water, SO_2 emissions in Asia, food production per capita, harvest of fisheries, etc. Simon refused to accept this bet, saying that he would only wager on direct measures of well-being, such as life expectancy.

As we saw in Chapter 2, human well-being has indeed improved over the decades, but ecosystems have also deteriorated. If Ehrlich and Simon had bet on environmental trends, Ehrlich would have won the bet. In fact, Ehrlich could have even won the original bet if the time period had been different, since metal prices tend to be quite volatile. However, what this bet indicates is that extrapolating environmental trends without considering human behavior and socioeconomic effects can be misleading. For resources that are included in the marketplace, the free market is a powerful mechanism to deal with potential scarcities. However, for resources that are outside the market, there are no prices to signal scarcities, so, in the absence of regulation by society or the government, significant deterioration is likely. Examples of such deterioration that we saw in Chapter 2 are for ecosystem services such as climate regulation, pollination, and water quality regulation. This bet also shows the tendency of ecologists to underestimate human adaptability and ability to find substitutes, and the tendency of economists to ignore biophysical limits and assume more substitutability than can be sustained.

Many of these projections have been at least partially correct, such as those of population growth and pollution, but the doomsday scenario of global food shortages that attracted the most attention has not materialized, mainly due to technological advances such as agricultural intensification by the development of high-yield crop varieties, and the use of artificial fertilizers and pesticides. Such human ingenuity has played and continues to play a crucial role in overcoming many environmental challenges. Some examples include the following:

- As wood became scarce in the UK in the 1700s due to its conversion to charcoal to obtain the high temperature required for making steel, it was replaced by coal. As domestic coal was depleted, it was replaced by imported coal. Ultimately, much of the steel industry moved to other parts of the world and the UK now relies on global trade to meet its demand for steel.
- As the easily accessible crude oil deposits have been depleted and demand has continued to grow, new technologies have been developed for more efficient oil extraction. Furthermore, oil exploration has been moving to increasingly remote and inhospitable locations, such as deep under the waters of the Gulf of Mexico and the Arctic. New technologies such as horizontal drilling, enhanced oil recovery, and hydraulic fracturing (fracking) have result in renewed production even from regions considered to be depleted, by unlocking previously inaccessible deposits such as those of shale gas and oil, as we saw in Figure 2.2.
- As urbanization continues and demands more fresh water, rivers are diverted and dams are built to satisfy this need. As discussed in Section 2.3, many of the major rivers have greatly reduced flows into the oceans, but human demand for fresh water is mostly satisfied.

Thus, human adaptation and ingenuity in exploiting new resources and developing more efficient technologies has been able to overcome challenges posed by ecological degradation and resource depletion. Science and engineering have played a critical role in such developments, so it is natural to wonder if, given the huge global spending on science and engineering, we will not be able to develop new technologies to overcome any problems we face in the future? Isn't technology development adequate for achieving sustainability? No one knows what the future holds, but the human experience of the last several centuries and scientific knowledge provide useful insight into this question. Let us now consider some experiences of technological efforts to address environmental challenges.

3.2.1   Energy-Efficient Lighting

The use of artificial lighting plays an essential role in our well-being. Over the last few centuries, many innovations have resulted in increasingly efficient technology for providing artificial light. These innovations include candles, kerosene lamps, gas lamps, incandescent bulbs, compact fluorescent bulbs, and light-emitting diode (LED) bulbs. The improvement in efficiency of these technologies is indeed remarkable. For example, the luminosity of a 100 W incandescent bulb can be provided by a 30 W compact fluorescent lamp or a 15 W LED bulb. Scientists and engineers have played an essential role in these efficiency improvements. They usually assume that higher efficiency is better for the environment since consuming less energy for accomplishing the same task will result in greater energy saving and a smaller environmental impact. As discussed below, for lighting technologies, these expectations have been incorrect.

The consumption of artificial light in the UK over the last three centuries is shown in Figure 3.3. We can see that, over time, various technologies such as candles, gas, and kerosene have been adopted and have then given way to newer alternatives. However, total lighting use, shown by the thick black line, has continued to increase. The total consumption in Figure 3.3 may seem to be flattening in recent years, but this figure does not include the recent increase due to the adoption of LED bulbs.

Trends in lighting efficiency, per capita use, and total energy use are shown in Figure 3.4. Note that the y-axis in these figures has a logarithmic scale. The figure depicts an exponential decrease in cost per lumen along with an exponential increase in all other trends: per capita light consumption, per capita luminous energy consumption, luminous efficacy, and total energy use for lighting. Thus, despite improvements in luminous efficacy due to technological breakthroughs, the energy consumption for lighting has only increased. Today, artificial lighting consumes 0.72 percent of global gross domestic product but uses 6.5 percent of global energy. This economic rebound is also called the Jevons paradox, owing to the English economist's observation in 1865 that technological advances that improved the efficiency of coal use encouraged more use of coal.

This experience with lighting shows that even though technology may become more efficient and have a smaller impact per unit, the overall impact may not

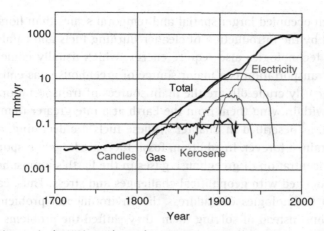

Figure 3.3 Lighting use in the UK in teralumen-hours per year over three centuries. Adapted from [11].

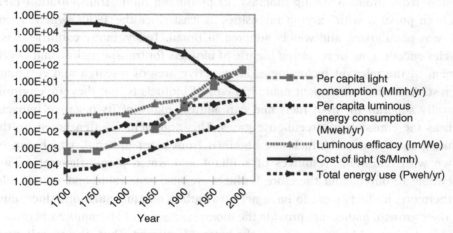

Figure 3.4 Trends associated with lighting in the UK.

decrease. In fact, greater efficiency per unit often encourages consumption since the more efficient technology usually costs less per unit of service delivered. Thus, environmental impact can shift from the domain of engineering to the domain of behavior and economics, since the benefits of engineering innovation can be negated by human behavior and the economics of increasing demand with lower cost.

3.2.2 Sustainable Transportation

From horses to horseless buggies. About 100 years ago, residents of New York and many other cities complained about horse dung in the streets. This was a local and short-lived environmental problem since the dung would decompose in a few days. It was solved by the introduction of the "horseless buggy" or automobile. However, as cars became more popular, they created their own environmental problem of smog, which occupied larger spatial and temporal scales than horse dung. This was addressed by the introduction of cleaner-burning fuels and catalytic converters, but these technologies also require energy, which usually comes from fossil fuels. Today, transportation is a major source of greenhouse gas emissions. Fossil resources, primarily crude oil, are the main sources of transportation fuels today. Since we are withdrawing them from the Earth at a rate greater than their rate of replenishment, as described in Chapter 2, these fuels are declining, and their use cannot be sustained forever. In addition, fossil fuels are largely responsible for the increasing concentration of greenhouse gases in the Earth's atmosphere, and they tend to be associated with geopolitical challenges and stress. Thus, cars and other transportation technologies did address the environmental problems they were designed for, but instead of solving them, they shifted the problems elsewhere in space and time.

Fuels from biomass. Using biomass for producing liquid transportation fuels has been pursued with varying intensities at least since the 1970s, when "gasohol" was popularized and widely adopted in Brazil. Today, many countries have policies encouraging or requiring the use of biofuels for transportation, and development of the relevant technologies is an active area of research and corporate interest. A common argument made in favor of biofuels is that they are environmentally superior to fossil fuels and can address many of its negative impacts, such as the emission of greenhouse gases. This argument is often based on the thinking that because biofuels are obtained from plant material, which can be grown with renewable resources of sunlight and water, plants themselves, are renewable resources and therefore biofuels are also renewable and sustainable. Furthermore, biofuels seem to have net zero carbon dioxide emissions since, during their growth, plants that provide the biomass raw materials capture the carbon dioxide that would be emitted when the biofuel is burned. Thus, the overall cycle seems to have zero net emissions of CO_2. These arguments were made in favor of policies to promote biodiesel in Europe and ethanol in the USA in the 1990s. The outcome of these policies, described below, provides a glimpse into the complex nature of environmental challenges.

- *Shifting of impacts along the supply chain.* Even though biomass itself is renewable, the resources needed for growing it by intensive agriculture are not. Crops such as corn or sugarcane for ethanol and soybean for biodiesel, which were most common in the 1990s, are usually grown by intensive industrial farming, which requires nonrenewable fossil resources for energy, pesticides, and fertilizers. As a result, at the life cycle scale, corn ethanol or soybean biodiesel are far from being fully renewable fuels. In addition, reliance on nonrenewable fuels results in the emission of CO_2, and the use of artificial fertilizers causes the emission of nitrous oxide, another greenhouse gas. There-fore, the net emission along the production and use networks of biofuels is positive. Depending on the nature of the farming practice, emissions from the life cycle of biofuels may be less than those from the life cycle of fossil-based fuels, but they are certainly not zero for fuels like corn ethanol and soybean biodiesel, which rely on intensive farming. Thus, if we consider processes from the life cycle of these fuels, we find that impressions about their complete renewability and carbon neutrality are incorrect since the use of nonrenew-able resources and CO_2 emissions simply shifts to other processes in the supply chain.

- *Shifting of impacts to other geographic regions.* Given the global market for agricultural raw materials, policies of encouraging biofuels provided a global incentive for planting relevant crops. Farmers in the USA planted as much corn as they could, including on land that was previously set aside for conservation. Farmers in Brazil also had incentives to increase land area under sugarcane (for ethanol) and soybean (for biodiesel). This caused them to push farther north into the Amazon rain forest, causing significant deforestation. Similarly, the increasing demand for palm oil to make biodiesel accelerated deforestation in Indonesia by converting rain forests into palm oil plantations. Consequently, carbon dioxide emissions from these land use changes may even have greatly exceeded any benefit from reducing the consumption of fossil fuels. Thus, the impact of policies to reduce greenhouse gas emissions ended up increasing emissions elsewhere, resulting in a perverse outcome since policies meant to be environmentally friendly had the opposite effect.

- *Shifting of impacts across disciplines.* Greater use of corn-based ethanol increased the demand for corn, causing a jump in prices of not just corn but also products made from corn, such as tortillas, meat, and various processed foods. This food versus fuel conflict caused hardship to poorer populations, mainly in Central and South America, for whom corn is a staple food. Thus, a technological solution, in addition to environmental impacts, also had economic and societal repercussions that could not be sustained.

- *Shifting of impacts between types of flows.* One of the main motivations for adopting biofuels was the expectation that carbon dioxide emissions would decrease. The environmental impact involves many different emissions and systems, and a narrow focus on a single type of emission or system is usually inadequate. For biofuels, even if a reduction in greenhouse gas emissions was possible, these products required more land area and water and caused greater disruption of the nitrogen cycle, owing to the use of artificial fertilizers, than impacts in these categories by fossil fuels. Thus, any reduction in CO_2 emis-sions also resulted in an increase in the emission of nitrates and greater use of water and land resources.

- *Ignoring supply and demand of ecosystem goods and services.* Biofuel policies were directed mainly toward technological enhancement and innovation, and little attention was paid to the demand placed by biofuels on ecosystem services and the capacity of ecosystems to supply these services. Thus, the effects of biofuel policies on the loss of carbon stored in the soil and forests due to deforestation and the conversion of land to farming were not considered. Farming practices that enhance the capacity of soil to sequester carbon and ways of preventing fertilizer runoff from exceeding the capacity of local bodies of water to absorb their impact also received little attention.

This experience of unintended consequences from well-intentioned transportation technologies and policies highlights the challenges in developing sustainable solutions. Many other efforts to reduce environmental impact have also resulted in unintended consequences. These include the development and use of refrigerants, as described in Box 2.1, and the use of artificial fertilizers, which we discussed in Section 2.8. These examples, along with the history of environmental and resource problems and their solutions, demonstrate that efforts to solve environmental problems often cause the problem to shift not only to larger temporal and spatial scales, as depicted in Figure 3.5, but also across disciplines, as illustrated for lighting, and across different types of physical flows. Therefore, approaches for solving environmental problems need to be capable of addressing such challenges.

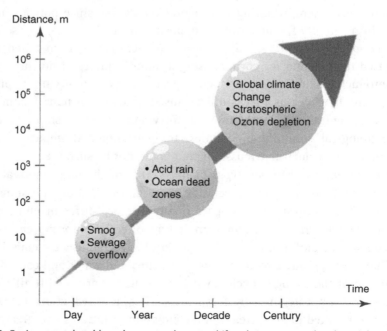

Figure 3.5 Environmental problems have a tendency to shift to larger temporal and spatial scales.

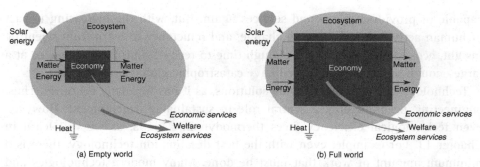

Figure 3.6 Thinking in most areas of human activities was developed when we lived in an "empty world" and nature seemed infinite. It needs to adapt to the "full world" we live in today. Adapted from [12].

3.3 Nature of the Sustainability Challenge

3.3.1 Need for Sustainable Engineering

The engineering we practice today was developed over the last two to three centuries, and continues to rely on some assumptions that were easier to justify at that time. As depicted in Figure 3.6a, the world of a few centuries ago, or even a few decades ago, was relatively "empty." In this empty world, human activities were a relatively small fraction of activities on the planet, and the biosphere occupied most of the planet. Thus, as shown in the figure, the production of economic goods and services was relatively small, while that of ecosystem goods and services was quite large. Under these conditions, goods and services from ecosystems not only seemed to be unlimited, but controlling and dominating nature seemed essential for human survival and prosperity. Maybe it is not surprising that thermodynamics assumed an "infinite sink" and economics considered ecosystems to provide "public goods," which are goods that do not deplete regardless of use and are freely available to everyone.

Unfortunately, these assumptions, which may have been reasonable in the "empty world," are not reasonable in the "full world," that we inhabit now. As depicted in Figure 3.6b, in the world of today, the footprint of human activities is much larger than in the past. The production of economic goods and services has grown significantly, while that of ecosystem goods and services has declined. The biosphere is under much more stress and is already encountering some fundamental limits. This is indicated by the degradation of ecosystems, which are exceeding their capacity at local, regional, and global scales, as we learned in Chapter 2. Ocean dead zones and the loss of stratospheric ozone are indeed examples of ecological collapse, and societies have been bearing the consequences and adapting with the help of science, technology, governance, migration, and global trade. Given enough time, degraded ecosystems are often able to recover and become capable of providing goods and services again. But, with the increasing impact of human activities, resistance to change, and reluctance to learn from scientific insight, ecosystems may not get enough time to recover and, if this happens at a large enough scale, it could indeed have catastrophic global consequences.

Technology can certainly provide solutions, as it has for many centuries. Thus, engineering needs to play a central role in sustainable development. However, even technological efficiency faces thermodynamic limits, as we will learn in Chapter 13. For example, even with the best desalination technology, there is a minimum amount of work that must be done. Many modern technologies and manufacturing processes are nearing this limit, leaving little room for further improvement. In addition, the use and impact of technology depend on the number and behavior of users and the per capita consumption of the technology.

So far, predictions of doom have spurred the development of new methods to avoid many of the predicted impacts. Thus, these predictions have themselves played a role in their being false. However, it is also clear that if we had maintained "business as usual" and ignored the indicators of environmental impact, we would increase human misery and even the likelihood of collapse. Just consider scenarios in which ozone-depleting chemicals are not constrained, and urban air pollution is not controlled. Thus, it is essential that we learn from past mistakes and actively protect and restore ecosystems, since they are the basis of our existence. Meeting such challenges requires engineering and ingenuity that truly solve problems as opposed to just shifting them to another domain, as illustrated in Section 3.2. It also requires deep insight into the dependence and impact of human activities on nature, and approaches that create synergistic harmonies between natural and human systems. Clearly, such engineering will be quite different from engineering of the past. It needs to adopt Einstein's words:

> The significant problems we face cannot be solved at the same level of thinking we were at when we created them.

The engineering that is required is what is being called *sustainable engineering*, and is the focus of this book.

3.3.2 Wicked Nature of Sustainability

The lighting and transportation fuel examples in the previous section convey the importance of considering temporal, spatial, and disciplinary aspects, and interactions with ecological and economic systems, in decisions for sustainable development. Of course, products also need to be profitable and societally acceptable. However, even after considering these and other such issues, it is usually not possible to know for sure whether the system being considered is truly sustainable. This is so because the interaction between the large number of subsystems makes the overall system complex and difficult to understand and predict with a reasonable degree of certainty. Interactions between different elements can result in unexpected surprises, and managing such systems requires continuous learning and adaptation. This is unlike typical problems in science and engineering, where it is possible to define a clear goal and know when it is achieved. Such examples include synthesizing a specific molecule, achieving the yield or quality of a desired product in a chemical process, building a bridge, going to Mars, or proving a mathematical property. Such problems, on which science and engineering usually focus, belong to the category of "tame" or "benign" problems, since the mission is clear and it is possible to tell when or whether the problem is solved [13].

Sustainable development, in contrast, does not possess these traits, and belongs to the class of "wicked" problems. For example, the end goal of sustainability is not clear and depends on the person or entity defining the problem: a person who values profit may be more willing to compromise on environmental quality and social equity. This explains why there are so many definitions of sustainability, each one being legitimate from the definer's point of view. Also, there is no definition of sustainability that can be stated in a rigorous mathematical manner, owing to the complex and multidisciplinary nature of the challenge. Determining with certainty whether a product or activity is sustainable is possible only in hindsight. Thus, after a system collapses it is possible to say that it was not sustainable, but while it persists it may not be possible to determine its sustainability with confidence [14]. Not being able to define a clear end goal also means that, unlike an engineering design problem where we can know when we have solved the problem, in the "design" of sustainable systems it is not possible to know when we have solved the problem. Therefore, such systems need to be managed adaptively, with continuous learning.

Wicked problems commonly occur in handling economic, environmental, and public policy issues. In fact, each one of us faces wicked problems just in making decisions as we go through life. The pursuit of happiness is one such problem and raising children is another. In both cases, the definition and approach for such goals varies from person to person, and knowing what will work for meeting the goal is difficult, except in hindsight. When we have to make decisions, we can only have an idea of which decision is better or worse with regard to our goal, but we cannot know whether it is optimal and whether it will get us to the desired goal. Whether the decision was correct can only be known in hindsight. Despite these difficulties, everyone goes through life and deals with its wickedness, some more successfully than others. Other examples of wicked problems include dealing with disease epidemics, climate change, and nuclear energy. Approaches for addressing such problems are discussed in the next section, in the context of sustainable development.

3.4   Requirements for Sustainability

Modern science and engineering have developed by dividing disciplines into specializations and by focusing on them to gain insight that often results in new technologies and solutions. As we will learn in more detail in Chapter 6, such a reductionist approach often misses the interaction between specializations and disciplines, and increases the chance of unexpected surprises when the results of such science are implemented, particularly on a large scale. This shifting of problems across scales is described by the examples in Section 3.2 and illustrated in Figure 3.5. In addition, despite the critical role of ecosystems, most disciplines, including engineering and economics, tend to ignore or undervalue their role. This amounts to assuming nature to be unlimited and free, implying that no matter how much we use a resource, it does not degrade or deplete. The fallacy of this assumption is clear from the examples of ecological degradation we saw in Chapter 2. We also learned in Chapter 1 and Figure 1.7 about the essential role of ecosystems in sustaining societal and economic systems.

Sustainability is not just about the environment, but also about societal and economic aspects. Therefore, for a system to be sustainable, it must

- operate within ecological limits;
- be acceptable in society; and
- contribute to economic prosperity.

These ecological, societal, and economic requirements are the triple values mentioned earlier since they go beyond the conventional single economic bottom line or value of business decisions. With regard to these three requirements, we learned in Chapter 1 that the economy is nested in society, which is nested in the biosphere. This is depicted in Figure 1.7. Thus, the three requirements are listed in order of their importance, which is why we can state a necessary but not sufficient meta-principle for sustainability [15]:

> Sustainable systems must not demand more from ecosystems than can be supplied without transgressing critical thresholds.

Violating this requirement for a long time results in fraying the foundation of human well-being by causing ecological degradation as was described in Chapter 2.

Efforts to satisfy these sustainability requirements need to ensure that they do not result in a shifting of the environmental problem, to cause unintended harm and create unexpected surprises. From the illustrations in Section 3.2, we know that environmental problems can shift across space, time, disciplines, and types of flows. All this insight may be converted into six requirements that need to be satisfied by methods for assessing sustainability. These requirements are listed in Box 3.3. We will consider these requirements when we learn about methods for assessing sustainability in Part III and about solutions toward sustainability in Part IV of this book.

BOX 3.3 Requirements for Sustainability Assessment Methods [15]

1. *Account for the demand of ecosystem goods and services.* The goods and services extracted by human activities from nature need to be quantified for sustainability assessment. Goods include water, minerals, and sunlight, while services include carbon sequestration, pollination, and air quality regulation.
2. *Account for the supply of ecosystem goods and services.* Accounting for the capacity of ecosystems to supply the goods and services that are demanded is needed to satisfy the requirement of staying within nature's carrying capacity as articulated in most definitions of sustainability.
3. *Consider multiple spatial scales.* The effect of decisions meant to enhance sustainability should be considered across spatial scales. For a technological system, this requires a consideration of the processes in its life cycle, all the way from extraction of resources, to manufacturing, use, and end of life. Effects on the largest scales, such as that of the planet, also need to be considered.
4. *Consider temporal interactions.* Over time, new technologies tend to become more efficient owing to advances in manufacturing and the benefits of large-scale production. The availability of resources may also change due to depletion, competing demands, etc., as may the capacity of ecosystems to absorb impact and provide resources. New policies may also be developed because of societal change. Thus, sustainability assessment needs to consider such changes over time and issues of intergenerational equity.
5. *Consider cross-disciplinary effects.* The introduction of new technologies is likely to affect the economy and society. For example, if newer technologies are cheaper, they may encourage greater consumption, which may reduce the overall savings due to higher efficiency. Ignoring such cross-disciplinary interactions is a common reason for unexpected surprises due to new technologies. Thus, methods are needed to quantify effects in multiple disciplines and interaction across disciplinary boundaries.
6. *Consider multiple flows.* Efforts for reducing one type of impact often cause another impact to increase. So the effect of decisions on multiple flows needs to be considered. This may be enabled by aggregating flows after conversion to a common unit, such as monetary value or energetic content. However, the resulting metrics measure "weak sustainability" because aggregation allows the depletion of one flow if another is enhanced to compensate. Such substitutability may not be practical for many resources

such as fresh water and biogeochemical cycles. Strong sustainability metrics avoid aggregation and consider each flow separately, along with the nexus between them.

The development of the field of sustainable engineering is motivated by the need for methods that satisfy these requirements, and because such work does not fit within any single conventional discipline. Among existing knowledge most relevant to meeting the above requirements is systems science, which focuses on understanding the behavior and interaction between systems at smaller scales, and their impact on systems at larger scales. It has the potential to develop methods that consider not just interactions between technological systems, but also effects across time, space, and disciplines.

Perhaps the greatest challenge facing the adoption of systems thinking is inadequate familiarity and appreciation among the reductionist thinkers who have dominated modern science, perhaps because of the relatively recent emergence of systems science as compared to reductionist science. Courses related to systems thinking and sustainability are still not required or even available in all curricula. Systems thinking also requires knowledge and collaboration across disciplines, which requires the overcoming of traditional disciplinary barriers.

3.5   Approaches Toward Sustainable Engineering

Given the wicked nature of the sustainable development task and the challenges and requirements discussed in this chapter, how to choose between alternatives for enhancing sustainability and developing new solutions that contribute to sustainable development are important questions. Some potential approaches for assessing alternatives in terms of their sustainability, and for designing sustainable systems, are summarized here, along with information about some existing methods that we will learn about in subsequent chapters.

- *Enhance efficiency.* Approaches for reducing emissions and resource use for a selected manufacturing process have been popular since at least the 1970s and are known by names such as "pollution prevention" and "waste minimization." They include methods for designing networks of resources such as heat, water, and materials in a manufacturing site. Such methods constitute a first step toward sustainability but are far from adequate since they ignore most of the requirements discussed in Section 3.4.

  More recent methods based on the concepts of footprints and life cycles consider efficiency at larger spatial scales such as that of direct inputs and the entire life cycle. We will learn about many of these methods in Parts III and IV of this book. The goal of these methods is to reduce the life cycle environmental impact by choosing between various alternatives.

- *Use renewable resources.* Many systems lack sustainability owing to their use of nonrenewable fossil resources. Switching to renewable resources such as biomass, sunlight, wind, and hydropower are considered to be promising ways of moving toward sustainability, and many efforts are directed toward meeting this goal. The challenges in such work are due to the dilute and intermittent nature of many renewable resources and their greater cost as compared to conventional fossil resources. Methods like life cycle assessment and exergy analysis are useful for developing this approach.

- *Emulate nature.* One way of tackling a wicked problem is by emulating the strategy used for addressing other wicked problems. For example, to achieve happiness (another wicked problem), it is common to emulate the strategies followed by others who seem to be happy. For achieving sustainability, we could learn from and emulate nature, which is a system that has sustained itself for millennia. Approaches based on this idea include biomimicry, industrial symbiosis, ecological engineering, and techno-ecological synergy. We will learn about these in Part IV of this book.

- *Adaptive management.* Sustainability is often thought of as a steady state that needs to be reached in the long run. In practice, for most real systems, such a state may not really exist, since the system is continually subjected to disturbances and perturbations. Therefore, a more practically meaningful goal could be that, rather than grappling with the difficulties of trying to determine and reach a desired sustainable state, we should focus on ensuring that society is resilient and can recover from perturbations to continually provide goods and services to maintain human well-being and to flourish.

These approaches and the requirements of sustainability that we discussed in Section 3.4 may be addressed by the general framework depicted in Figure 3.7. This framework may be used for assessing and developing technologies that contribute to sustainable development. As shown in the figure, the first step relies on methods for conventional engineering design and analysis such as cost analysis to develop a novel engineering intervention, depicted by the white circle in Step 1. This intervention could be due to a new technology such as more efficient light bulbs or driverless cars. This engineering intervention would be introduced into society, which is indicated by the dark circle surrounding it, which is in the ecosystem, indicated by the gray outermost circle in Step 1. With wider use, the technology has effects across spatial scales through its life cycle, as depicted in Step 2. Methods such as life cycle assessment and footprint analysis are relevant for assessing effects across spatial scales. With wide use, the technology becomes

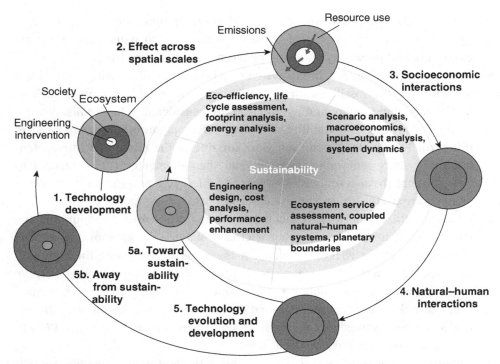

Figure 3.7 Steps in the development of technologies that contribute to sustainable development. Reproduced with permission from [15].

a part of society and may have positive and negative impacts on society and the environment. Such socioeconomic interactions may be evaluated by economic analysis. The assessment of environmental impacts requires consideration of the coupling between human and natural systems and the role of ecosystem services. Over time, the technology is likely to evolve toward or away from sustainability, as shown in Steps 5a and 5b. This conveys the need for continuous learning and adaptive management.

3.6   Summary

Many definitions of sustainable development have been proposed. All of them consider present and future generations, the limit of ecological systems to absorb human impact, and the importance of economic, environmental, and societal aspects. However, defining sustainability in a way that can be used to make decisions has been difficult owing to the wicked nature of this problem. Previous efforts to reduce environmental impact have shown that impacts can shift across temporal, spatial, and disciplinary boundaries and between types of flows. Because of these characteristics, meeting the sustainability challenge is not possible by means of a single existing discipline, but requires a new discipline, which is emerging under the name of sustainable engineering. Among various approaches that may help in achieving sustainability are enhancing efficiency, emulating nature, and adaptively managing complex human–nature interactions. We will learn about many of these methods in future chapters.

Key Ideas and Concepts

- Sustainable development
- Anthropocentric approach
- Emulation of nature
- Adaptive management
- Demand and supply of ecosystem goods and services

- Wicked problems
- Shifting of impacts in time, space, across disciplines, and between flows
- Meta-principle for sustainability

3.7 Review Questions

1. State three requirements that are common to most definitions of sustainability.
2. What is the Jevons paradox?
3. What is the food versus fuel conflict that was caused by the widespread adoption of biofuels?
4. State two characteristics of wicked problems.
5. State the six requirements for the sustainability assessment of a new technology.

Problems

3.1 The green revolution of the 1970s played a crucial role in satisfying global food requirements. It involved the development and adoption of new crop varieties and intensive agriculture.
  1. Discuss how solving the problem of providing food for everyone resulted in other unanticipated problems across temporal, spatial, and organizational domains.
  2. The introduction of genetically modified crops is considered by many to be the second green revolution. What are some of the "unanticipated" problems that this technology is creating?

3.2 Explain the challenges in determining whether an existing activity is sustainable. Why can sustainability only be known in retrospect or after the fact? Give illustrative examples to support your points.

3.3 Reducing energy consumption per lumen of light is a tame problem, but lighting that reduces total energy consumption is a wicked problem. Explain why this statement is true.

3.4 On the basis of the history of chlorofluorocarbons described in Box 2.1, discuss how this technology resulted in unexpected impacts and shifting of problems from one domain to another.

3.5 Answer the following questions about the definitions of sustainability given in Box 3.1.

1. Jefferson wrote, "Then, no generation can contract debts greater than may be paid during the course of its own existence." Provide one example with explanation of how your parents' generation has violated this requirement. Is your generation doing better than your parents'?

2. Gandhi suggests that "we are thieves in a way." In what way? Explain with examples.

3.6 For the emerging technologies listed below, discuss whether they will result in any shifting of impacts or in exceeding nature's capacity to supply goods and services.

1. Electric cars. These cars have no tailpipe emissions and are considered to be pollution-free.

2. 3D printing. Unlike conventional machining to make a part by removing and wasting the material that is removed, this technology builds a part by constructing layers and reducing waste.

3. Ionic liquids. These are nonvolatile salts that can replace volatile organic solvents, thus eliminating fugitive emissions.

3.7 Since ecosystems are self-sustaining, many efforts are being made to develop industrial systems that emulate ecosystems. Describe how such mimicry may be developed, and what could be some of the key similarities between industrial and ecological systems.

3.8 A popular thermodynamic reference state used for many decades involves the composition of the atmosphere, the ocean, and the top few meters of the earth's crust. This reference state was devised when the world was empty. Now that the world is full, have the compositions in this reference state changed in any way?

3.9 The generation of thermal power requires a large quantity of water for cooling. In arid parts of the world, it is being suggested that power plants could use dry cooling or air cooling technology. Knowing that air cooling involves a smaller heat transfer coefficient, will adoption of this technology have unintended consequences on the efficiency of power generation? Which requirement for sustainability addresses this issue?

3.10 Conventional engineering focuses on developing technologies for improving efficiency. Explain how efforts toward more efficient light bulbs can result in unintended harm by failing to satisfy the requirements for sustainability.

References

[1] A. C. Revkin. Climate, "not the story of our time." *New York Times,* December 2008.

[2] United Nations. World population prospects: The 2017 revision. https://esa.un.org/unpd/wpp/Graphs/Probabilistic/POP/TOT/, 2017, accessed November 22, 2018.

[3] World Commission on Environment and Development. *Our Common Future.* Oxford University Press, 1987.

[4] B. F. Woods, editor. *Thomas Jefferson: Thoughts on War and Revolution.* Algora Publishing, 2009.

[5] M. K. Gandhi. *Trusteeship.* 4th edition. Jitendra T. Desai, Navajivan Mudranalaya, 1960.

[6] J. Ehrenfeld. *Sustainability by Design.* Yale University Press, 2009.

[7] B. R. Bakshi and J. Fiksel. The quest for sustainability: challenges for process systems engineering. *AIChE Journal,* 49(6):1350–1358, 2003.

[8] T. R. Malthus. *An Essay on the Principle of Population* Pickering, 1986. First edition was published in 1798.

[9] P. R. Ehrlich. *The Population Bomb.* Ballantine Books, 1971.

[10] D. H. Meadows, D. L. Meadows, J. Randers, and W. W. Behrens III. *The Limits to Growth: A Report for the Club of Rome's Project on the Predicament of Mankind.* Universe Books, 1972.

[11] J. Y. Tsao, H. D. Saunders, J. R. Creighton, M. E. Coltrin, and J. A. Simmons. Solid-state lighting: an energy-economics perspective. *Journal of Physics D: Applied Physics,* 43(35):354001, 2010.

[12] H. Daly. Economics for a full world. www.greattransition.org/publication/economics-for-a-full-world, 2015, accessed November 22, 2018.

[13] H. W. J. Rittel and M. M. Webber. Dilemmas in a general theory of planning. *Policy Sciences,* 4(2):155–169, 1973.

[14] R. Costanza and B. C. Patten. Defining and predicting sustainability. *Ecological Economics,* 15(3):193–196, 1995.

[15] B. R. Bakshi, T. G. Gutowski, and D. P. Sekulic. Claiming sustainability: requirements and challenges. *ACS Sustainable Chemistry and Engineering,* 6(3):3632–3639, 2018.

PART II
Reasons for Unsustainability

In this part, we will learn about some of the underlying reasons for the conflict between human and environmental well-being. A simple but insightful way of understanding the causes underlying the environmental impact of human activities is provided by the IPAT equation:

$$I = P \times A \times T$$

Here, I stands for the total environmental impact, P is population, A is affluence measured as consumption per capita, and T is technology in terms of environmental impact per unit of consumption. This equation conveys the relevance of diverse cross-disciplinary factors in determining environmental impact. The relevance of engineering is captured by the T term, but the A and P terms are influenced by other disciplines such as economics, business, and social sciences. The chapters in this part of the book describe how various disciplines have contributed to unsustainability. Understanding this role of each discipline is essential for devising meaningful solutions toward sustainable engineering.

4 Economics and the Environment

Nature did not appear much in twentieth century economics, and it
doesn't do so in current economic modelling. When asked,
economists acknowledge nature's existence, but most deny that she
is worth much.

Partha Dasgupta [1]

An important component of human well-being is economic well-being, which is
usually measured in monetary terms. Today, economic activities in most parts of
the world are determined by markets. Therefore, markets play an important role
in determining the production and consumption of economic goods and services.
Since economic activity relies on goods and services from nature, as shown in
Figures 1.6 and 1.7, economic activity can strongly influence their availability
and status. In this chapter we will learn about the following ways by which economic activities, methods, and policies can contribute to increasing ecological
degradation and unsustainability of human activities:

- ignoring the role of ecosystems in supporting economic activities by keeping
  nature outside the economic system or market;
- valuing the present more than the future, which makes it difficult to justify
  decisions that have long-term environmental benefits;
- assuming the substitutability of ecological and economic resources; and
- not accounting fully for the physical basis of the economy.

This understanding is essential for developing market-based solutions that
enhance sustainability, as will be discussed in Chapter 21. Before digging deeper
into these four issues, we will learn some basics about the free market economy.
Specialized textbooks on environmental economics and ecological economics
provide more detail [2, 3].

4.1 The Free Market Economy

Components. Ideally, a competitive market consists of buyers and sellers interacting with each other according to the prices of goods and services. In general,
if the price increases, the demand for that product decreases, as illustrated by the

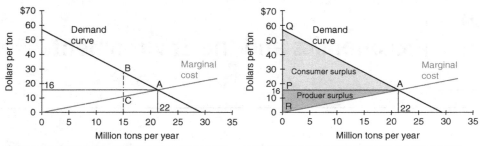

Figure 4.1 Elements of a free market. Adapted from [3].

demand curve in Figure 4.1a. Thus, if iron ore becomes more expensive, firms and people tend to use less of it. The other line in Figure 4.1a is called the marginal cost or supply curve. It represents the effect of producing and selling goods in the marketplace. For the iron ore illustration, the marginal cost curve indicates the cost of each additional ton of ore that is extracted. Thus, the first few tons that are extracted by suppliers are the ones that are the least expensive to extract. As more ore is extracted, the cost of each additional ton increases since the new ore may be more difficult to extract, transport, or process into products. Thus, the marginal cost curve represents this additional cost, making it the first derivative of the total cost of ore extraction versus the number of tons extracted.

A competitive market involves interaction between the demand and supply (marginal cost) curves, and tends to reach equilibrium at the intersection of both curves. This equilibrium point is shown by Point A in Figure 4.1a. An important feature of this intersection point is that even if there is a perturbation from this point due to changes in price or production, the market brings the system back to equilibrium. To understand this, consider a situation to the left of the equilibrium point in Figure 4.1a, where the consumption and production of ore are less than that at equilibrium. In this situation, Point B on the demand curve indicates that there are buyers willing to pay $30 per ton, while Point C indicates that extracting an additional ton of ore will cost about $10. This opportunity for profit will encourage the extraction of additional ore. Consequently, more ore will be extracted and sold in the market, causing points B and C to move toward each other, until the equilibrium point, A, is reached. Similarly, if the current situation is to the right of Point A, where the cost of extracting each additional ton of ore exceeds what people are willing to pay, extractors will reduce production because they would incur a loss for the additional tons extracted, and the system will move toward Point A.

Characteristics. One of the most attractive features of competitive markets is their ability to find solutions that balance trade-offs. For the illustration in Figure 4.1b, the triangle APQ represents the benefit of the market to consumers.

The reason is that the market price of ore of $16 per ton is much less than what many consumers are willing to pay for it. As the demand curve indicates, there are people willing to pay a range of values between $16 and $56 per ton. Thus, the area of triangle APQ represents the "consumer surplus" or the monetary benefit to society due to the free market equilibrium. Similarly, the free market is beneficial for producers as well, since producers spend an amount less than or equal to the market price for extracting a ton of ore. As shown by the marginal cost curve, all tons of ore until the 22nd cost producers less than the market price of $16 per ton, allowing them to make a profit. The total profit to all producers is indicated by triangle APR in Figure 4.1b, which is called "producer surplus."

The market equilibrium represents a "win–win" solution for producers and consumers since each surplus is maximized without hurting others. This optimally efficient solution is also called a Pareto solution. The validity of this win–win claim can be appreciated by considering the situations when prices are larger or smaller than the equilibrium value. In each situation, as discussed in the previous paragraph, both consumers and producers lose money, which is why the market moves toward equilibrium. Thus, if the supply and demand curves capture consumer and producer behavior and all costs, then the market can find the optimal balance between satisfying demand and the impact of economic activities. This is a very attractive property of free markets, since it means that the market can find the optimal solution that is best for society completely by itself, without any external intervention by the government, individual, or any other entity. The metaphor of an "invisible hand" introduced by Adam Smith [4] refers to this ability of markets to convert individual actions that are driven by personal benefit into benefits to society. This property of free markets is commonly used to argue against government intervention through mechanisms such as regulation and taxes.

Assumptions. For the free market to result in an optimally efficient solution, the following conditions need to be satisfied:

1. all industries are competitive;
2. consumers and producers have economic knowledge about the costs and benefits of their actions;
3. everyone in the market is driven by the desire for maximum financial gain;
4. there is no benefit of increasing the scale of an activity, that is, there are no economies of scale; and
5. there are no external or ignored social costs such as damage to society, employees, or the environment.

As we will see in the rest of this section, these assumptions of free markets are commonly violated, owing to which they may not operate in a manner that maximizes benefits to consumers and producers. Some violations may encourage human actions that cause excessive ecological degradation and are unsustainable.

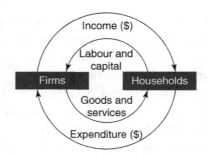

Figure 4.2 Typical view of the economy in most neoclassical economics texts.

The tragedy of the commons is one such example, which we will learn about in Section 4.2.

   Macroeconomy. Our discussion so far has focused on the market for a single product. In practice, the macroeconomic system involves multiple markets interacting with each other, which is commonly depicted in mainstream or "neoclassical" economics as shown in Figure 4.2. The overall economy consists of firms and households; the firms supply goods and services to households, for which households pay firms. The lower loop is the goods market, which is where prices are determined by interaction between producers and consumers, as described earlier in this section. Similarly, households supply factors of production (land, labor, capital) to firms, for which firms pay households. The upper loop is the factors market. According to this worldview, the goal of the economy is to maximize the gross domestic product (GDP), which is an indicator of economic activity in a given period. The GDP may be calculated as the total income or the total expenditure in the economy. These are in the outer loop of Figure 4.2, and are equal under the assumption of a steady state in the selected time period. This is a reasonable assumption since expenditure from one activity must be another's income. According to this view, greater circulation of money in the economy is good since it increases the GDP, which is considered to translate into greater prosperity and economic well-being. Recollect that we discussed the historical trend of the GDP in Chapter 1 and saw in Figure 1.2 that per capita GDP has increased dramatically over the centuries.

4.2   Environmental Externalities

If the price of an economic activity does not include its impact on all the relevant resources, then the market will be blind to the impact of this activity on the ignored resources. For such resources, the marginal cost curve in Figure 4.1 does not change as the ignored resources are affected and altered. Such resources or services that are not included in the market are called externalities. The market is unable to balance trade-offs between economic activities and externalities, and fails to find the economically efficient win–win solution. For example, if transportation is available at no cost, then the role of this activity is not included in the price of ore, and the marginal cost of ore is not affected by changes in the transportation network. Thus, a deterioration of the transportation system will not affect the price of ore, so there will be no feedback that maintains the feasibility of the transportation system or prefers mines that are easier to access. The importance of transportation may be felt only after a significant disruption that could have severe repercussions for the mining industry.

Externalities can be positive or negative. Examples of positive externalities are the benefits of a nice garden to the neighbors. This is a positive externality because the neighbors do not pay for the garden but benefit from it: the garden adds value to the neighborhood and may even increase home prices without incurring costs. Another example is the use of natural wetlands to treat industrial or municipal wastewater. Here, the cost of maintaining the wetland is borne by the company or municipality, but benefits such as an increase in biodiversity, aesthetic value, carbon sequestration, and maintenance of the water table are available to society, which does not pay for the benefits.

An undesirable environmental impact due to economic activities is often due to negative externalities. As an example, consider a community in which each residence pays for water in proportion to the quantity of water used. The costs incurred to each residence cover the resources used to withdraw the water, such as electricity and pumps, to treat the water, such as chemicals, to transport the water, such as pipes and electricity, and to clean the water at the sewage treatment facility. These costs do not include the cost of the water itself or the natural processes in the water cycle that make the water available. That is, water and the ecosystem services that make it available are considered to be free to use at will for everyone, and constitute a negative externality. In this case, consumers reap the benefits of water and its ecological cycle, but nature and society may pay a price that is not reflected in decisions or the market. Similarly, the market price of oil usually excludes costs such as those due to the impact of carbon dioxide emissions, ground-level ozone pollution, and the contribution of national defense to ensuring the safety of oil-carrying ships.

In terms of demand and supply curves, this situation may be depicted as shown in Figure 4.3. Including the cost of externalities, or external social cost, in the marginal cost causes this curve to shift upward, as indicated by the dashed line of the true marginal cost curve in the figure. For the given demand curve, the market equilibrium with the true marginal cost will be at point A′. Thus, keeping

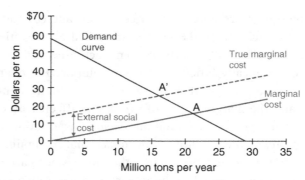

Figure 4.3 Effect of negative externalities in a free market.

the full cost of water outside the market results in the equilibrium at point A which encourages overconsumption, since point A is to the right of point A'. This overconsumption often results in ecological degradation and unsustainable economic growth, which are not good for society. However, economic activities that are not internalizing the externalities may benefit, at least in the short run, owing to the lower price and higher consumption of their products.

We learned in Chapter 1 that our well-being depends on the availability of goods and services from ecosystems, and in Chapter 2 about the degradation of ecosystems. One way in which economics has contributed to this degradation is that conventional markets have ignored the role of nature, and have made ecosystem goods and services negative externalities since their contribution has not been included in market prices. The result has been overconsumption and ecological deterioration. Such a situation is also referred to as a "tragedy of the commons," a term popularized by Hardin [5]. Here, "commons" refers to common property or public goods. These are goods such as air, public land, oceans, rivers, biogeochemical cycles, etc. that belong to everyone and are usually outside the market. Each user of such goods benefits from access to them but the use also causes a negative impact. While the benefit is to the user, the impact is spread across a larger area or population, including other users and non-users. For example, the benefit of commuting by car is to the driver, but the impact of tailpipe emissions is borne not just by the driver but by a much larger population in the region where the emissions spread, including those who do not commute by car. In this situation, since benefit is privatized while impact is socialized, the outcome is that more individuals choose to commute by car, while the common property of clean air gets polluted, resulting in a loss to society at large. Many such tragedies of the commons exist today, including those involving the ecological degradation described in Chapter 2. Some specific examples are described in Box 4.1.

BOX 4.1 Tragedy of the Commons: Some Examples

The **Aral Sea** is an inland sea in Central Asia between Kazakhstan and Uzbekistan. It used to be among the largest lakes in the world, but today is 10 percent of its original size, as shown in Figure 4.4. The reason is a classic tragedy of the commons. The common resource here was the fresh water flowing into the Aral Sea from the Amu Darya and Syr Darya rivers. This water was diverted by the Soviet Union for the irrigation of crops such as rice and cotton. As the inflow decreased, the sea started disappearing, and the sand that was exposed was contaminated by pesticides and fertilizers brought by the agricultural runoff from irrigated fields. Local fishing, shipping, farming, and other industries that used to thrive around the lake disappeared, which along with the polluted sand caused significant hardship to the local population. Thus, the benefits of the fresh water were reaped by the state (Soviet Union) and those who benefited from the irrigation and canals, but the lake ecosystem and population around it had to bear the consequences.

Lake Erie is one of the five great lakes in North America and has severe water quality problems due to harmful algal blooms. Here, the common resource is the water quality, which deteriorates mainly due to fertilizer and pesticides in runoff from farms and other surfaces within the lake's watershed, mainly in the vicinity of the city of Toledo and the Maumee river basin. Even with advanced soil and fertilizer management practices, the nitrogen and phosphorous fertilizers applied on corn and soybean farms and on lawns run off during rains, and introduce excessive nutrients into the lake, causing harmful algal blooms. In this case, the beneficiaries are the fertilizer users, but the harm is borne by all in the vicinity of the lake. In an extreme case, residents of the city of Toledo on the lake's southwestern coast were without drinking water for several days in 2014 owing to this pollution. An image of the lake with algal blooms is shown in Figure 4.5.

Plastic in the oceans. Plastics constitute a large category of human-made materials that seem indispensable in modern society. They form an essential part of a large number of consumer products, particularly as packaging material. Reasons for their popularity include their durability and low cost. Unfortunately, the life cycle of these materials is mostly linear, since only a small percentage of plastics are recycled. A large amount of plastic products end up in the ocean owing to the direct dumping of waste and in the form of microparticles from the washing of synthetic clothing, cosmetic products, etc. These materials accumulate in and around oceans, as shown in Figures 2.7 and 2.8. It is estimated that, at the current rate, by 2050 oceans will have a larger mass of plastic than marine life. Whales have died and been washed ashore as a result of having consumed many kilograms of plastic bags, ropes, and even drums. In this case, the beneficiaries are the people who generate plastic trash: they benefit from the convenience of plastic products and getting rid of the trash. Pollution of the ocean commons results in harm to the global community.

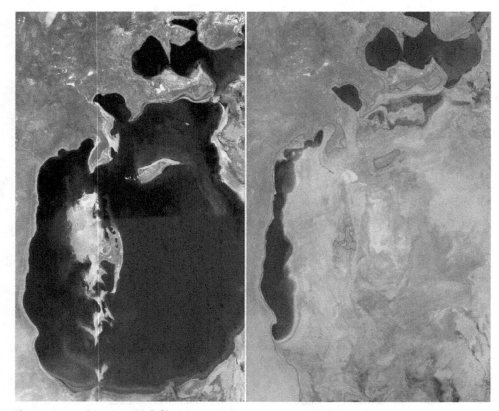

Figure 4.4 Aral Sea in 1989 (left) and 2014 (right). Image courtesy of NASA.

The understanding of environmental externalities and approaches for addressing them have been available in economics for many decades. Potential solutions include regulations, taxes, and emissions trading, and are the subject of Chapter 21. As we will learn in that chapter, implementing these solutions faces challenges due to social, political, and cultural factors.

4.3   Discounting and Benefit–Cost Analysis

Making decisions with monetary information often involves costs and benefits over different time periods. For example, the capital cost of setting up a manufacturing process is incurred at the beginning of the project, while the benefits are obtained every year over the life of the facility. The value of money changes over time, which should be considered when comparing or combining monetary flows at different times. The concept of discounting is used for dealing with the changing value of money over time. In this section, we will learn about this method with the

Figure 4.5 Lake Erie with algal blooms (gray water) on September 24, 2017. Image courtesy of NASA.

help of some simple examples, and then discuss its use for benefit–cost analysis, and its implications for environmental degradation. Additional details and more advanced methods in the context of sustainable engineering decisions and designs are presented in Chapter 17.

Let us start by asking the question, "Why discount?" The simple answer is that usually we value money and the things it can buy more today than in the future. Discounting reflects this human preference. Therefore, if you deposit a dollar in the bank, the bank can lend it to someone who needs it. The bank charges interest to the borrower and pays you interest. The basic equation used for such calculations is

$$P = \frac{F}{(1 + r)^n} \tag{4.1}$$

where P and F are the present and future values, respectively, r is the interest or discount rate, and n is the number of years. Thus, $15 invested today ($P$) will be worth $15 \times (1 + 0.1)^{20} = \$100$ (F) in 20 years (n) with a 10 percent (r) discount rate. With a 3 percent discount rate, it will grow to $27. Conversely, if an investment today is going to result in a benefit of $100 ($F$) in 20 years, then the value of that investment today is $15 ($P$) given a 10 percent discount rate. With a 3 percent discount rate, the value today is $55. This shows that the higher the discount rate, the greater the value of the present versus the future. Thus, a higher discount rate prefers the present over the future, and would give less importance to expenses that have a benefit far into the future. Addressing many environmental challenges requires expenses today for benefits in the future.

Example 4.1   Consider a scenario in which a significant catastrophe of $500 billion is predicted 50 years from now as a result of greenhouse gas emissions. This could be avoided by spending $10 billion today. Does it make sense to spend the money today? You may consider discount rates of 10 percent and 3 percent.

Solution

Whether it makes sense to spend this money may be determined by calculating the present value of the future catastrophe, and comparing this present value of the future benefit with the present cost. The two cases are considered below.

Case 1. Discount rate $r = 10$ percent; future value $F = \$500 \times 10^9$.

Using Equation 4.1, the present value of the future catastrophe is

$$P = \frac{\$500 \times 10^9}{(1.1)^{50}} = \$4.26 \times 10^9 \tag{4.2}$$

The present value of the future benefits (avoiding future costs) of this catastrophe, with a 10 percent discounting rate, is $4.26 billion. This is less than the present cost of $10 billion. Thus, in terms of present values, spending $10 billion for a benefit (to prevent a cost) of $4.26 billion is not justifiable.

Case 2. Discount rate $r = 3$ percent; future value $F = \$500 \times 10^9$.

Using Equation 4.1, the present value of the benefit of avoiding the future catastrophe is $114 billion. Now, spending $10 billion today to prevent a catastrophe that will result in a future loss of $114 billion in today's dollars is justified.

This example demonstrates that the decision to invest in preventing a future catastrophe depends on the selected value of the discount rate. Often, expenses are incurred annually, in which case the present value changes over time. Then the total present value is calculated as

$$P = \frac{1 - (1 + r)^{-n}}{r} A \tag{4.3}$$

Here, A is the annual cost. More details about this equation and its use in engineering design are provided in Chapter 17.

Example 4.2   American National Parks attract thousands of visitors from all over the world for their views and recreation opportunities. However, visibility in many parks is often poor, as shown in Figure 4.6. One reason for the loss of visibility is identified to be the presence of a local coal-burning power plant. For the Grand

Figure 4.6 Good and poor visibility in Sequoia and Kings Canyon national park [6]

Canyon, the US National Park Service analyzed the cost of reducing visibility-impairing pollution from the power plant and the benefit of improved visibility in the park [6].

- Costs of pollution control at power plant:
  - capital cost of sulfur removal equipment: $330 million at beginning of project
  - operation and maintenance cost of equipment: $75 million per year
- Benefits of better visibility:
  - visitors' willingness to pay for better visibility: $210 million per year

Does it make economic sense to implement this project?

Solution

Using Equation 4.3, the present value of the costs in millions of dollars for a 30-year life and 10 percent discount rate is

$$P_{costs} = 330 + \left(\frac{1 - 1.1^{-30}}{0.1} \right) 75$$

$$= \$1037 \text{ million}$$

The present value of the benefits is

$$P_{benefits} = \left(\frac{1 - 1.1^{-30}}{0.1} \right) 210$$

$$= \$1980 \text{ million}$$

Since $P_{benefits} > P_{costs}$, controlling power plant emissions makes economic sense.

From these examples we can see that a higher discount rate implies greater preference for the present than the future. This preference makes sense for manufactured goods, but is more difficult to justify for critical goods and services such as those from nature. Since they are essential to our well-being, as we saw in Chapter 1, societies would like to ensure that ecosystem services are available to our children and grandchildren as well. This is a basic requirement of sustainability, as we learned in Chapter 3. Thus, for environmental benefits and impacts, it makes sense to use a smaller discount rate than the one used for other goods and services. A discount rate of 3 percent is often recommended, although some economists argue in favor of a 0 percent discount rate (no discounting) for some critical goods and services. This is the emerging consensus among environmental economists but is still a topic of some debate. One example of the controversy is the debate between Lord Stern and Professor Nordhaus about the economic impacts of climate change, as discussed in Box 4.2.

BOX 4.2 Stern–Nordhaus Debate about Discounting Climate Change

A famous debate took place between two well-known economists, Lord Nicholas Stern of the London School of Economics and Professor William Nordhaus of Yale University, about the actions that society should take to tackle climate change. Lord Stern, in his *Stern Review on the Economics of Climate Change*, prepared for the UK government, argued in favor of government action in the form of an environmental tax. However, Professor Nordhaus, with the help of his integrated assessment model, argued against immediate action. Both economists used cost–benefit analysis, which we learned about in Section 4.3. They reached opposite conclusions owing to the use of different discount rates. Lord Stern used a discount rate of 1.6 percent, while Professor Nordhaus used a rate of 6 percent. Thus, Stern's smaller discount rate assumed that the future benefit of spending a dollar on mitigating climate change today is roughly equal to the value of today's dollar, whereas Nordhaus assumed that the future benefit of spending a dollar will be less than the value of the dollar spent today. In other words, Stern assumed that when it comes to the effects of climate change, people value the present and future equally, while Nordhaus assumed that people value the present more than the future. The latter assumption is tantamount to assuming that future generations will be wealthier, so that spending the money now rather than saving it for the future makes sense. However, no one knows what life will be like on a planet that is warmer by five degrees. Maybe the assumption that future generations will be richer will not hold.

4.4 Substitutability

Economics represents all activities and goods in terms of monetary value. This common unit is extremely appealing owing to its simplicity, familiarity, and ability to account for a large number of factors such as human preference, the supply of raw materials, the cost of labor, environmental impact, etc. An implicit assumption in using a common unit to represent diverse flows is that of substitutability between them. This assumption is common in many economic models, and it does make sense for many economic flows. For example, substitutable sources of energy include coal, natural gas, crude oil, sunlight, wind, etc. For building houses, substitutable materials include stone, wood, and cement. Modes of transportation are substitutable, as are sources of heat.

Going beyond products, economists also usually assume substitutability between the factors required for economic activity such as labor and capital. This assumption is captured by the following equation, called the production function [7]:

$$Y = F(K, L) \tag{4.4}$$

Here, K represents capital, and L stands for labor. This equation implies that many combinations of K and L can result in the same value of Y, that is, capital and labor are assumed to be substitutable. Capital in the form of machinery and automation often replaces human and animal labor, just as tractors have replaced animals for ploughing fields and automated checkout counters at supermarkets are replacing human cashiers. Thus, the assumption of substitutability seems quite reasonable for many products. However, this assumption has its limits, particularly when dealing with ecosystem goods and services. Nevertheless, economic theory based on the assumption of perfect substitutability has resulted in statements such as the following [8]:

> If it is very easy to substitute other factors for natural resources, then there is, in principle, no problem. The world can, in effect, get along without natural resources, so exhaustion is just an event, not a catastrophe.

Such a statement about resources being unlimited could be valid in a world where substitutes are always available and reversibly convertible into desired resources. Unfortunately, the world we live in is certainly not like that. As discussed in Chapter 1, human well-being depends on ecosystem goods and services that do impose limits. All ecosystem goods and services are required, and, beyond a point, lack substitutes. For example, developing substitutes for services such as climate regulation, biogeochemical cycles of carbon and nitrogen, and the water cycle is not possible with current technology and may never be possible. Even when ecosystems can be regenerated, the process is usually very slow: much slower than the production of economic goods and services. Thus, exhausting many natural resources could indeed be a catastrophe, not just an event! The assumption, "if it is very easy to substitute other factors for natural resources" in the statement above is itself wrong, particularly in the "full" world we live in today. Therefore, the rest of the statement is irrelevant. In addition, many ecosystems can undergo sudden changes such as collapse without prior warning.

In addition, considering economic activities to be functions of only labor and capital, as conveyed by Equation 4.4, and assuming substitutability between them, can violate the laws of mass and energy conservation. As stated by Daly and Farley [2],

> according to this logic, if our cook is making a 5-pound cake, he can increase to a 1000-pound cake with no extra ingredients – just by stirring harder and baking longer in a bigger oven!

Making a 1000-pound cake must have the additional mass of ingredients. No amount of energy or oven volume can act as a substitute.

Furthermore, the assumption of substitutability also fails to take into account the second law of thermodynamics or "arrow of time," which says that resources can flow spontaneously in only one direction. This limits substitutability since, for example, fuels cannot be made from their products of combustion without an input of energy. Even if natural resources are included in the production function, as was done until a few decades ago, the implied substitutability between labor, capital, and natural resources is also incorrect. In fact, labor and capital are transformation agents that convert natural resources into economic goods and services, and not necessarily substitutable. Many nature-made products or ecosystem goods and services are not substitutable. This includes the fact that there are fish in the ocean, fresh water, and biogeochemical cycles, since these services need to be maintained to ensure the generation and availability of other goods and services that support economic activities. As described by Daly [9], perfect substitutability would imply that a lack of fish in the ocean could be handled by building more boats. However, this does not make physical sense since, unlike fish, boats by themselves are not food.

Modern economics does realize these flaws in the assumption of substitutability, but the assumption continues to persist. One effect is that sustainability metrics derived by economists rely on this assumption to aggregate diverse economic and ecological flows. The resulting metrics can quantify only weak sustainability. We were introduced to weak sustainability in Box 3.3 and will learn about such metrics in Chapter 21. Such a worldview also causes many economists to be overly optimistic that the challenges of ecological degradation can be overcome, by implicitly assuming that substitutes will be found. One example of such optimism is the attitude of the cornucopians led by economist Julian Simon, as described in Box 3.2.

4.5   A Scientific View of the Economy

In the previous sections, we learned about three ways in which decisions based on the currently practiced "neoclassical" economics can hurt the environment. Figure 4.2 shows the conventional view of the economy. Let us now look at this neoclassical economist worldview from a scientific point of view. What is flowing in this diagram in the inner loop are physical goods, including items produced by firms and consumed by households, such as cars, food, fuel, etc. Money is also flowing in this diagram in the outer loop, in the opposite direction. Look at this diagram from a physical point of view as you would for a manufacturing process. You will realize that the diagram looks strange since goods are flowing in the system and enabling economic activity without any external inputs or outputs. We know from thermodynamics that for any system to maintain itself in a state of low entropy, as is the state of the economy owing to the order desired in economic goods and services, external inputs are required. Without any external inputs, Figure 4.2 looks like a perpetual motion machine, which is impossible!

In reality, the economy involves many flows in addition to those that are considered in neoclassical economics and shown in Figure 4.2. A more complete diagram of the actual system is shown in Figure 4.7 [10]. As can be seen, this system includes the environment, which is the source of resources to firms and which absorbs the wastes produced by economic activities. In this view, the economy is nested within the environment and is dependent on it and governed by all the laws of nature. As is clear from this figure, the economy is not a perpetual motion

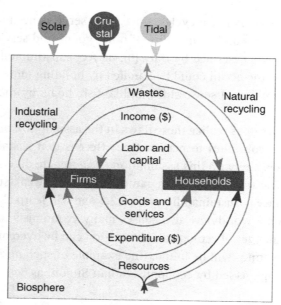

Figure 4.7 Interaction between the biosphere and the economy.

machine. This view of the economy is not new, but is still not widely accepted in making economic policies. Note that this figure bears similarities to the triple value model of Figure 1.7.

The conventional neoclassical view of Figure 4.2 was developed nearly three centuries ago, when the world was relatively "empty" since humanity did not have a large impact and available nature was vast and seemingly infinite. This is illustrated in Figure 3.6. However, today's world is much more "full" and human activities are so dominant that they are causing severe ecological degradation. Any discipline that ignores this reality and takes nature for granted is likely to contribute to ecological degradation and encourage unsustainable activities. New developments in economics, such as environmental economics and ecological economics, are addressing these issues [2, 3]. We will learn in Chapter 21 about how these developments can contribute to sustainable development.

4.6 Summary

In this chapter we learned how conventional economics could contribute to unsustainable activities. We identified four main ways by which conventional or neoclassical economics can hurt the environment. These are: ignoring external social costs or negative environmental externalities due to use of and impact on ecosystem goods and services; the use of discounting factors that make it difficult to justify investments that involve benefits after a long time; the assumption of perfect substitutability between products and factors of production; and the lack of a physical basis for economic models, resulting in assumptions that can violate the laws of nature. In recent years, many researchers have been working on addressing these issues. We will learn about these efforts and potential solutions in Chapter 21.

Key Ideas and Concepts

- Free market
- Marginal cost
- Producer surplus
- Neoclassical economics
- Tragedy of the commons
- Cost–benefit analysis

- Demand curve
- Consumer surplus
- Economic equilibrium
- Environmental externalities
- Discount rate
- Substitutability

4.7 Review Questions

1. Explain the terms demand curve, marginal cost curve, and market equilibrium.
2. State three conditions that need to be satisfied for a free market to be optimally efficient.
3. Give one example each of a positive and negative economic externality.
4. Which discount rate will result in a higher present value of a future benefit, 3 percent or 10 percent?
5. Provide one example of a resource that lacks perfect substitutes.

Problems

4.1 What is the present value of $100 of benefit available 50 years from now? You may use discount rates of 1 percent and 3 percent.

4.2 Connect the concepts of economic externalities and tragedy of the commons to the problem of carbon emissions. How is the approach of pressure from shareholders on carbon-emitting companies a way of internalizing the externalities? One example of such pressure is at http://nyti.ms/2isxILM.

4.3 Restoration of wetlands can help reduce nutrient runoff and harmful algal blooms like those that occur on Lake Erie (Figure 4.5). A year-long restoration project is estimated to cost $10 million. It is estimated to provide benefits worth $750,000 per year for the next 20 years. Is this a worthwhile investment if the discount rate is 8 percent or 2 percent? For what value of discount rate will this project be feasible?

4.4 Contaminated soil in a region that used to be an industrial site is leaching pollution into the local environment, including the ground water. Treating this soil is estimated to cost $750 million. The benefit of this program is estimated to be $50 million per year for 50 years. Is the investment justified?

4.5 In a free market, the marginal cost (M) of a commodity in $/kg may be modeled as $M = 7Q$, where Q is the quantity produced in kilograms. The demand for this commodity is related to its price (P) as $P = 3Q + 100$. Answer the following questions:

1. Determine the equilibrium point for this market.
2. If an environmental tax of $25/kg is imposed on this system, what is the new equilibrium point?

4.6 One of the assumptions of free market economics is perfect substitutability.

1. Is it possible to find substitutes for all types of goods and services? Illustrate your answer with examples and a detailed discussion.
2. It has been argued that the assumption of perfect substitutability violates physical laws such as those of thermodynamics. Discuss with the help of examples whether this argument makes sense.

4.7 Read the article at `https://nyti.ms/2jKdgKK` and answer the following questions.

1. Which businesses are approaching the challenge of addressing climate change as a benefit to their economic bottom line, and which businesses treat it as a loss? Explain the reason for this disparity.
2. The article says,

   "The bottom line is that the policies will increase the cost of carbon and electricity," said Roger Bezdek, an economist who produced a report for the coal lobby that was released this week. "Even the most conservative estimates peg the social benefit of carbon-based fuels as 50 times greater than its supposed social cost."

   Discuss the possible role that discounting may play in this conclusion. To help you in answering this question, read about the Stern–Nordhaus debate in Box 4.2.

4.8 Read the article available at `http://nyti.ms/1p45jNW`. Connect the concepts of economic externalities and tragedy of the commons to the issues discussed in this article. On the basis of this connection, suggest a solution to this problem.

4.9 With bacteria becoming resistant to antibiotics, there is an urgent need to develop new antibiotics. However, most drug companies are not working in this area because to maintain the effectiveness of existing antibiotics, new ones are used only after older ones stop working. This discourages drug companies from developing new antibiotics since they may not be able to make money by selling them for some time. Thus, what is better for public health is not good for the corporation. What is the external social cost or tragedy of the commons in this situation? Suggest approaches that could address this problem.

4.10 In February 2015, the city of São Paulo, Brazil, home to 20 million people, was running out of water. According to an article at `https://nyti .ms/2jKb99w`, one of the reasons for this dire situation in a country often called the "Saudi Arabia of Water" is "the destruction of surrounding forests and wetlands that have historically soaked up rain and released it into reservoirs." Discuss in detail the role traditional (neoclassical) economics may have played in contributing to this situation.

References

[1] P. Dasgupta. Nature in economics. *Environmental and Resource Economics*, 39(1):1–7, 2008.

[2] H. E. Daly and J. Farley. *Ecological Economics: Principles and Applications*. Island Press, 2004.

[3] D. Chapman. *Environmental Economics: Theory, Application, and Policy*. Addison-Wesley, 2000.

[4] A. Smith. *The Wealth of Nations*. Thrifty Books, 2009.

[5] G. Hardin. The tragedy of the commons. *Science*, 162(3859):1243–1248, 1968.

[6] National Park Service. Park Air Quality, Sequoia and Kings Canyon National Park, August 31, 2016. `www.nps.gov/seki/learn/nature/airqualitymon .htm`, accessed November 23, 2018.

[7] N. Gregory Mankiw. *Macroeconomics*, 7th edition. Worth Publishers, 2009.

[8] R. M. Solow. The economics of resources or the resources of economics. *American Economic Review*, 64(2):1–14, 1974.

[9] H. E. Daly. Economics in a full world. *Scientific American*, September, 100–107, 2005.

[10] J. M. Harris and A. M. Codur. Macroeconomics and the environment. `www.ase .tufts.edu/gdae/education_materials/modules/Macroeconomics_ and_the_Environment.pdf`, 2004, accessed November 23, 2018.

5 Business and the Environment

> The days of measuring business success through financial metrics alone are over. Our vision is that by 2050, all companies will measure, value and report their true value, true costs and true profits. To do this, companies need to go beyond just accounting for financial capital. They need to take an integrated approach, via a better understanding of how to incorporate and account for their natural and social capital as well.
>
> World Business Council for Sustainable Development

Business activities constitute the foundation of our economy, but they can also be significant contributors to the generation of wastes, consumption of resources, and degradation of ecosystems. Businesses can also have a positive environmental impact by developing and implementing new technologies that are much cleaner than existing approaches, and adopting corporate strategies that influence the environmental impact of their entire supply and demand chains. In this chapter, we will learn about the role of business activities in contributing to unsustainability. We will focus on how the attitude of businesses toward the environment has evolved over the last few decades, from treating environmental protection as a threat to treating it as a business opportunity. We will get a brief overview of business efforts toward sustainability, with more details about specific efforts in subsequent chapters of the book.

5.1 Pre-1980s: Environmental Protection as a Threat

While providing goods and services for enhancing human well-being, industrial activities have had a large impact on all compartments of the environment: land, air, and water. In this section, we will first focus on some typical examples of this impact, followed by a general insight into the attitude of industry before the 1980s.

Mining and refining activities were among the early causes of soil contamination. In Japan, in the late 1800s, copper and cadmium entered the human foodchain through rice due to the deposition and uptake of emissions from local smelters. This killed hundreds and sickened thousands. Similar pollution was common around metal refining regions such as Sudbury, Ontario, Canada and Hesse, Germany. Another source of soil pollution was the dumping of industrial waste in pits and landfills. Exporting of hazardous waste to developing countries was also common in the 1970s until its phaseout over 20 years. Mining causes environmental disruption by moving massive quantities of earth. The following quote about mining for nickel in New Caledonia, a small Pacific island that was annexed by France, conveys the magnitude of the operation [1]:

Between 1890 and 1990, miners employed by the Société le Nickel (SLN) moved half a billion tons of rock to get 100 million tons of ore and 2.5 million tons of nickel. This took the form of opencast mining, which after World War II involved beheading mountains.

The environmental and social effects proved profound. To get at the nickel, miners decapitated the ridges. Streams filled with silt and debris, making fishing and navigation impossible. Floods and landslides destroyed lowlands, dumping gravel on arable land and shearing away coconut groves. Silt smothered offshore corals in one of the world's largest lagoons. Many Kanaks (as the Melanesians of New Caledonia are called) lost their livelihoods, their homes and their lands.

Examples of severe and large-scale industrial air pollution are most common after 1900, and started with the burning of coal. Although the use of coal caused severe smog and deaths in London in the late 1800s, there seemed to be little political will to do anything about it. Smog in London had killed hundreds in 1873, but when it killed 4000 in 1952, the public outcry finally resulted in the Clean Air Act of 1956. Similar situations were common in cities across the world that used coal, such as in Pittsburgh, USA, due to its steel mills, the Ruhr region of Germany, and around Osaka, Japan. The trend of air pollutants in many regions of the world is shown in Figure 2.17. The impact of air pollution was severe, and some estimates are that the toll from air pollution in the twentieth century was 25–40 million, comparable to the toll of both World Wars combined. [1] As we saw in Chapter 2, challenges due to air pollution have now moved to China, India, and other developing countries.

The earliest examples of intense pollution by human activity include the Thames and Seine rivers in London and Paris, respectively, in the late 1800s. However, these and other early examples were of pollution due to human waste and untreated sewage. The Rhine river, which flows from the Swiss Alps to the North Sea, became increasingly polluted owing to industrial activities since about 1880. The Mississippi river and others in the agricultural regions of the USA suffered fish kills due to industrial pollution and pesticide runoff from farms. The effects of dichlorodiphenyltrichloroethane (DDT) insecticide on bald eagle populations became the basis for Rachel Carson's book, *Silent Spring*. Pollution was often so severe that the Cuyahoga river in Cleveland, Ohio even caught fire several times. An image from a fire in 1952 is shown in Figure 5.1.

The attitude in the late nineteenth century and most of the twentieth century was that pollution was inevitable and just had to be accepted. For example, despite

Figure 5.1 Cuyahoga river on fire, November 3, 1952. Courtesy of Cleveland Press Collection, Cleveland State University Library. http://web.ulib.csuohio.edu/SpecColl/croe/accidx.html

the severe air pollution, many Londoners defended the coal-burning hearth as their birthright and sign of freedom. As problems due to industrial pollution grew and as its harmful effects became more apparent, public pressure against pollution became more intense, particularly after World War II. However, the attitude of industry toward protecting the environment was mostly negative and confrontational. For example, at a hearing on April 10, 1964 [2] on the agricultural use of the pesticides aldrin, eldrin, and dieldrin, a representative of the manufacturer, Shell Chemical Company, argued that there was no evidence of any harm to public health and the benefits of these pest killers should be considered before any potential harm. Industry argued that fish kills from unknown causes were common in the Mississippi River, and there was no evidence of the pesticide runoff from farms causing them. The president of the National Agricultural Chemicals Association argued that to address the challenge of malnourishment, pesticides were essential to grow food in many areas of the world. These pesticides are now known to be persistent organic pollutants and are banned in most countries.

This corporate attitude, that environmental protection is an imposition and an expense, was encouraged and justified by writings such as those of Nobel Laureate economist Milton Friedman, who said that corporations should not make pollution control expenditures "beyond the amount that is in the best interests of the corporation or that is required by law." Any additional efforts "in order to contribute to the social objective of improving the environment" were labeled as "pure and unadulterated socialism" [3]. This argument was based on the understanding that markets should be left to themselves to obtain maximum benefit to producers and consumers. We learned about this ability of free markets to obtain the "win–win" solution in Section 4.1. Deliberately trying to do social good by other means such as altruism or regulation was considered to result in market inefficiencies and more harm to society than good. This argument can be supported by free market theory, but only if the underlying assumptions listed in Section 4.1 are satisfied. However, as we learned in Section 4.2, these assumptions are routinely violated and can cause environmental degradation owing to negative environmental externalities. Such an understanding was not widely accepted at that time.

In summary, for most of the period since the industrial revolution, the attitude of business activities with regard to the environment has been negative and confrontational, and the environmental impact of industrial activities has been ignored, denied, or both. A specific example that exemplifies this attitude is provided in Box 5.1.

BOX 5.1 The Ozone Hole and DuPont: From Denial to Business Opportunity

The history of chlorofluorocarbons (CFCs) was introduced in Box 2.1. Here we focus on industry's reaction to the theory of ozone depletion and the subsequent discovery of the ozone hole. A scientific article predicting stratospheric ozone depletion due to CFCs appeared in 1974 [4]. The environmental organization, the Natural Resources Defense Council, presented a press release based on this finding later that year. It blamed chemical companies and their customers for environmental destruction. Their focus was on the use of CFCs in aerosol sprays for deodorants, hair spray, and cosmetics. Manufacturers of CFCs saw these reports as threats to their industry and attacked the science. As written in [5],

Chemical Week (11 June 1975), report Molina and Rowland had "conjured up [their] now-famous theory that the ozone layer in the upper atmosphere may be shrinking under the action of free chlorine atoms originating from chlorofluorocarbon propellants" and (16 July 1975) quoted British meteorologist Richard S. Scorer as referring to the ozone-depletion theory as "a science fiction tale … a load of rubbish … utter nonsense," and characterizing the Molina-Rowland computer model as a simplistic representation of "exceedingly complex chemical and meteorological processes."

DuPont, a major manufacturer, sponsored tours by the British ozone skeptic cited above and delayed action until adequate evidence implicating CFCs in ozone depletion was found. They also attacked the "integrity and scientific acumen" of Molina and Rowland in the media [6], through a group called the Alliance for Responsible CFC Policy. The ozone hole over the Antarctic was discovered by British and Japanese scientists in 1985–1986. With the controversy resolved, the Montreal Protocol was signed in 1987, which phased out CFCs. After a decade of blocking change, DuPont now played an important role in developing hydrochlorofluorocarbon (HCFC) refrigerants that have a smaller ozone depletion potential than CFCs. Today, these compounds are also being phased out and replaced by hydrofluorocarbons (HFCs), which have no ozone depletion potential. However, these compounds have a high global warming potential and alternatives with zero ozone depletion and zero global warming potentials are now available and are gradually being adopted. Meanwhile, as shown in Figure 2.6, the ozone hole persists.

5.2   Post-1980s: Environmental Protection as an Opportunity

Increasing public pressure, along with clear evidence of the negative impacts of many industrial activities, such as the ozone hole over the Antarctic, due to CFCs, severe contamination and health impacts near many manufacturing sites, and industrial disasters such as the accident in Bhopal, India that killed thousands, resulted in gradual realization and acceptance of the harmful environmental impacts of many industrial activities. Over time, this resulted in a dramatically different attitude of industry toward environmental protection. Business strategists such as Michael Porter declared that "environmental protection was not a threat to the business enterprise but rather an opportunity" and that "strict environmental regulations do not inevitably hinder competitive advantage against rivals; indeed, they often enhance it" [7, 8]. Examples in support of this hypothesis include improvement in resource efficiency achieved by reducing pollution, and the spurring of innovation by policies such as the trading of SOx emission permits. These arguments were in effect saying that "companies that control pollution beyond what was required are practicing pure and unadulterated capitalism" [9], which is the stark opposite of Friedman's statements in the previous section. These ideas were further developed toward the approach of creating shared value. In this strategy, companies and corporations consider the creation of value not just for themselves but also for society and the environment: components of the triple value model in Figure 1.7. Examples of this change in attitude include 3M's 3P program (pollution prevention pays), described in more detail in Box 5.2, and the chemical industry's Responsible Care program.

BOX 5.2 Environmental Protection as an Opportunity: 3M's 3P Program

3M's pollution prevention pays (3P) program was initiated in 1975 as a way to prevent pollution rather than treat it after it is created. This was based on the insight that once pollutants are created, treating them does not eliminate their impact since the impact often shifts to other systems, as we learned in Chapter 3. Preventing pollution requires technological innovation and changes to eliminate the use of pollutants, such as product reformulation, changes in manufacturing and operation, and recovery and reuse of wastes. The initial motivation behind this program was that reducing pollution could be done in a manner that saved money. For recognition of a project in the 3P program, it was required to meet the following guidelines:

- it must eliminate or reduce a current or potential pollutant;
- in addition to preventing pollution, the project must also reduce the use of energy, raw materials, or other resources;
- it should involve technical solutions or innovation;
- the project should be monetarily attractive.

In the period from 1975 to 1994, the 3P program resulted in savings of at least $575 million in the USA. If the global facilities of 3M are included, the savings were over $710 million. The environmental benefits of this program have been substantial. Toxic air emissions reduced by 70 percent from 1975 to 1993, and were expected to go down by a total of 90 percent by 2000. Toxic water emissions reduced by 96 percent and waste production decreased by 21 percent from 1975 to 1993. [10]

Several innovations developed at 3M are attributed to the 3P program. Some examples include recyclable Post-it® notes made from recycled materials, solventless processes for adhesives, and reusable and returnable packaging. Perhaps the biggest success of the 3P program is that pollution prevention became part of the 3M culture, and has evolved toward life cycle management and sustainable engineering. This success has also inspired many other corporations to adopt similar strategies.

Despite this transformation toward considering environmental protection as an opportunity, debate about the general validity of this hypothesis continues since examples in support of and against the hypothesis are not difficult to find. The traditional perspective of businesses toward the environment considers environmental protection and economic growth to have a "win–lose" relation. In this perspective, depicted in Figure 5.2a, increased environmental protection reduces profit and economic growth. This purely confrontational perspective considers environmental matters as those of regulatory compliance. Examples of

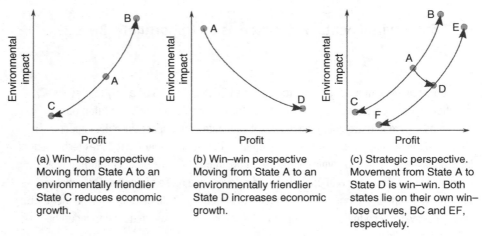

(a) Win–lose perspective Moving from State A to an environmentally friendlier State C reduces economic growth.

(b) Win–win perspective Moving from State A to an environmentally friendlier State D increases economic growth.

(c) Strategic perspective. Movement from State A to State D is win–win. Both states lie on their own win–lose curves, BC and EF, respectively.

Figure 5.2 Perspectives on business and the environment. Adapted from [9].

this perspective include industry opposition to reducing mercury emissions from coal-burning power plants and resistance from fossil-fuel burning industries to reducing carbon dioxide emissions. From an economic point of view, this resistance often translates into an unwillingness to pay the external social cost due to the negative externalities imposed by industrial activity on society. As we learned in Chapter 4, such negative externalities result in larger than optimal consumption in society and a larger profit to businesses than if the impacts were internalized. This is an example of a situation in which profit is privatized but impact is socialized. That is, the business makes the profit, while society bears the impact.

The perspective conveyed by Porter considers environmental protection and profit to be mutually beneficial, as depicted in Figure 5.2b. In such situations, the costs of environmental protection may be eliminated by innovation. Examples of this win–win perspective include the development of gas–electric hybrid engines that reduce fuel use and emissions while being more profitable, the use of wetlands instead of conventional methods for treating industrial waste water, and the use of green chemistry technologies such as process intensification and greener solvents to replace toxic substances. Even DuPont's development of alternatives to CFCs, after their initial opposition, turned out to be a win–win solution for them that was economically and environmentally superior to CFCs. Developing such win–win solutions requires innovation to overcome the shortcomings of existing technologies. Such innovation requires smart and creative engineers who are able to see beyond the current state-of-the-art, and intellectual capital in the organization and society. A corporate culture of environmental protection also seems to help, as described in Box 5.2.

Environmentally friendlier processes can often be win–win because, by reducing emissions or resource use, the process also becomes more efficient. If the cost of reducing emissions is less than the benefit of higher efficiency, then the innovation is likely to be win–win. Note that, as we learned in Chapter 3, such a solution, which is more profitable to a business and reduces emissions, would make sense from a business standpoint but it need not enhance sustainability. This is so because the solution may cause the impact to shift across spatial or disciplinary boundaries and still exceed the capacity of ecosystems to provide needed services. Thus, sustainability requires solutions that are win–win for a business and have a smaller global environmental impact. Finding such solutions requires methods like those that we will cover in Parts III and IV of this book.

In practice, it is likely that a combination of the win–lose and win–win perspectives is prevalent. In this view, shown in Figure 5.2c, win–lose is valid in the short run, as shown by line BC. Innovation can cause a shift from point A to point D, resulting in a win–win solution, followed by a new win–lose situation shown by line EF. An example of such innovation in the automobile industry is provided in Box 5.3.

BOX 5.3 Win–Lose and Win–Win Perspectives in the US Automobile Industry

Let us consider the US automobile industry for examples of the win–lose and win–win perspectives. Enhancing efficiency of the conventional internal combustion engine by strategies such as manufacturing smaller and lighter cars is usually win–lose since American consumers do not prefer such cars. This could be represented by line BC in Figure 5.2a. Development of the gas–electric hybrid engine may be represented by line AD in Figure 5.2b, since it has allowed companies like Toyota to enhance their profitability while reducing emissions and fuel use without compromising on features that consumers desire, such as safety and power. This technology also has its own win–lose trade-off, shown by line EF in Figure 5.2c. A new technology such as plug-in hybrid or electric cars may represent the next win–win automotive innovation.

5.3 Modern View: Corporate Sustainability

Today, most corporations claim to be embracing sustainability and working toward it. A survey of 4000 managers from 113 countries showed that, even in 2011, more than half of them had sustainability on their management agenda. This number has been growing exponentially, especially since 2004. Sixty-seven percent of those surveyed considered sustainability-related strategies to be necessary for being competitive, and another 27 percent considered that such strategies would

become necessary in the future [11]. This survey also identified factors that motivate corporations to include sustainability considerations in their business models.
The top factors were identified to be the following:

1. customer preference for sustainable products/services;
2. legislative/political pressure;
3. resource scarcity;
4. competitors' increasing commitment to sustainability;
5. stricter requirements from partners along the value chain;
6. owners' demands for broader value-creation (i.e., more than just profits);
7. competing for new talent;
8. customers willing to pay a premium;
9. meeting demands of existing employees;
10. maintaining a "license to operate."

Businesses have realized that environmental protection contributes to enhancing shareholder value. According to the above survey, 89 percent of survey
respondents consider pursuing sustainability to be necessary for being competitive. Activities that are commonly undertaken in efforts toward sustainability
include emissions reduction, resource conservation, life cycle system design,
engaging stakeholders, and strategic philanthropic activities. As shown in Figure 5.3, these activities help enhance shareholder value in tangible and intangible

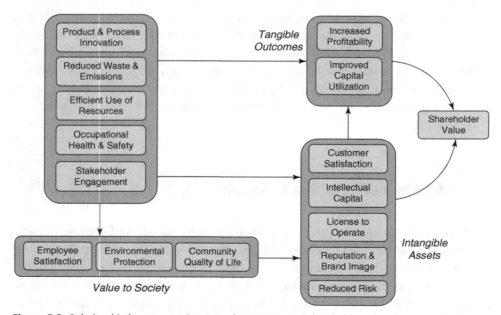

Figure 5.3 Relationship between environmental protection and shareholder value. Reproduced with
permission from [12].

BOX 5.4 Innovation due to Efforts Toward Sustainability

Among corporations, Procter and Gamble was an early adopter of life cycle thinking. Results of a life cycle energy analysis of many of its consumer goods showed that for their laundry detergent, the highest energy use was in the clothes-washing phase owing to the use of warm water. In fact, the energy used to heat the water was about three-quarters of the energy used in doing a load of laundry. This encouraged P&G to consider ways of reducing this energy use and it resulted in the development of "Coldwater Tide," a detergent that is able to clean as well as regular Tide, but in cold water. Thus, this product can reduce the life cycle energy use of their detergent, while saving consumers money and reducing energy use and resulting emissions. Even though the direct monetary benefit was to consumers, P&G expected to benefit from the resulting consumer loyalty. Since then, further innovations in surfactants and enzymes have resulted in other coldwater detergents. The success of a Danish product, Care Coldwash, is not that consumers save money, since the detergent costs more as the company charges for the new chemistry. The company claims that consumers use their product since the smaller energy use translates into a smaller carbon footprint for the consumer [13].

ways. Tangible benefits include direct monetary savings because there is less waste and more efficient utilization of equipment, and lower insurance premiums as a result of the improved risk profile caused by a reduction in the use and emission of toxic compounds. Intangible benefits include leadership, reputation, innovation, etc., as shown in the figure and listed above. In addition, sustainable business practices also help the corporation in providing higher value to stakeholders such as customers, employees, and neighborhoods around manufacturing sites. These contribute further to enhancing intangible value drivers. The benefit of the 3P program to 3M's innovation was described in Box 5.2. Another such example of innovation due to more advanced sustainability efforts is provided in Box 5.4.

Today, organizations such as the World Business Council for Sustainable Development and Business Sustainability Roundtable enable the development and adoption of sustainable business practices. Many companies are assessed by the Dow Jones Sustainability Index and vie to be included in it. It is now common for environmental organizations to work with industry for finding solutions that reduce environmental impact. This is very different from the past, when the relationship between these entities was mostly confrontational, such as that between DuPont and the Natural Resources Defense Council with regards to CFCs, described in Box 5.1. Through efforts such as the Carbon Disclosure Project, corporations are reporting their greenhouse gas emissions and goals for reducing them.

The change in corporate attitude toward the environment and claims by an increasing number of corporations about striving toward sustainability have decreased the impact per unit of a large number of products. Of course, as we discussed in Chapter 3, corporate action by itself will not be enough to achieve sustainable development, since other aspects such as consumer behavior and governance also matter. In addition, the primary goal of a business is to make money. Therefore, as long as environmental impacts remain outside the market, economic signals to corporations may cause sustainable practices to become win–lose, thus discouraging their adoption. We discussed some examples of such negative environmental externalities in Chapter 4. In addition, it is likely that at least some corporations are simply paying "lip service" and are not really committed to global sustainability. One example of such "greenwashing" may be when a corporation claims to be working toward sustainability, but the trade organization to which it belongs is opposing government regulations directed toward sustainability. The challenges facing corporate sustainability are summarized in the following quote at the end of a study describing the high corporate interest in sustainability [11]:

> This rosy picture must be balanced against another set of data. While sustainability has made it onto many management agendas, responses indicate it ranks just eighth in importance among other agenda items. Meanwhile, economic growth continues to deplete the planet's stocks of natural capital, despite the efforts of many companies to minimize their impacts through activities such as decreasing their carbon footprints and cultivating closed-loop production systems.

Corporate efforts toward sustainability utilize many of the methods of sustainable engineering that we will learn about in future chapters. A glimpse into the future of corporate sustainability is provided in the next section by a leading thinker in the field.

5.4   The Future of Corporate Sustainability, by Joseph Fiksel

As described in previous sections, over the last several decades, the concept of sustainability has been steadily integrated into the business practices of leading global corporations. They have demonstrated a commitment to environmental and social responsibility throughout their supply chains, while finding innovative ways to build shareholder value. They have engaged with diverse stakeholder groups, developed sophisticated management systems, set aggressive goals, implemented comprehensive reporting schemes, and communicated their accomplishments broadly.

It is ironic that while the global economy still faces enormous challenges – climate change, poverty, conflict, pollution – the practice of enterprise sustainability seems to be maturing. Most companies have learned to shrink their "footprint" by minimizing waste and using resources more efficiently, but they are approaching a plateau owing to diminishing returns. Some, such as Dow Chemical, have established new goals that are more outward-facing, aimed at making a positive impact and bettering people's lives. Dow's sustainability goals for 2025 include "creating a blueprint for a sustainable society" – certainly a challenging aspiration. However, there are limits to what any company can accomplish through incremental progress.

Industry leaders are beginning to step back and reconsider enterprise sustainability in a more holistic fashion. The success and continuity of every business depends on the effective functioning of social, economic, and environmental systems that are inherently complex, dynamic, and often unpredictable. Conventional indicators of sustainability – emissions, energy, water, waste – are easily measurable but conceal important interdependencies that may produce unexpected consequences. For example, a shift to renewable resources may reduce a company's carbon footprint, but may also lead to increased pressure on water and land resources, while further destabilizing the nitrogen cycle. How can a company establish the right incentives and guidelines for its employees, suppliers, and customers?

A key emerging issue is the challenge of enterprise resilience – the capacity to adapt continuously to a turbulent business environment. Supply chains are exposed to disruptions ranging from natural disasters to political upheavals, which are difficult to address through traditional risk management. Meanwhile, sustainability analysis tools provide a static snapshot and do not account for potential changes due to factors such as climate volatility or human behavior. For example, economists point to a paradox called the "rebound effect" (discussed in Section 3.2): decreasing waste will increase economic efficiency, which results in more goods and services being consumed, which in turn causes a net increase in pollution and waste. How can the long-term goals of sustainability be aligned with the need for day-to-day resilience?

To answer these questions will require fundamental rethinking of our approach to measuring, modeling, and managing sustainability. But individual companies must remain focused on competitiveness and shareholder returns, and cannot afford the luxury of pursuing this type of research. We can only achieve a major advance in enterprise sustainability through an effective, multilateral partnership among businesses, non-profits, and universities that have access to the necessary resources. Today we can envision a new approach to enterprise sustainability based on a holistic understanding of global megatrends and the dynamic linkages among industry, society, and the environment. It will take time – at least another decade – to fully realize this type of breakthrough.

In view of the changing landscape described above, we can anticipate a major leap forward in enterprise sustainability, based on the following premises:

- Businesses thrive in a flourishing society that drives improved and equitable human well-being.
- The well-being of human societies, including health, prosperity, and dignity, depends upon a continued balance of natural, economic, and social capital.
- While some progress has been made, greater trust and collaboration among industry, civil society, and government is essential to understanding this balance and driving further progress.

We envision a society in which businesses routinely and voluntarily perform an integrated assessment of significant risks and opportunities based on comprehensive analysis of the potential benefits or trade-offs in terms of human well-being. This process considers both the short- and long-term impacts of business decisions, including potential hidden consequences for natural, economic, and social capital.

There are a number of fundamental advances needed to realize this vision and to align both policy makers and enterprise sustainability leaders around a common view of sustainability:

- *Adopting systems thinking.* Business strategy needs to consider present and future value creation through a holistic lens that embraces different stakeholder perspectives, from global to local. Systems thinking will identify business innovations that lead to a low-carbon, zero-waste, resilient, and equitable economy in which economic growth is decoupled from resource extraction.
- *Measuring human well-being.* Businesses and government agencies alike need to move beyond GDP and adopt a new paradigm based on "comprehensive wealth," including economic, social, and natural resource assets, in order to clarify the value proposition for sustainability initiatives.
- *Extending ethical frameworks.* Principles of ethical business conduct need to evolve in order to keep pace with social and technological change, and to address human rights issues such as diversity, human dignity, gender equality, access to clean water, and availability of medical care.
- *Earning stakeholder trust.* The business community needs to be viewed by regulators and the public as a constructive force for positive impact and to be trusted to address sustainability issues that are material to the prosperity and well-being of both business and society.

On a practical level, putting the above insights into practice will depend on the ability of progressive companies to drive and execute business solutions. This will require collaborative dialogue among companies to explore the value and feasibility of new initiatives such as the following:

- *Embracing transformational change.* Innovative companies can work toward building investor confidence and stakeholder acceptance for disruptive technologies and regenerative business models (e.g., renewable energy, distributed manufacturing) that help to improve human well-being, protect natural capital, and increase supply chain resilience.
- *Data-driven decision making.* Analysts can utilize big data analytics, including massive data acquisition and predictive modeling tools, to help identify investments and business decisions that promise "shared value" in terms of positive impacts for both shareholders and society.
- *Collaborating at scale.* To drive meaningful progress in key focus areas (e.g., food, energy, water), the private sector can engage supply chain partners, communities, and non-governmental organizations (NGOs), while existing forums can be consolidated to develop multi-stakeholder alignment around shared goals.
- *Enabling public policy solutions.* Public–private partnerships can help to design effective policies and regulations, and government agencies can expand their role by convening stakeholder dialogues, adopting systems thinking, and striving for global harmonization; for example, by establishing commodity prices that reflect the true value of natural resources.
- *Influencing consumer behavior.* Businesses can develop improved communication tools and appropriate metrics to influence consumer perceptions and motivate sustainable choices.
- *Extreme transparency.* Sustainability-related communication and reporting can evolve toward the disclosure and investigation of issues that are material to both businesses and their stakeholders, including projected outcomes of current global trends and potential interventions.
- *Educating the next generation.* The business community can collaborate with universities to design curricula that integrate sustainability, and systems thinking sustainability into science, engineering, business, and humanities education for future customers and employees.

5.5   Summary

Over the last century, environmental protection by businesses has moved from being unnecessary, to a necessary evil, to a source of business advantage. Businesses have gradually moved away from their earlier confrontational attitude to an attitude in which efforts toward environmental protection are ways of discovering win–win solutions through innovation, retaining the license to operate, and attracting the best talent. Today, industry is increasingly active in reducing its environmental impact to enhance sustainability of its activities. Many of the techniques described in this book are actively used by many corporations. However, to make meaningful progress toward sustainability, in addition to continuing and growing corporate efforts societal pressure and governmental involvement are also needed.

Key Ideas and Concepts

- Win–lose perspective
- Corporate sustainability
- Tangible and intangible benefits
- Win–win perspective
- Environmental protection and shareholder value

5.6 Review Questions

1. How did industry react to the environmental impact of DDT?
2. What was Friedman's view about environmental protection?
3. What view did Porter put forth about environmental protection?
4. Give two examples of "win–win" industrial innovation.
5. What fraction of companies consider sustainability strategies essential for being competitive?

Problems

5.1 Visit the web pages of some corporations that describe their efforts toward sustainable development. Are these efforts addressing all the requirements for sustainability presented in Section 3.4?

5.2 A strategy followed by some corporations for reducing their corporate carbon dioxide emissions has been to sell the divisions that emitted most CO_2 per unit of product and buy the raw materials previously produced by these divisions from other manufacturers. Discuss whether such strategies constitute genuine steps toward sustainability. Assess this approach by using the requirements for claiming sustainability from Section 3.4.

5.3 Are the following innovations examples of win–win or win–lose situations? Answer from the point of view of the specified entities.
   1. Catalytic converters in cars, from the perspectives of automotive companies and catalyst manufacturers.
   2. Carbon dioxide removal technology, from the perspective of coal-burning electricity generators.

5.4 If you visit the sustainability pages of any automotive manufacturer's website or browse through their sustainability reports, you will see many indications of the company's seriousness toward enabling sustainable development. At the same time, trade groups representing this industry such as the Alliance of Automobile Manufacturers actively lobby politicians to relax fuel efficiency standards. Knowing that the largest impact of a vehicle is during the use or driving phase, can you explain the efforts of individual companies to contribute to sustainability versus the efforts of their trade organization to oppose fuel efficiency? Does this raise doubts about industry's commitment to sustainability?

5.5 Perfluorooctanoic acid (PFOA) is a chemical that does not occur naturally but has many industrial uses. Its derivative, perfluorooctane sulfonate (PFOS), was used in the product ScotchGuard manufactured by 3M. As the properties of PFOA and its derivatives became better known, 3M withdrew the product, reformulated it to eliminate PFOS and reintroduced it. Among the worrisome properties of PFOA/PFOS were: their persistence in the environment; their wide presence, not just in people but also in seals and polar bears in the Arctic; and their association with developmental disorders in laboratory animals. 3M acted in the absence of regulations or lawsuits in a way that would fall under the "pure and unadulterated socialism" category of Milton Friedman. Explain the reasoning that may have been behind 3M's decision to eliminate PFOA/PFOS from ScotchGuard. In what way did their decision make business sense? Discuss both the tangible and intangible factors.

5.6 From the history of chlorofluorocarbon refrigerants, the ozone hole, and industry's reaction to it, do you see parallels between this and the reaction of some industries to the science of climate change? When did it become possible to get industry to buy-in to fighting the ozone hole by signing the global treaty of the Montreal Protocol? What is needed for a similar outcome for addressing climate change? Compare the similarities and differences between the challenges of stratospheric ozone depletion and climate change. How will the approach used for addressing ozone depletion have to be modified for addressing climate change?

5.7 Many corporations are striving toward "net positive impact" of their activities. One such example is that of beverage manufacturers who have been encouraging practices that reduce water use and enhance the water table. Such practices include encouraging dryland rice farming, which uses much less water than traditional rice farming, and rainwater harvesting. Companies take credit for the water saved by these practices and, if this amount is larger than the water they withdraw from the watershed, they are said to achieve a positive water balance. From the point of view of corporate strategy, do you think it makes sense for a company to strive toward net positive impact? Discuss some advantages and disadvantages to the company from pursuing such activities.

5.8 Based on the material covered in Chapter 4 on the free market, answer the following questions.

1. Explain why Milton Friedman recommended that companies should not try to reduce emissions beyond the minimum regulatory requirement. What must he have assumed about negative externalities due to business activities?
2. Explain the opposite recommendation from Michael Porter about businesses and environmental protection.
3. Who do you think is correct? Justify your answer.

5.9 Describe the role of corporations in the problem of plastic pollution. Suggest actions that corporations or industry groups could take to address this scourge. Describe examples of such efforts and discuss whether corporate efforts alone can solve this problem or at least prevent it from getting worse.

5.10 Characteristics identified as essential in future corporate efforts toward sustainability include "collaborating at scale" and "extreme transparency." Describe these approaches and the barriers facing their wide application. Do you know of specific corporate efforts in these directions?

References

[1] J. R. McNeill. *Something New Under the Sun*. Norton, 2000.

[2] *New York Times*. Use of pesticides backed by maker. https://nyti.ms/2uof0yP, April 10, 1964, accessed November 23, 2018.

[3] M. Friedman. The social responsibility of business is to increase its profits. *New York Times*, September 13, 1970.

[4] M. J. Molina and F. S. Rowland. Stratospheric sink for chlorofluoromethanes: chlorine atom-catalysed destruction of ozone. *Nature*, 249:810–812, 1974.

[5] C. S. Zerefos, G. Contopoulos, and G. Skalkeas, editors. *Twenty Years of Ozone Decline: Proceedings of the Symposium for the 20th Anniversary of the Montreal Protocol*. Springer, 2009.

[6] S. H. Schneider. *Science as a Contact Sport: Inside the Battle to Save Earth's Climate*. National Geographic Books, 2009.

[7] M. Porter. America's green strategy. *Scientific American*, 264(4):168, 1991.

[8] M. Porter and C. van der Linde. Toward a new conception of the environment–competitiveness relationship. *Journal of Economic Perspective*, 9(4):97–118, 1995.

[9] A. J. Hoffman. *From Heresy to Dogma: An Institutional History of Corporate Environmentalism*. Stanford Business Books, 2001.

[10] M. Ochsner, C. Chess, and M. Greenberg. Pollution prevention at the 3M corporation: case study insights into organizational incentives, resources, and strategies. *Waste Management*, 15(8):663–672, 1995.

[11] D. Kiron, N. Kruschwitz, K. Haanaes, M. Reeves, and I. von Streng Velken. Sustainability nears a tipping point. *MIT Sloan Management Review*, 53(2):69–74, 2012.

[12] J. Fiksel, J. Low, and J. Thomas. Linking sustainability to shareholder value. *Environmental Management*, June 2004.

[13] M. McCoy. Energy impact: small Danish firm has big hit with cold-water formula. *Chemical and Engineering News*, 86(3):16, 2008.

6 Science, Engineering, and the Environment

Science discovers and understands truths about the greater world, from the human genome to the expanding universe. Engineering, for its part, solves problems for people and society.

C. D. Mote, Jr., President, US National Academy of Engineering

Confidence in the ability of science and technology to solve problems has been high, and continues to be high, so you may be surprised to see this chapter on how science and engineering contribute to the unsustainability of human activities. Like the roles of economics and business activities that we covered in the previous two chapters, the role of science and engineering is also mixed. On the one hand, advances in science and engineering have played a significant role in enhancing human well-being and in addressing environmental challenges, as we saw in Chapter 1. However, on the other hand, many scientific and engineering advances have been responsible for substantial environmental harm and unsustainable activities, and have been major contributors to the trends covered in Chapter 2. In this chapter we will focus on understanding some of the underlying reasons for the negative impacts of science and engineering. This insight will include:

- the attitude of science and engineering toward the environment;
- the resulting approach; and
- the outcomes of this approach.

Such an understanding plays a critical role in methods to evaluate and ensure that science and engineering contribute to sustainable development. We will learn about these methods in the rest of the book.

6.1 The Attitude

The aim of science is to understand the natural world, while that of engineering is to use this understanding to develop solutions for satisfying human needs. Thus, as conveyed by the quote at the beginning of this chapter, science understands or discovers what is already there and engineering innovates or invents new ideas and technologies that may never have existed before. These goals are hallmarks of human activity and have most likely been pursued since the beginning of our species. Before the advent of modern science and engineering, powered by fossil fuels, these activities tended to work with nature. This is still the case in many traditional societies, as we will learn in Chapter 20.

However, modern science has developed with an attitude of dominating nature and improving upon it, and modern engineering is the way of making this happen. This attitude is conveyed in the following quote from a speech delivered over 100 years ago by Rossiter Raymond, a decorated member of the Society of Chemical Industry, Chemists' Club [1].

> What is Engineering? The control of nature by man. Its motto is the primal one – "Replenish the earth and subdue it" ... Is there a barren desert – irrigate it; is there a mountain barrier – pierce it; is there a rushing torrent – harness it. Bridge the rivers; sail the seas; apply the force by which all things fall, so that it shall lift things ... Nay, be "more than conqueror" as he is more who does not merely slay or capture, but makes loyal allies of those whom he has overcome! Appropriate, annex, absorb, the powers of physical nature into human nature!

With such an attitude, much of the science and engineering of the last few centuries has developed with the intention of conquering nature and improving it. Examples are easy to find and include the following. For controlling the flow and availability of water, we build dams and canals, straighten rivers, and drain wetlands. For controlling pests and maximizing the yield of desired farm products, we develop pesticides, weedicides, and genetically modified crops. For ensuring the supply of nutrients to plants, we have developed commercial nitrogen fixation by the Haber–Bosch process. There is no doubt that such advances have resulted in a tremendous improvement in our understanding of how the natural world works and in controlling our environment, which have played central roles in enhancing our well-being. For example, the Haber–Bosch process is considered to be the most important invention of the twentieth century since it "detonated the population explosion" [2], and the development of vaccines and medicines has saved and enhanced millions of lives.

Since this attitude has been deeply ingrained over several decades, most scientists and engineers ignore nature and take it for granted as being limitless and free. As we saw in Chapter 1, industrial activities cannot be sustained without ecosystem goods and services, but with this attitude of ignoring or undervaluing the role of nature, engineers rarely consider the importance of these services and their status. Keeping part of an interconnected system outside the decision boundary usually means that this ignored part of the system may deteriorate, and this deterioration is not recognized until it is too late. For example, if the role of groundwater recharge is not included in the decision boundary, aquifers are likely to be overused, which may not be realized before there is a significant decline in the groundwater level. Similarly, ignoring the capacity of the biosphere to mitigate CO_2 emissions means that the ecosystem service of climate regulation may be disrupted, as has already happened. This attitude of ignoring nature may be explained because, as discussed in Section 3.3, the basics of modern science and engineering were developed in an "empty" world. These basic principles and the underlying paradigm need to be modified for today's "full" world.

In addition to taking nature for granted, scientists, engineers, and society at large also often assume that technology can always be developed to solve problems, including the problem of deteriorating ecosystems. Without any doubt, science and engineering have been amazingly successful at developing all kinds of technologies to address the depletion of various resources, and to enable their efficient use. However, as we learned in Section 4.4, most ecosystem goods and services cannot be substituted at all or in any reasonable manner by means of technology. These include fundamental ecological processes that sustain our well-being. Therefore, finding technological solutions to replace degraded ecosystems and the services that they provide seems highly unlikely.

A possible side-effect of this attitude of science and engineering is the relatively low ecological literacy of many scientists and engineers. This lack of appreciation among engineers of the services from nature means that deterioration of ecosystems and loss of their ability to provide essential goods and services are not sources of much worry for many mainstream scientists and engineers. The result is that, except for ecologists, most other scientists and engineers pay little attention to and have limited knowledge of the complexities of the very systems that sustain us on Earth. Thus, an ignorance of nature combined with the arrogance of being able to conquer, control, and replace it are common underlying reasons behind how science and engineering can encourage unsustainable activities. Note that ecological ignorance in taking nature for granted was also the reason, identified in Chapter 4, why economics contributes to unsustainability. Thus, taking nature for granted is emerging as a common reason for unsustainable human activities. We will revisit this many times in this book, including when we learn about solutions for sustainability in Part IV.

With the increasing impact of human activities, the fallacy of the attitude of taking nature for granted and as an infinite source and sink is slowly becoming clear to many. Gradually, the attitude of science and engineering is moving from "nature the adversary" to "nature the mentor." This transformation poses many challenges and opportunities, and is giving rise to unique approaches for sustainable engineering that we will learn about in later chapters of this book.

6.2 The Approach

6.2.1 Reductionism

Many advances in science have arisen because researchers have specialized their studies to consider smaller and smaller components of a system, understanding individual components and then manipulating them to affect the larger system. This philosophy of scientific reductionism is pursued with the expectation that the behavior of things can be explained by knowing how their constitutive components behave and interact. This philosophy considers the whole to be equal to the sum of its parts. Examples of reductionist science include studying the behavior of individual particles or molecules and their interaction using Newton's laws of motion to determine the macroscopic properties of a substance made up of these particles or molecules. This approach is commonly used in methods such as molecular thermodynamics to understand and develop new materials. Another example is the effort to decipher the structure of DNA in each organism, and then to look for correlations with specific characteristics of the organism. The resulting insight is the basis of genetically modified organisms such as Roundup-ready soybean, which is produced by inserting a gene for resistance to the weedicide glyphosate (N-(phosphonomethyl) glycine) in the soybean genome. This gene is obtained from microorganisms found to be resistant to glyphosate, whose commercial name is Roundup. This makes the soybean plant immune to glyphosate weedicide, while the weeds remain susceptible. Such reductionist thinking has resulted in groundbreaking technological advances that have contributed to enhancing human well-being and material wealth.

However, such thinking also encourages extremely narrow specialization, and a lack of the ability to see the "big picture" that is required for reducing the chance of unintended consequences, and for satisfying the requirements for sustainability discussed in Chapter 3. The popular fable of Six Blind Men and the Elephant conveys the shortcomings of reductionism, and is presented in Box 6.1. The history of chlorofluorocarbons (CFCs) described in Box 2.1 also demonstrates the use of reductionist thinking that focuses mainly on the chemical, physical, and toxicological properties of this class of compounds. The ozone hole was due to factors beyond those considered at that time: the effect of the widespread and dissipative use of these compounds, their interaction with the ozone layer, and the scale of their use were not considered. The green revolution of the 1970s, which led to the adoption of high-yield crop varieties and increases in global food production, was also based on reductionist research. These crops required high fertilizer and pesticide use and have increased food production, but have also resulted in many negative and unintended side-effects such as water contamination, deterioration of soil quality, depletion of groundwater, aquatic dead zones, etc. The scale of

BOX 6.1 Six Blind Men and the Elephant

Six blind men encounter an elephant. Being blind, they don't know what it is and so try to figure it out. One man pulls at the elephant's tail and thinks that what they have encountered is a rope. Another man pushes on the legs and thinks they have found tree trunks. The third feels the pointed tusks and thinks of them as spears. The trunk seems to be a snake, and the body is a brick wall. The six men reach very different conclusions about what they have encountered and all of them fail to see the "big" picture and cannot figure out that they have encountered an elephant.

This fable illustrates the folly of reductionist thinking. The six blind men are like different reductionist thinkers, research areas, or disciplines, who don't communicate with each other and are unable to see the broader implications of their activities. This fable was captured in verse by John Godfrey Saxe, part of which is reproduced below.

And so these men of Indostan
Disputed loud and long,
Each in his own opinion
Exceeding stiff and strong,
Though each was partly in the right,
And all were in the wrong!

these negative impacts is so large, as we saw in Chapter 2, that even some of its key proponents now consider it to be unsustainable.

Another more recent example of reductionist engineering that may be running into trouble is related to the introduction of genetically modified organisms (GMOs). With genetic engineering it has become possible to engineer the genome of species to introduce new properties. Examples include rice with higher vitamin content, tomatoes with a longer shelf-life, and crops resistant to specific herbicides. A widely adopted GMO is Bt-cotton. This is a form of cotton in which a toxin-producing gene from the bacterium *Bacillus thuringiensis* (Bt) is incorporated into the cotton genome. The resulting Bt-cotton naturally produces the toxin that kills the larvae of its common pests. Thus, Bt-cotton is often claimed to be environmentally friendlier than conventional cotton as it requires less use of pesticides, and it has been widely adopted all over the world. Side-effects of this large-scale adoption of Bt-cotton that are now being detected include an increase in the population of pests that are not affected by the Bt toxin. This requires the use of some pesticides, but less than what would have been used in conventional cotton. Field studies are showing the evolution of resistance in pests to the Bt toxin, raising fears of unexpected side-effects [3]. This approach is reductionist because it combines genes from different organisms to produce transgenic organisms that are expected to have the combined properties of both genetic codes. As discussed in Box 6.2, this need not be the case. We will learn in Section 6.2.2 that the non-reductionist or holistic approach would be quite different. Unfortunately, such reductionist research and decision making and its unintended harm continue to be commonplace.

The limits of reductionism are most apparent for systems that demonstrate properties such as nonlinearity, complexity, and emergence. These are systems where the "whole is greater than the sum of the parts." Examples of such systems include the national or global economy, ecological systems such as a tropical rain forest, and behavior of groups such as crowds of people or swarms of birds. Many such systems possess so many parts that knowing all the parts and understanding them in a manner that permits the modeling of macroscopic behavior is impossible.

6.2.2 Holism

To address the shortcomings of reductionist thinking, the last few decades have witnessed the growth of systems thinking, which focuses on the whole by understanding the interaction and links between its components. Such holistic thinking is represented in engineering by approaches used to design and operate manufacturing processes. Here, for proper design and operation, it is necessary not just to understand and design individual equipment, but also to account for its interaction with other equipment so that the manufacturing process produces products while meeting multiple goals such as quality, safety, and profitability.

A reductionist approach to process design may focus on maximizing reactor conversion without considering the challenges of the separation problem and recycling of unreacted components. However, a holistic approach accounts for the presence of recycle loops that permit reuse of the reactants, and such an approach may prefer operation with lower conversion. A reductionist approach for designing a building would choose technologies to ensure comfort, energy efficiency, and aesthetics. A more holistic approach would consider the performance of the entire building and also consider the role of the surroundings such as local trees, wetlands, and other ecological components in improving heating and cooling efficiency, air quality, and well-being of the building occupants. Similarly, a holistic approach for new product development would consider the broader implications of the product, such as its environmental impact, social acceptance, and ability to satisfy human needs, in addition to its ability to make money for the company. The use of such an approach for the development of new refrigerants described in Box 2.1 could have considered the loss of CFC molecules and their interaction with different parts of the environment. However, given the

BOX 6.2 Reductionist versus Holistic Approaches in Developing New Crops

The approach of breeding plants to enhance desired traits was the basis of the green revolution of the 1970s. More recent efforts are developing genetically modified plants that are often claimed to constitute a second green revolution. These efforts have been of a reductionist nature, since the goals of the plant breeder community have been to increase crop yield per plant, increase resistance to pests, or change the nutrition content. Such approaches have been highly successful in developing hybrid varieties and, more recently, genetically modified varieties of many species. A more holistic view of these efforts from the thermodynamic point of view can easily reveal that the creation of unexpected and sometimes undesirable side-effects is inevitable. This is so because, regardless of its genetic makeup, the ability of the plants to metabolize resources is limited by its photosynthetic efficiency, which is determined by the underlying biochemistry and physiology [4]. Plant breeding changes how the plant uses material and energy resources, but it cannot violate the laws of thermodynamics, which means that no new material or energy can be created. Thus, enhancing one property often means giving up on some other traits. Plants that have a high crop yield are less capable of resisting pests or competing for nutrients from the soil. In the first green revolution, farmers had to take over the responsibility for these activities. This was done by using artificial fertilizers and pesticides. Developing plants with high yield, but less demand for pesticides, fertilizers, and weedicides, as is often claimed for the second green revolution, means that some other plant characteristic will be reduced because of the first law.

Continuing with systems thinking, we would also consider how plants and other species would evolve in an agroecosystem. Under evolutionary pressures, pests and weeds developed resistance to pesticides and weedicides, respectively, during the first green revolution, and that is also happening with second-generation plants. Industry response so far has been to use a stronger cocktail of weedicides, thus eliminating the claimed benefits of genetic modification, or relying on further genetic modification that makes the plants tolerate other weedicides. For example, with increasing resistance to glyphosate, new genetically modified plants are resistant to glyphosate and dicamba (3,6-dichloro-2-methoxybenzoic acid). Thus, even the newer plants seem to involve the toxic treadmill that was pursued by farmers with the earlier hybrid varieties. Holistic thinking indicates that the solution for increasing yield along with pest and weed resistance may require going away from monocultures to polycultures, where multiple species are planted together in a manner that emulates a natural ecosystem. Such methods were practiced in traditional societies, as we will learn in Chapter 20. Adapting such techniques to the modern world while still meeting global food requirements is a formidable challenge.

knowledge at the time of their development, it is likely that even a holistic approach would not have resulted in predictions of stratospheric ozone depletion. However, with a holistic view, scientists and engineers might have paid more attention to the release of these brand new molecules into the environment, monitored their effect, been more willing to accept the possibility of unexpected surprises, and been better prepared to respond.

In the case of genetically modified organisms, discussed in Section 6.2.1, the holistic approach would be quite different. It would assess the results of reductionist research, such as the development of Bt-cotton, for its effects other than just increasing the yield of cotton or reducing the use of pesticides, as described in Box 6.2. It may also consider more holistic farming practices that overcome the shortcomings of intensive monoculture farming.

An important advantage of holistic approaches is their ability to identify emergent properties or behavior. Emergent properties cannot be determined just by understanding the behavior of individual components. An example of such properties is that water molecules have very different properties from those of its constituent atoms, hydrogen and oxygen. Another example is that the behaviors of crowds of people, traffic, flocks of birds, and swarms of fish are often very different from those of individuals. Many such systems exhibit patterns such as traveling waves and threshold effects. A further example of such a phenomenon involving

Figure 6.1 Starling murmuration [5]. Image credit: Walter Baxter / A murmuration of starlings at Gretna / CC BY-SA 2.0.

emergent behavior is that of the murmuration of birds, illustrated in Figure 6.1. The inability of reductionist methods to model or predict emergent properties can be a significant bottleneck in their ability to determine the broader implications of reductionist science and technology. Models of holistic complex systems can be developed, but they often lack the accuracy of the models produced by reductionist systems, and can only provide general insight into the behavior of the system. Such insight can still be very useful, and these methods are increasingly common in the management of complex systems such as business organizations, the stock market, and socioecological systems.

Many traditional methods used in societies before the advent of Western science were holistic in nature, as we will learn in Chapter 19. However, modern science has been mostly reductionist for the last few centuries. The importance of holistic methods is being rediscovered, mainly since World War II, and has resulted in the development of systems science and systems engineering, and now sustainable engineering.

6.3   The Outcome

The attitude of science and engineering, of taking nature for granted and wanting to dominate it, combined with an approach based on a narrow reductionist perspective, has resulted in many outcomes that are unsustainable. These outcomes are usually justified or encouraged by the goal of businesses and governments to satisfy economic objectives, which also take nature for granted as we saw in Chapter 4. Examples of such outcomes include the heavy reliance on nonrenewable resources, the accumulation of greenhouse gases in the atmosphere, and degradation and the erosion of fertile soil. We discussed many such issues in Chapter 2, and most of them can be traced to some contributions from science and engineering. A question that should arise is this: if the shortcomings of reductionist methods are known, how can we overcome them? Why do we still continue to develop "solutions" based on reductionist thinking? The rise and fall of many reductionist technologies tends to involve the following steps, as articulated by Lovins [6].

1. Scientific and technical challenges attract scientists to explore and develop the technology. With research effort and government funding, the ability of the technology to solve a problem is demonstrated in the laboratory. Once the technology attracts venture capital, large government grants, and promoters, the promoters promise its societal benefits and often shield the technology from dissent and independent assessment.
2. As people find out more about the technology, questions are raised. Public relations efforts attempt to deflect these questions.

3. As the negative side-effects become better known or as more questions are raised, public relations efforts become ineffective and are replaced by efforts to hide the use of the technology.

4. Either alternatives are developed or people realize the benefits of earlier technologies and prefer them. The new technology becomes economically infeasible or gets regulated, causing it to gradually fade away. However, its negative side-effects linger in the environment and society, while the positive benefits, usually in the form of greater wealth, accrue to a few. This may even result in a tragedy of the commons such as the ozone hole due to chlorofluorocarbons.

Well-known examples of the above steps include pesticides in the 1960s, chloro-fluorocarbons in the 1970s and 1980s, first-generation biofuels, and possibly some GMOs and nonrenewable fuels now.

Fortunately, the fallacy of the attitude and approach of science and engineering is being realized, and new methods and efforts that utilize the principles of systems science to address the challenges of sustainable development are being developed. One outcome is the field of sustainability science and engineering that is the subject of this book. Many approaches from this field are described in the rest of the book, from Part III onwards. As we learned in Chapter 5, such methods are being adopted by industry, government, and investors.

In practice, reductionist and holistic approaches are both needed and should work together, as illustrated in Figure 6.2. Reductionist approaches are needed to explore and develop new alternatives, while holistic approaches are needed to assess the alternatives from a sustainability standpoint and to provide feedback to reductionist researchers about the shortcomings of the alternatives, in order to spur more reductionist research to overcome them.

Figure 6.2 Desired interaction between reductionist and holistic research.

6.4   Summary

Science and engineering have played a critical role in meeting human needs but have also been contributors to unsustainable activities. The attitude of being better than nature and ignoring it, combined with the approach of reductionism, are key factors that have contributed to unsustainable technological developments. The lack of sustainability has often come as an unexpected surprise in seemingly promising developments. This understanding is now being used to develop methods that seek harmony with nature and attempt to prevent unintended harm by integrating reductionist and holistic science and considering interactions with systems beyond the conventional boundaries of science and engineering.

Key Ideas and Concepts

- Dominating nature
- Holism
- Reductionism
- Emergent behavior

6.5 Review Questions

1. What has been the traditional attitude of engineering toward ecosystems?
2. Give two examples each of reductionist and holistic thinking.
3. Provide some advantages and disadvantages of reductionist research.
4. Give an example of emergent behavior.
5. What did the six blind men consider the elephant's trunk to be? Why?

Problems

6.1 The development of high-yielding crop varieties has been an active area of research for many years. It has resulted in various hybrid varieties that have played an important role in the green revolution for meeting global food demand. However, it has also contributed to the contamination and depletion of aquifers, as discussed in Chapter 2.

    1. Is the development of high-yielding crops a reductionist or holistic activity?

    2. How could the development of new seeds be done to reduce its negative side-effects?

6.2 The Haber–Bosch process produces ammonia (NH_3) by means of the following reaction: $N_2 + 3H_2 \rightarrow 2NH_3$. For a process to be economically feasible, 98 percent of the hydrogen moles fed to the entire system need to be converted into ammonia. Also, the least expensive reactor design converts 10 percent of the hydrogen into ammonia. The following questions address the effect of the system boundary on reactor design.

1. Consider only a single reactor in which this reaction takes place. To produce 100 moles per hour of ammonia, calculate the amount of hydrogen and the minimum amount of nitrogen that must be fed to the reactor to obtain the specified overall conversion of 98 percent.

2. Now consider a reactor followed by a separator, and recycle streams that return unreacted hydrogen and nitrogen back to the reactor. Is it possible to operate this reactor at 10 percent efficiency while achieving 98 percent overall conversion and 100 moles per hour of ammonia production?

3. What does the difference in reactor efficiencies for these two cases tell us about the disadvantages of considering only part of a system in the design?

6.3 The "Game of Life" developed by John Conway consists of live or dead squares on a grid. It is implemented by some simple rules, such as the following.

- For a square that is alive or populated,
  - a cell with one or no neighbors dies due to loneliness
  - a cell with four or more neighbors dies due to overpopulation
  - a cell with two or three neighbors survives.
- For a square that is dead or unpopulated,
  - a cell with three neighbors becomes populated.

Several online implementations of this game are available. With the help of one of these implementations, answer the following questions.

1. Is each rule reductionist or holistic?

2. Run the game for some of the patterns. You may also create your own patterns. Identify some stable patterns that appear over time.

3. Why are these patterns examples of emergent behavior?

This simple game has led to many sophisticated tools for modeling complex systems by methods such as agent-based modeling.

6.4 The Twelve Principles of Green Chemistry shown in Box 17.2 were developed to incorporate sustainability considerations in chemistry. Discuss at least three of these principles in terms of whether they are encouraging chemistry toward more holistic thinking.

6.5 The Everglades in Florida used to be a large, slow-moving body of water between Lake Okechobee and the Atlantic Ocean. These wetlands were drained, starting from about 1850, to develop farmland and for urban development. This was done by building canals that sped up the flow of water. The common attitude during the period of its draining was that the swamp did not have much use and caused pestilence. However, since 1990, the Everglades restoration project is working toward removing the canals and slowing down the flow of water to its natural state. Explain how the engineering attitude may have contributed to the draining of this swamp, and how holistic thinking may have contributed to recent efforts for its restoration.

6.6 Coastal regions of the world where rivers meet the ocean are constantly undergoing gain and loss of land due to the deposition of sediment carried by rivers and to erosion, respectively. The loss of land in coastal Louisiana has been greatly accelerated by human activities, as conveyed by the following quote from the website of the Coastal Wetlands Planning, Protection and Restoration Act:

> with construction of extensive levee systems along the Mississippi River to maintain navigation and reduce flooding of adjacent homes and businesses, the Mississippi River has been confined to a small portion of its original flood plain. The levees have prevented coastal wetlands from receiving the regular nourishment of riverine water, nutrients and sediment that are critical to coastal wetland survival. In addition, the declining sediment load in the Mississippi River, due to upstream dams on the river and its tributaries, results in less sediment available for coastal marsh nourishment to compensate for subsidence (Kesel 1988). The amount of sediment currently being carried by the Mississippi River is only 50% of that carried during historic delta building conditions (Kesel 1988, Kesel 1989, Kesel et al. 1992, Mossa 1996).

Discuss the role of engineering in contributing to this loss by building levees, dams, and other infrastructure. How may the engineering attitude and approach have contributed to the development and implementation of such designs? Suggest an approach that is likely to be less harmful to the environment but can still meet at least some of the human needs.

6.7 Mosquitoes are responsible for spreading many deadly diseases in the world, such as malaria, dengue fever, and the Zika virus. One technology proposed to address the spread of such mosquito-borne diseases is to release genetically modified male mosquitoes into the wild. These mosquitoes would breed with females and weaken their offspring due to their modified DNA. The result would be a decrease in the population of mosquitoes, and a lower prevalence of the diseases. Field tests in Brazil have shown a decline of 82 percent in the mosquito population. Also, this technology is claimed to have fewer side-effects than the current use of insecticides for controlling the mosquito population. It may even wipe out mosquitoes entirely and make them extinct, resulting in a significant reduction of disease among the poorest populations. Is this a holistic or reductionist approach? Discuss the pros and cons if this technology is adopted more widely than in a few controlled field trials.

6.8 In Chapter 4, we learned about free markets and their equilibria at the intersection of the demand and supply curves. Is the concept of market equilibrium reductionist or holistic?

6.9 An unintended outcome of the use of plastics is ocean pollution and its impact on marine life. Would it be fair to argue that this problem is the result of reductionist thinking? Suggest a holistic approach that could have prevented this problem. Also suggest approaches that can address this problem now. Are your suggestions reductionist or holistic?

6.10 Reductionist research has contributed tremendously to improving the efficiency of lighting technologies. However, the total energy consumed for lighting has not decreased. Details about this are provided in Section 3.2. Suggest ways in which reductionist research on improving technological efficiency could be combined with holistic research, to make sure that efficiency improvements per lighting device translate into similar improvements in total energy consumption for lighting.

References

[1] R. W. Raymond. The new age. *The Journal of Industrial and Engineering Chemistry*, March, 249–251, 1913.

[2] V. Smil. *Enriching the Earth: Fritz Haber, Carl Bosch, and the Transformation of World Food Production*. MIT Press, 2004.

[3] B. E. Tabashnik, T. Brevault, and Y. Carriere. Insect resistance to Bt crops: lessons from the first billion acres. *Nature Biotechnology*, 31(6):510–521, 2013.

[4] C. F. Jordan. *Environmental Challenges and Solutions: An Ecosystem Approach to Sustainable Agriculture*. Springer, 2013.

[5] B. Keim. The startling science of starling murmuration. *Wired Magazine*. www.wired.com/2011/11/starling-flock, November 2011, accessed November 23, 2018.

[6] A. B. Lovins and J. H. Lovins. Replacing Nature's wisdom with human cleverness, *St. Louis Post Dispatch*, August 1, 1999

7 Society and the Environment

No amount of factual information will tell us what we ought to do. For that, we need moral convictions – ideas about what it is to act rightly in the world, what it is to be good or just, and the determination to do what is right. Facts and moral convictions together can help us understand what we ought to do – something neither alone can do.

Moore and Nelson [1]

In the previous chapters of Part II, we have learned about how economics, business activities, science, and engineering can contribute to ecological degradation and unsustainable human activities. Ultimately, all these activities and decisions are controlled and determined by people. Thus, no matter how useful or revolutionary a technology or policy might be, if society does not adopt it then it will have little effect. Clearly, societal and human behavior play pivotal roles in determining the sustainability of our activities. In this chapter, we will learn how societal aspects such as

- cultural narrative,
- contact with and knowledge about nature,
- politics,
- morals, ethics, and religion

can contribute to the unsustainability of human activities. Even though such aspects are not the main focus of this book or of most engineering activities, knowledge about their role is important for understanding the challenges and for formulating holistic solutions that do not cause unintended harm by shifting problems to the societal domain.

7.1 Cultural Narrative

A cultural narrative describes the attitude and behavior of a group of people. It can have a large effect on sustainability since it determines society's impact on the environment and its willingness to accept and respond to these impacts. Today, the dominant cultural narrative across the world revolves around economic growth and consumption of material goods, without being too concerned about the negative side-effects of these activities. The lack of full monetary accounting that includes the total external social cost in prices reflects societal preference to consider the environment as an unlimited source and sink, and to value the present more than the future even for critical and degraded ecosystem services.

Awareness about ecological degradation and the effects of human activities has been increasing. For a few decades, scientists have put forth warnings like those listed in Box 7.1. While not all such warnings have been completely accurate, as was described in Box 3.2, the main warning has been about ecological degradation, for which there is plenty of incontrovertible evidence, which we learned about in Chapter 2. However, the common cultural narrative does not seem to be affected by such information, since trends such as the atmospheric concentration of CO_2 and the loss of habitats and species continue to worsen. There could be many reasons for this disconnect: (1) the inability of market forces to provide feedback to society about environmental harm; (2) absence of an appropriate moral compass, as conveyed by the quote at the beginning of this chapter and discussed further in Section 7.4; (3) apathy or ignorance exemplified by the thinking that environmental impact is inevitable so there isn't much point in trying to reduce it. Not only is the cultural narrative unaffected, it also becomes more difficult for many individuals to become motivated to change their behavior to do the right thing. Even for individuals who are aware and wish to reduce environmental impact, it is common for them to wonder if their individual actions will have any benefit at all when no one else seems to care.

BOX 7.1 Examples of Warnings from Scientists about Ecological Degradation

1992 statement of the Union of Concerned Scientists [2]:

We the undersigned, senior members of the world's scientific community, hereby warn all humanity of what lies ahead. A great change in our stewardship of the Earth and the life on it is required if vast human misery is to be avoided and our global home on this planet is not to be irretrievably mutilated.

The Millennium Ecosystem Assessment, a group of over 1000 scientists, issued a similar warning in 2005 [3]:

At the heart of this assessment is a stark warning. Human activity is putting such a strain on the natural functions of the Earth that the ability of the planet's ecosystems to sustain future generations can no longer be taken for granted.

The Fifth Assessment Synthesis Report of the Intergovernmental Panel on Climate Change made the following statements in 2014 [4]:

Anthropogenic greenhouse gas emissions have increased since the pre-industrial era, driven largely by economic and population growth, and are now higher than ever. This has led to atmospheric concentrations of carbon dioxide, methane and nitrous oxide that are unprecedented in at least the last 800,000 years. Their effects, together with those of other anthropogenic drivers, have been detected throughout the climate system and are extremely likely to have been the dominant cause of the observed warming since the mid-20th century.

Changes in many extreme weather and climate events have been observed since about 1950. Some of these changes have been linked to human influences, including a decrease in cold temperature extremes, an increase in warm temperature extremes, an increase in extreme high sea levels and an increase in the number of heavy precipitation events in a number of regions.

Continued emission of greenhouse gases will cause further warming and long-lasting changes in all components of the climate system, increasing the likelihood of severe, pervasive and irreversible impacts for people and ecosystems. Limiting climate change would require substantial and sustained reductions in greenhouse gas emissions which, together with adaptation, can limit climate change risks.

If we think of the cultural narrative in engineering and business then, not surprisingly it has many similarities with the cultural narrative in society at large. Engineering values innovation and problem solving, but until the 1980s had relatively little regard for the ecological impact of many activities and treated environmental protection as a hindrance, as we learned in Chapter 5. Despite the increasing interest in sustainable development among engineering firms, academic disciplines, and professional bodies, activities with a strong dependence on non-renewable resources and hazardous materials still continue to be treated by many as environmental externalities. As we discussed in several earlier chapters, despite evidence of ecological degradation the role and capacity of ecosystems is still mostly ignored or undervalued, and the attitude of controlling and dominating nature is changing only slowly, if at all.

Despite the large knowledge base of science, awareness and appreciation about the role of ecosystems is still highly limited. Over the last few centuries, virtually every academic discipline and most activities of modern humans have taken nature for granted. Owing to engineering's success in dominating and controlling nature, many individuals tend to be techno-optimists: they expect that technology will be able to solve the problems associated with the unsustainability of human activities. However, as we have seen in previous chapters, technology is not likely to provide a solution by itself, and, rather than dominating nature, working with it is likely to be better for human, societal, and environmental well-being. However, modern society does not seem to know how to work with nature instead of against it.

While many have taken the warnings in Box 7.1 to heart and are making efforts to respond, the continued degradation of ecosystems that we learned about in

Chapter 2, and the inability to address the challenge of climate change indicates the limited success of these scientific warnings and efforts. Clearly, society is not paying adequate attention to these statements even though they are based on the best research and are from some of the brightest scientists in the world. This may seem strange since society has been very willing to accept many scientific predictions, like the effect of life style risk factors on health, the effect of vaccines in preventing diseases, predictions about celestial events such as eclipses, etc. Often, society readily accepts even unproven technologies that result from scientific advances, but seems less willing to accept warnings that go against the cultural narrative or involve some discomfort. This apparent disconnect between environmental impact and human behavior is articulated in Box 7.2. A similar disconnect exists in some countries with the theory of evolution, despite its wide acceptance in science for several decades. For example, in the USA a 2013 survey found that 33 percent of the population rejected the idea of evolution and consider humans to have existed in their present form since the beginning of time.

Society's "cognitive dissonance" and "collective denial" may have arisen because, like all species, humans also have inherently expansionist tendencies. This has been essential for our success as a species, and was very effective in a relatively "empty" world. This may explain why, even now, warnings that should rein in our expansionist tendencies encounter societal resistance. It seems that as the world is becoming more "full," such tendencies need to be controlled. Here, the foresight that humans possess, unlike other species, seems essential but does not appear to be helping, at least not yet.

BOX 7.2 Quotes Articulating the Disconnect Between Human Activities and the Environment

The following statement by William Rees describes the apparent cognitive dissonance of modern society:

Ours is allegedly a science-based culture. For decades, our best science has suggested that staying on our present growth-based path to global development implies catastrophe for billions of people and undermines the possibility of maintaining a complex global civilization. Yet there is scant evidence that national governments, the United Nations, or other official international organizations have begun seriously to contemplate the implications for humanity of the scientists' warnings, let alone articulate the kind of policy responses the science evokes. The modern world remains mired in a swamp of cognitive dissonance and collective denial seemingly dedicated to maintaining the status quo. [5].

A major reason for societal unwillingness to change is deeply rooted in human history, as summarized in this quote by E. O. Wilson:

According to archaeological evidence, we strayed from Nature with the beginning of civilization roughly ten thousand years ago. That quantum leap beguiled us with an illusion of freedom from the world that had given us birth. It nourished the belief that the human spirit can be molded into something new to fit changes in the environment and culture, and as a result the timetables of history desynchronized. A wiser intelligence might now truthfully say of us at this point: here is a chimera, a new and very odd species come shambling into our universe, a mix of Stone Age emotion, medieval self-image, and godlike technology. The combination makes the species unresponsive to the forces that count most for its own long-term survival. [6]

For successful transformation toward sustainable development, the cultural narrative needs to change toward one in which harmony between human activities and the environment is of paramount importance. While such a change is certainly formidable, society has indeed undergone large shifts in its cultural narrative. For example, practices such as slavery and apartheid were widely accepted and commonly practiced at one time, but are now illegal or considered taboo. A more recent change in the cultural narrative is with regard to smoking. Less than 25 years ago, airplanes had a smoking section, something that is unthinkable now to many. Today, smoking is severely restricted in most parts of the world. Thus, such changes are certainly possible. They can take time but, once in motion, change can be very rapid. No one truly understands the nature and mechanism of such changes, but they seem to involve the combined effect of many smaller actions to get over a threshold or tipping point. Some such efforts for bringing about change in society toward sustainability will be discussed in Chapter 22.

7.2   Ecological Literacy

Scientific knowledge about nature has increased by leaps and bounds, but this knowledge is restricted to specialists. In the general public many studies indicate that ecological literacy, measured by metrics such as knowledge about species in their own backyards, is declining. For example, a study of British eight-year-old children showed that they could identify 78 percent of Pokémon characters but only 53 percent of common British wildlife species. A study of cultural products in English such as stories, songs, and films found that words about the natural environment have steadily declined since the 1950s, while words about the human-made environment have increased. Another study has shown that as societies become more prosperous, they tend to lose their connection with nature and their ecological literacy diminishes. Figure 7.1 shows how knowledge about local plant species has decreased with increasing gross domestic product and human

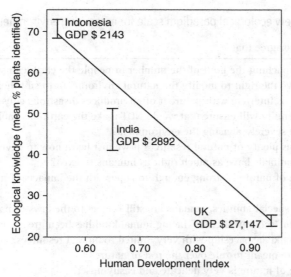

Figure 7.1 Deterioration of ecological knowledge with increasing economic prosperity. Reproduced with permission from [8].

development index. This may seem counterintuitive, since scientific knowledge about ecosystems has grown tremendously. However, such knowledge is limited to scientists and researchers, and most of us know less about nature than our ancestors. Studies about ecological attitudes among college students in various disciplines have shown that students in technological disciplines and economics have a more negative view toward the environment and are less ecologically literate than students in other disciplines [7]. Questions in such a survey are listed in Table 7.1.

These observations and surveys indicate that the human–nature gap has been widening. This gap contributes to unsustainability since, if individuals and societies take nature for granted and do not appreciate its essential role in supporting human well-being, then efforts to protect nature, especially in a proactive manner, are more likely to be resisted. Efforts initiated after realizing the effects of ecological degradation, as was done with the ozone hole (Box 2.1), are often too late since reversing damage is much more difficult than preventing it.

7.3   Political Aspects

Like any change, societal transformation toward sustainability will result in winners and losers. If, in a society, a large fraction of wealth and power are concentrated among those who benefit from maintaining the status quo, then bringing about change away from the status quo can be very difficult, even when it may be

Table 7.1 The new ecological paradigm scale for measuring environmental attitudes [9].

Do you agree or disagree that:

1.   We are approaching the limit of the number of people the earth can support?
2.   Humans have the right to modify the natural environment to suit their needs?
3.   When humans interfere with nature it often produces disastrous consequences?
4.   Human ingenuity will ensure that we do NOT make the earth unlivable?
5.   Humans are severely abusing the environment?
6.   The earth has plenty of natural resources if we just learn how to develop them?
7.   Plants and animals have as much right as humans to exist?
8.   The balance of nature is strong enough to cope with the impacts of modern industrial nations?
9.   Despite our special abilities, humans are still subject to the laws of nature?
10. The so-called "ecological crisis" facing humankind has been greatly exaggerated?
11. The Earth is like a spaceship with very limited room and resources?
12. Humans were meant to rule over the rest of nature?
13. The balance of nature is very delicate and easily upset?
14. Humans will eventually learn enough about how nature works to be able to control it?
15. If things continue on their present course, we will soon experience a major ecological catastrophe?

beneficial to society at large. This is so because the ample resources of those who wish to prevent change can influence the political system and decision makers.

Furthermore, support or opposition to change is often influenced by political ideology and can be a knee-jerk reaction based on the perceived political orientation of those pushing for change. Such reactions may seem irrational, and proper communication is essential for bringing various parts of society together, as illustrated by the examples based on psychological studies in Box 7.3.

BOX 7.3 How Political Ideology Affects Attitudes Toward Pro-environmental Options

Bringing about societal change toward sustainability can be helped by understanding human attitudes toward technologies and policies that are likely to be more sustainable. Here we describe two psychological studies that demonstrate how communication can result in different outcomes depending on political ideology.

Energy-efficient lighting. Technologies such as compact fluorescent lamps (CFLs) and light emitting diodes (LEDs) are much more efficient than conventional incandescent bulbs. Therefore, many governments and power companies are making efforts to encourage consumers to switch to these more efficient lighting alternatives. A study to understand consumer psychology toward energy-efficient lighting evaluated the effect of product labeling on its acceptance among people of different political ideologies in the USA [10]. A label on the bulbs either conveyed their environment-friendliness, or was blank. The efficient CFL bulbs either cost the same as incandescent bulbs or were three times more expensive. What this study found was that if both bulb types cost the same, political leaning had no effect on who bought which bulb. However, if the CFL bulb was more expensive, consumers with a conservative political ideology were more likely to buy a CFL bulb if it was not labeled versus if it was labeled with a message of being environmentally friendly. Thus, the environmental label turned off conservatives from the energy-efficient bulb. This showed that politically conservative individuals are likely to be less inclined toward energy-efficient options owing to their placing less importance on the resulting environmental benefits such as lower CO_2 emissions. This insight may be used to develop appropriate labeling schemes for environmentally superior consumer technology.

Focus on past versus present. This study focused on the difference among conservative and liberal political ideologies in terms of their focus on the past versus the future [11]. Conservatives are more focused on the past and therefore are more likely to be influenced by arguments that compare the present with the past. In contrast, liberals are more focused on the future and therefore more affected by comparison of the present with the future. This study found that comparing today's climate with that of the past was more likely to elicit action from conservatives, while comparing what the climate is likely to be like in the future as compared to now had more traction with liberals. Developing a message based on understanding this psychology was found to eliminate the effect of political leaning on action. Thus, both conservatives and liberals were equally likely to act on addressing climate change if the message was conveyed to them in an appropriate manner.

7.4 Ethics, Morals, and Religion

Like other disciplines, modern ethics and philosophy have developed while taking nature for granted. Their focus has been on guiding the behavior of people with other people, and not with the natural world. Ethics and moral principles on how to treat nature are relatively recent, and in some cases are being rediscovered from much older literature.

Even though all religions teach about protecting the Earth and the environment, this message seems to have been forgotten or is just ignored. In the last few decades, many scholars have thought about how religion contributes to environmental degradation. The following reasons have been identified [12]:

- Anthropocentrism considers humans to be separate from nature and superior to it. This makes it easier to justify actions that destroy nature by human domination.
- Many religions strip nature of any sacrality, thus making way for its exploitation.

- Technological progress has been identified with religious virtue regardless of the environmental impacts of the technology.
- The notion that allows a nonhuman as a relative has been rejected. This is seen in the opposition to the theory of evolution. If nonhuman relatives were accepted, then hurting them would be more difficult to justify.
- Wilderness has been viewed as a cursed land, thus encouraging its domestication. This attitude is seen through concepts such as the Garden of Eden being desirable, as opposed to the unmanaged wilderness.
- The pervasiveness of otherworldliness is due to the belief that the promised land is another world different from the Earth. Thus, degrading the planet seems acceptable since that is not our final and desired destination.

These attitudes were used in the past to justify the eradication of many cultures and eco-friendly attitudes.

In older, often Eastern and polytheistic religions, such as Confucianism, Taoism, Buddhism, Hinduism, and Shinto, nature has an important role and is considered to be on an equal footing with human beings. Environmental components such as trees, rivers, mountains, reptiles, birds, and animals are considered to be sacred. In newer, usually monotheistic, religions such as Judaism, Christianity, and Islam, people are considered to be at a higher level than nature, but are still required to be stewards of nature. Unfortunately, these teachings in all religions seem to have little effect on human actions today, and the only global "religion" that the majority follows seems to be that of economic growth. Today, even the most "sacred" components of the environment are heavily polluted, and environmental stewardship has become secondary to rampant consumerism. Fortunately, there are movements among most religions to rediscover their connection with the environment and to encourage their adherents to protect nature. We will learn more about these in Chapter 22 when we focus on solutions.

7.5 Summary

The societal aspects are perhaps the most important in determining the sustainability of human activities. These aspects in fact contribute to unsustainability owing to the present cultural narrative of consumerism, the increasing gap between human and natural systems, political efforts to maintain the status quo by those who fear that sustainable development will hurt their profits, and the inability of morals, ethics, and religion as practiced today to address the relationship of humans with nature.

Key Ideas and Concepts

- Cultural narrative
- Cognitive dissonance
- Ecological literacy
- Political ideology and communication
- Anthropocentrism
- Polytheism

7.6 Review Questions

1. Explain the cognitive dissonance in today's society.
2. Give two examples of cases in which the cultural narrative has undergone dramatic change.
3. Describe a communication strategy that will enable the acceptance of energy-efficient technologies across the political spectrum.
4. Explain the attitude of viewing wilderness as cursed land and its impact on ecological degradation.
5. Is it fair to consider modern ethics and philosophy as anthropocentric? Justify your response.

Problems

7.1 Have the warnings about unsustainability of human activities like those in Box 7.1 affected the cultural narrative? Why or why not?

7.2 How does E. O. Wilson explain human unwillingness to respond to unsustainable activities?

7.3 The cultural narrative has changed quite drastically in most parts of the world with regards to some human activities. How did this change occur for smoking? Was there a tipping point or threshold that had to be overcome before the change could be accepted and spread across societies? How long did the effort take? What insight does this experience provide with regard to changing the cultural narrative toward sustainable development?

7.4 The finding that British children were more familiar with Pokémon species than natural species alarmed many. Learn about the game Phlo (http://phylogame.org), to answer the following questions.

   1. How was this game influenced by the finding that children knew more Pokémon characters than natural species?

   2. How does it attempt to teach players about species in nature?

   3. Do you think such an approach will be effective in reducing the human–nature gap?

7.5 Discuss in brief the relationship between the Jevons paradox (see Chapter 3) and the present cultural narrative. Can the Jevons paradox be affected by changing the cultural narrative?

7.6 Explain how (a) viewing wilderness as a cursed land and (b) stripping nature of any sacrality can contribute to decisions that are not sustainable.

7.7 Discuss how environmentally beneficial farming activities would need to be communicated to conservative and to liberal farmers.

7.8 A history of chlorofluorocarbon (CFC) compounds and the ozone hole is provided in Chapters 2 and 5. What role did the cultural narrative and political aspects play in the opposition to addressing this problem? What was the tipping point that resulted in the Montreal Protocol?

7.9 A large number of decisions about changes in land use due to the construction of large dams, long pipelines, and roads have resulted in conflicts between indigenous populations and businesses. Examples include the North Dakota oil pipeline in the USA, the Belo Monte dam in Brazil, and the Sardar Sarovar project in India. Can these conflicts be understood on the basis of differences in ethical and moral values of the different stakeholder groups? Choose one such conflict and explain in detail.

7.10 As an effort toward addressing plastic pollution, many cities and communities are proposing bans of plastic bags, disposable food packaging, and other products. What political challenges is such a ban likely to encounter? Determine whether such bans have been successfully implemented anywhere in the world.

References

[1] K. D. Moore and M. P. Nelson, editors. *Moral Ground: Ethical Action for a Planet in Peril*. Trinity University Press, 2012.

[2] Union of Concerned Scientists. 1992 world scientists' warning to humanity. www.ucsusa.org/about/1992-world-scientists.html, 1992, accessed January 11, 2019.

[3] Millennium Ecosystem Assessment. *Ecosystems and Human Well-Being: A Framework for Assessment*. Island Press, 2003.

[4] Inter-Governmental Panel on Climate Change. IPCC fifth assessment report: summary for policy makers, 2014.

[5] W. Rees. What's blocking sustainability? Human nature, cognition, and denial. *Sustainability: Science, Practice & Policy*, 2:13–25, 2010.

[6] E. O. Wilson. *The Creation: An Appeal to Save Life on Earth*. Norton, 2006.

[7] P. M. Tikka, M. T. Kuitunen, and S. M. Tynys. Effects of educational background on students' attitudes, activity levels, and knowledge concerning the environment. *The Journal of Environmental Education*, 31(3):12–19, 2000.

[8] S. E. Pilgrim, L. C. Cullen, D. J. Smith, and J. Pretty. Ecological knowledge is lost in wealthier communities and countries. *Environmental Science and Technology*, 42(4):1004–1009, 2008.

[9] R. E. Dunlap, K. D. Van Liere, A. G. Mertig, and R. E. Jones. Measuring endorsement of the new ecological paradigm: a revised NEP scale. *Journal of Social Issues*, 56(3):425–442, 2000.

[10] D. M. Gromet, H. Kunreuther, and R. P. Larrick. Political ideology affects energy-efficiency attitudes and choices. *Proceedings of the National Academy of Sciences*, 110(23):9314–9319, 2013.

[11] M. Baldwin and J. Lammers. Past-focused environmental comparisons promote proenvironmental outcomes for conservatives. *Proceedings of the National Academy of Sciences*, 113(52):14953–14957, 2016.

[12] D. L. Haberman. *River of Love in an Age of Pollution*. University of California Press, 2006.

PART III
Sustainability Assessment

In Part III, we will learn about several methods that are commonly used for assessing the sustainability of engineering and other human activities. These methods expand the boundary of conventional engineering to prevent the shifting of impacts. Many also consider multiple flows such as resources and emissions. All the methods we will cover rely on a common theoretical framework. We will also learn how well these methods satisfy the requirements for claiming sustainability that we learned about in Box 3.3.

8 Goal Definition and Scope

> When we try to pick out anything by itself, we find it hitched to
> everything else in the Universe.
>
> John Muir [1]

Most sustainability assessment methods require consideration of the life cycle of a selected product or process. Before applying any such method, this life cycle network needs to be determined. In this chapter, we will understand some characteristics of life cycle networks and learn that there are standardized steps in sustainability assessment. We will investigate the details of each step in various chapters of this part of this book. In this chapter, we will consider the first step, which is to define the goal and scope of the problem. This involves determining the functional unit for comparing alternatives and the boundary of the life cycle network.

8.1 Nature of Life Cycle Networks

Consider the problem of choosing an environmentally friendlier transportation fuel that reduces reliance on fossil energy and the resulting CO_2 emissions. Ethanol from corn seems like a good option instead of gasoline, since, unlike gasoline, ethanol is not made from a fossil feedstock. In fact, its renewable plant feedstock uses CO_2 for its growth, thus absorbing the emissions from ethanol combustion. This closed CO_2 loop should cause lower net greenhouse gas emissions from using ethanol instead of gasoline. From this relatively narrow point of view, using corn ethanol instead of gasoline seems to be a way of reducing fossil energy use and greenhouse gas emissions. However, if we adopt a broader and more holistic point of view, and consider the fossil energy used in the supply network for making ethanol, we find that most of the steps rely on fossil energy and emit greenhouse gases, including the following:

- the transportation of ethanol to the gas station by trucks burning diesel;
- the process energy derived from the coal or natural gas needed to convert corn into ethanol;
- the farming machinery running on diesel and used for planting and harvesting the corn;

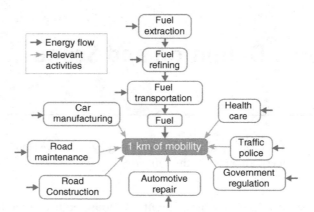

Figure 8.1 Contributors to the life cycle energy used to obtain 1 km of mobility.

- the natural gas and other fossil resources needed to produce the fertilizers and pesticides that are applied to grow the corn;
- the fossil resources needed to manufacture the equipment used for farming, transportation, and manufacture.

Thus, a holistic approach that considers activities in the life cycle indicates that there could be substantial fossil energy use over the life cycle of corn ethanol, and the reductionist conclusion that corn ethanol is independent of fossil energy may be naive and incorrect.

Another example about the importance of life cycle thinking considers direct and indirect energy use in a vehicle for a kilometer of mobility. This total energy use includes the following [2], as illustrated in Figure 8.1:

- the fuel burned;
- plus the energy required to extract, refine, and transport the fuel;
- plus the energy to manufacture the car (kilometer equivalent);
- plus the energy to build and maintain roads;
- plus the energy to maintain auto repair shops, government regulation, registration services, traffic police, etc.;
- plus the energy to produce and maintain that portion of health system used to care for the consequences of auto accidents and auto-related health problems;
- plus ...

Like the ethanol example, the system we need to consider to determine the total energy used to drive 1 km is large and complicated. The indirect energy use, that is energy use other than that due to the fuel burned in the car, is likely to be significant.

In addition to the use of energy over a product's life cycle, we are often interested in the life cycle flow of other resources and emissions as well. Commonly

Table 8.1 Different types of sustainability assessments.

What is being assessed?	System and its characteristics
Product	All activities needed to make, use, and dispose of the product. May need to allocate flows between multiple co-products.
Process	All activities that contribute to design and operation of the process.
Region or country	All activities in the selected geographical region, including imports and exports.

considered resources include water, land, and minerals, while emissions of interest include carbon dioxide, reactive nitrogen, toxic chemicals, and others. The total quantity of a resource used or pollutants emitted is often referred to as the footprint or life cycle flow, and is required for a holistic analysis that reduces the chance of unexpected surprises by shifting the impact to other stages of the life cycle.

Example 8.1   For the case of the energy used in driving considered above, identify the step in the car's life cycle that is likely to have the largest energy use. You may consider the steps to be broadly divided into resource extraction, manufacturing, transportation, use, and end of life.

Solution

The life cycle of a car involves extraction of resources that go into the car itself, manufacture of the car, transportation of the car to the dealer, use of the car by driving it, and managing the car's end of life by processes such as recycling and landfilling. In the life cycle of an automobile, the largest energy consumption is in the use phase. This is when the fuel is burned to power the car. This step overwhelms the energy use in any other stage, including crude oil extraction, refining, oil and gas transportation, and end-of-life waste processing of the vehicle. This can be confirmed quantitatively by using the methods we will cover in subsequent chapters.

For assessing sustainability, we will consider various types of life cycle networks and perspectives, some of which are summarized in Table 8.1.

8.2   Steps in Assessing Life Cycle Networks

Assessing the life cycle of products and processes requires proper definition of the system being analyzed and the alternatives being compared. The four steps in such assessments are shown in Figure 8.2 and summarized below.

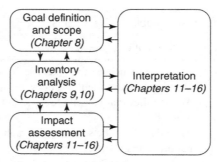

Figure 8.2 Steps in assessing life cycle networks and chapters in which they are described.

1. *Goal definition and scope.* This step involves defining the goal of the sustainability assessment study, selecting the functional unit, and defining the system boundary. We will learn about it in Section 8.3.
2. *Inventory analysis.* This step involves finding data for all the processes in the specified boundary. As we will see in Chapter 9, many sources of data are available for developing the life cycle inventory. The approach for doing calculations with these data is described in Chapters 9 and 10.
3. *Impact assessment.* This step utilizes the results of inventory analysis to determine the impact of the selected system. Various sustainability assessment methods have their own approaches for impact assessment. These approaches are covered in Chapters 11–16.
4. *Interpretation.* Here we glean insight about the pros and cons of the selected system over its life cycle. This step is also described in Chapters 11–16. This information is used to identify opportunities for improvement, which will be the focus of Part IV of this book.

The steps described in this subsection are essential for a fair comparison of alternatives. Steps 1 and 2 are identical in all the sustainability assessment methods covered in Part III of this book. Steps 3 and 4 differ for various methods, and are covered in Chapters 11–16.

8.3   Goal Definition and Scope

The goals of sustainability assessment are often to choose between alternative options and/or to identify opportunities for improving a product or steps in a life cycle.

8.3.1   Functional Unit

Understanding and defining the goal of a life cycle study is essential for its proper formulation and execution. For example, the goal in comparing gasoline with ethanol may be to choose between them from the perspective of mobility, while the goal in comparing grocery sacks could be to determine the best way of transporting groceries. After determining such a common functional unit for the comparison, the quantity of each item may be calculated for a fair comparison. This is illustrated by the following examples.

Example 8.2 Determine an appropriate functional unit for comparing the use of paper, foam, and ceramic cups for drinking hot beverages. The paper and foam cups are used once, while a ceramic mug is typically used 500 times.

Solution
In this problem, the goal is to choose a cup and the function of the cups is to carry a hot beverage. Let the functional unit be one drink in a paper cup. Then the number of foam cups required for the same function is also one, while the number of ceramic mugs required is $\frac{1}{500}$. Thus, for the functional unit of one hot beverage served in a paper cup, we get

$$1 \text{ paper cup} \equiv 1 \text{ foam cup} \equiv 0.002 \text{ ceramic mug}$$

Alternatively, we may choose the functional unit to be 500 drinks in the ceramic mug, which is the typical use of a single such mug. Then,

$$500 \text{ paper cups} \equiv 500 \text{ foam cups} \equiv 1 \text{ ceramic mug}$$

Example 8.3 The fuel E10 is 10 percent ethanol and 90 percent gasoline. It is the composition of gasoline used in many countries. The fuel E85 contains 85 percent ethanol and 15 percent gasoline. It covers 20 percent less distance per volume of fuel than E10. This is due to the lower fuel value of ethanol than gasoline and the differences in the engines that run on the two fuels. Determine the functional unit for comparing the life cycles of these two fuels when used in otherwise identical vehicles. How much ethanol and gasoline will each vehicle use for the selected functional unit if the fuel economy of the E10 vehicle is 15 km/L?

Solution
The function of the E10 and E85 fuels is to provide mobility. Therefore the functional unit may be the same distance traveled with either fuel. Let us choose this distance to be 100 km.

Now we determine the amounts of ethanol and gasoline needed for the selected functional unit and the given fuel economy for E10, $\eta_{E10} = 15$ km/L. Since the E85 powered vehicle travels 20 percent less distance per liter of fuel, we get

$$\eta_{E85} = 0.8 \times 15 \text{ km/L}$$
$$= 12 \text{ km/L}$$

To drive 100 km with E10 requires

$$V_{E10} = \frac{100 \text{ km}}{15 \text{ km/L}}$$
$$= 6.67 \text{ L}$$

Since 10 percent of this is ethanol and 90 percent is gasoline, we get for the volumes of the two fuels

$$V_{E10,gas} = 6 \text{ L}$$
$$V_{E10,EtOH} = 0.67 \text{ L}$$

Similarly, for E85,

$$V_{E85} = \frac{100 \text{ km}}{12 \text{ km/L}}$$
$$= 8.33 \text{ L}$$

and so

$$V_{E85,gas} = 1.25 \text{ L}$$
$$V_{E85,EtOH} = 7.08 \text{ L}$$

8.3.2   Life Cycle Boundary

Another decision required for a life cycle study is about the scope of the study as defined by the boundary of the network. As we learned in Section 8.1, to capture all the direct and indirect resource flows in a life cycle, the network may need to be very large. The quote at the beginning of this chapter by John Muir was most likely to have been about ecological systems, but it also applies to the life cycle of technological systems. Capturing such a large network may not be practically feasible owing to challenges in obtaining data about each process in the network and to the resulting cost and computational intractability. We will consider different ways of approximating the analysis boundary. These approaches are illustrated in Figure 8.3 and described below. Here, the full life cycle without any approximation consists of all the nodes in Figure 8.3a.

- *Process network*. This approach selects the processes from the life cycle that are considered to be the most important. They are represented by dark circles in Figure 8.3a, and identified as P_1 to P_8. The rest of the life cycle, denoted by lighter circles, is ignored.
- *Input–output network*. Here, similar processes are lumped together to result in a finite number of aggregate sectors. As depicted in Figure 8.3b, 25 processes are aggregated into five sectors. For example, the aggregate sectors may be defined by combining all foam product manufacturing processes, all processes

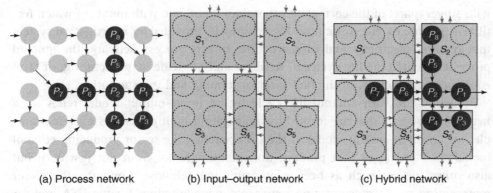

Figure 8.3 Types of life cycle networks.

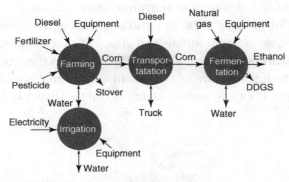

Figure 8.4 Selected processes in the life cycle of corn ethanol.

that generate electricity, and farming activities that grow any type of grain. Such aggregation makes it easier to define a life cycle boundary while reducing computational challenges in life cycle calculations.

- *Hybrid network.* This model combines the detailed models of the process network with the completeness of the input–output network, as shown in Figure 8.3c. Here, the input–output model captures the "background" flows for the selected processes P_1 to P_8, which would otherwise be ignored in a process network model like the one in Figure 8.3a.

Sources of data for such network models are described in Section 9.1.

To gain further insight into these three ways of defining life cycle networks, let us consider how the corn ethanol life cycle discussed in Section 8.1 may be modeled by each approach. A partial process network is shown in Figure 8.4. Here, ethanol is produced by fermentation of corn, which also requires other inputs such as water, fuel, and equipment. A byproduct from this process is distillers' dried grain solids (DDGS). The corn needs to be transported to the factory, which requires trucks, roads, diesel, etc. The farming process produces corn grain along with stover (parts of the corn plant other than grain) and with inputs of water, fertilizers, pesticides, etc. Agricultural water may be provided by an irrigation system that requires electricity and infrastructure. In this process network, the ignored processes include the life cycles of fertilizer and pesticide manufacturing, of farm equipment production, of the fuel used for farming and transportation, etc.

Modeling the corn ethanol life cycle by an input–output model relies on a network of sectors that are formed by aggregating multiple processes. For example, for an input–output model of the US economy, the sector producing ethanol includes all processes that produce nonpotable ethanol from not just corn but also other sources such as beet, sugarcane, and cellulose. Similarly, the sector representing truck transportation represents average values for the USA and for the transportation of many types of freight, not just corn. The sector of fertilizer manufacturing includes not just anhydrous ammonia, which is the commonly applied fertilizer for growing corn, but also other nitrogenous fertilizers such as ammonium nitrate, ammonium sulfate, and urea. With such aggregated sectors, an input–output network will include all economic activities that contribute directly or indirectly to the manufacture of ethanol in the selected region or country. A partial input–output life cycle with some key sectors is shown in Figure 8.5.

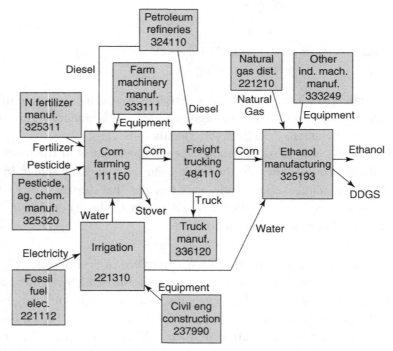

Figure 8.5 Selected economic sectors in the input–output life cycle network of corn ethanol. Economic sectors are indicated by their North American Industrial Classification System (NAICS) codes.

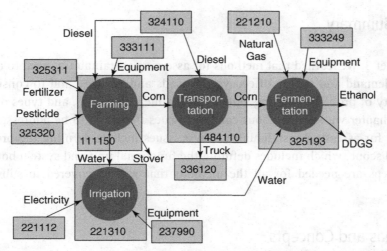

Figure 8.6 Hybrid network model for corn ethanol by combining the process network in Figure 8.4 with the input–output network in Figure 8.5.

A hybrid network for the ethanol life cycle will combine the process network in Figure 8.4 with the input–output network in Figure 8.5 to result in a hybrid network, as shown in Figure 8.6. Here, inputs to the ethanol process network whose life cycles are ignored, such as diesel, equipment, and fertilizer, are included by means of the corresponding economic sectors. In this figure, the economic sectors are represented by the North American Industrial Classification System (NAICS). Thus, the corn farming process belongs to Sector 111150, which is "Corn Farming," pesticide is obtained from Sector 325320, which is "Pesticide and Other Agricultural Chemical Manufacturing," and diesel is bought from Sector 324110, which is "Petroleum Refining." Practical and mathematical details about developing such network models will be covered in Chapter 10.

The network models described above are commonly used in sustainability assessment with two types of system boundaries:

- *Cradle-to-grave.* This boundary attempts to consider all processes starting from the beginning to the end of the activities and products being assessed. For a product such as a fuel, this boundary will include the extraction, manufacturing, use, and disposal phases.
- *Cradle-to-gate.* For some studies, part of the system may be ignored for convenience or because it is common to the activities or products being compared. For example, in comparing different ways of generating electricity, the use phase of the generated power is the same, regardless of how it is produced. Therefore, the use phase may be ignored and the boundary may be truncated at the power plant output gate.

8.4   Summary

In Chapter 3, we learned that methods for assessing sustainability need to account for the demand and supply of ecosystem goods and services, and to consider the possibility of problems shifting across space, time, disciplines, and types of flows. In this chapter we learned about common issues that need to be addressed in all methods for assessing sustainability. These issues include defining the problem's goal and scope, which includes defining the functional unit and system boundary. These steps are needed for all the methods that will be covered in subsequent chapters.

Key Ideas and Concepts

- Process network
- Hybrid network
- Life cycle inventory

- Input–output network
- Functional unit
- Cradle to grave

8.5 Review Questions

1. State the goal of a sustainability assessment of grocery sacks.
2. What are the four steps in assessing life cycle networks?
3. Explain the concept of functional units, with the help of an example.
4. Which are the three methods that can be used to determine the life cycle boundary?
5. What is a cradle-to-gate life cycle boundary?

Problems

8.1 A common quandary for the environmentally conscious shopper is about the type of bag that should be used to carry groceries. Identify the process network to approximate the life cycles of grocery sacks made of (a) paper and (b) polyethylene. Suggest a functional unit to compare the life cycle impact of the two types of sacks.

8.2 The average projected lifespans of 60-W-equivalent incandescent, CFL, and LED bulbs are 1200 hours, 10,000 hours, and 50,000 hours. Determine the functional unit for comparing the life cycles of these bulbs.

8.3 Determine the functional unit for comparing the life cycle impact of a textbook printed on paper versus one that is delivered electronically. Identify the most important processes in both life cycles. Which step of each life cycle is likely to have the highest energy use?

8.4 Many studies identify the importance of changing global diets for addressing issues related to environmental sustainability and health. For a comparison of the life cycle impacts of different diets, suggest appropriate functional units. Which steps are likely to have the largest emissions of greenhouse gases in the life cycles of vegetables, fish, and red meat?

8.5 In its efforts toward sustainability, an American university campus is replacing paper towel dispensers in restrooms with electric hand dryers based on air drying. Another hand-drying system popular in Europe uses a roll of cloth towel. Many restrooms in other parts of the world lack a hand-drying system, so people carry their own cloth handkerchief for drying their hands. What information will you need to determine the functional unit for comparing these four systems in terms of their life cycle environmental impact? On the basis of your own research about the functioning of these systems and their typical use, determine the functional unit for each system.

8.6 Consider six common products: cars, shoes, laundry detergent, fleece jackets, beer, and milk, and the following five steps in their life cycles: resource extraction, manufacturing, transportation, use, and end of life. Considering the carbon dioxide emissions in the life cycle of each product, identify two stages in the life cycle that have the largest CO_2 emissions for each product. (Provided by Professor Tim Gutowski and based on an article in the *Wall Street Journal* [3].)

8.7 You wish to conduct a sustainability assessment to compare the impacts of commuting by a personal vehicle versus by car pooling. Define the goal of this study and determine the appropriate functional unit. Select the most important activities in this life cycle and draw the corresponding process network.

8.8 A life cycle assessment of disposable cold cups conducted for Starbucks Coffee Company and NatureWorks LLC [4] compared cups and lids made out of polypropylene (PP), polyethylene terephthalate (PET), and polylactic acid (PLA). The report states: "The life cycle phases of the beverage cup and lid product systems that were studied include:
- Cradle-to-polymer factory gate pellet production for PET, Ingeo, and PP,
- Pellet transportation to converter,
- Conversion of pellets into cups and lids,
- Transport of cups and lids to Starbucks shops,
- Waste disposal in landfill."

What type of life cycle boundary does this statement imply? Discuss its benefits and shortcomings.

8.9 Answer the following questions for assessing a polyethylene terephthalate (PET) bottle.

1. Identify the processes in the life cycle of this plastic bottle that may be the most important. Draw the corresponding process network and show the important inputs and outputs for each process in your network.
2. Identify the economic sectors that provide inputs to the process network developed in the previous question and develop a hybrid network model for a PET bottle. You may use the North American Industrial Classification System for determining the appropriate economic sectors. Repeat this exercise for a glass bottle.

8.10 In today's globalized economy, it is common for supply chains to extend across multiple countries. Draw the process network for the following products.

1. A shirt sold in the USA says "Made in Honduras" but the synthetic fabric was produced in China with raw materials from Saudi Arabia.
2. Consider an electronic item such as a computer or phone. On the back, it says "Assembled in Mexico." By searching on the internet, draw a "cradle to gate" process network diagram that includes constituent parts such as the chip and display, and indicate possible countries of origin.

References

[1] J. Muir. *My First Summer in the Sierra*. Houghton Mifflin, 1911.
[2] R. A. Herendeen. *Ecological Numeracy: Quantitative Analysis of Environmental Issues*. John Wiley & Sons, 1998.
[3] J. Ball. Six products, six carbon footprints. *Wall Street Journal*, October 2008.
[4] PE Americas. Comparative life cycle assessment Ingeo™ biopolymer, PET, and PP drinking cups. www.natureworksllc.com/~/media/The_Ingeo_Journey/ EcoProfile_LCA/LCA/PEA_Cup_Lid_LCA_FullReport_ ReviewStatement_121209_pdf.pdf, 2009, accessed January 12, 2019.

9 Inventory Analysis

We can manage what we can measure.

Peter Drucker

In Chapter 8, we learned about the four steps that are common to all sustainability assessment methods and discussed the first step, that of goal definition. In this chapter, we will focus on the second step, which involves obtaining data for doing the assessment, and then using it in the calculations. We will consider various sources of data for building process and input–output networks, and address practical issues such as allocating the resource use and emissions between multiple products. Our approach in this chapter will rely on relatively simple mathematics, with a more general, matrix-based framework provided in Chapter 10. We will also cover approaches for assessing the effects of uncertainties in the inventory data.

9.1 Sources of Data

Calculations with life cycle network models like those described in Section 8.3.2 require data about the selected activities and their flows. These data are referred to as the life cycle inventory (LCI). Obtaining data for a large life cycle is a formidable task owing to the diversity of processes and the need for quantifying inputs and outputs of many types of raw materials, fuels, products, byproducts, and wastes. Consider the ethanol life cycle shown in Figure 8.4. The inventory requires data from diverse areas such as engineering, agriculture, and transportation. Using expertise in each area it is certainly possible to determine inputs and outputs for each process. This could rely on fundamental knowledge and models about the underlying processes, or on practical knowledge about the flows in functioning systems. Traditionally, efforts toward process modeling or industrial data measurement have focused mainly on economically valuable flows such as raw materials and products. However, sustainability assessment requires data about wastes and emissions, which may not be readily available from traditional models and databases.

Over the last several years, LCIs for a large variety of systems have been compiled and have become available as free and commercial databases. These inventories are usually based on data from relevant activities. To address corporate

Table 9.1 Typical life cycle inventory: partial inventory data for electricity generation from natural gas in the USA [1].

Flow	Category	Type	Unit	Amount
Outputs				
Benzene	Air/unspecified	Elementary flow	kg	1.00E−08
Carbon dioxide, fossil	Air/unspecified	Elementary flow	kg	5.85E−01
Carbon monoxide, fossil	Air/unspecified	Elementary flow	kg	4.01E−04
Cobalt	Air/unspecified	Elementary flow	kg	4.01E−10
Dinitrogen monoxide	Air/unspecified	Elementary flow	kg	1.07E−05
Electricity, natural gas, at power plant	None	Product flow	kWh	1.00E+00
Formaldehyde	Air/unspecified	Elementary flow	kg	3.58E−07
Lead	Air/unspecified	Elementary flow	kg	2.39E−09
Mercury	Air/unspecified	Elementary flow	kg	1.24E−09
Methane, fossil	Air/unspecified	Elementary flow	kg	1.07E−05
Nitrogen oxides	Air/unspecified	Elementary flow	kg	4.78E−04
Particulates, >2.5 μm, and < 10 μm	Air/unspecified	Elementary flow	kg	3.61E−05
Sulfur oxides	Air/unspecified	Elementary flow	kg	3.02E−06
VOCs (volatile organic compounds)	Air/unspecified	Elementary flow	kg	2.64E−05
Inputs				
Dummy-transport, pipeline, unspecified	None	Product flow	tkm	3.54E−01
Natural gas, processed, at plant	None	Product flow	m^3	2.98E−01
Transport, combination truck, average fuel mix	None	Product flow	tkm	5.92E−02
Transport, train, diesel powered	None	Product flow	tkm	3.54E−03

proprietary concerns, data are collected from multiple processes of the same activity, and only their average values are included in the inventory databases. Typical inventory data for a process are shown in Table 9.1. Note the comprehensive nature of this data since it contains even relatively small flows, such as those of lead, mercury, and dinitrogen monoxide, and relevant activities such as transportation. Such modules for various processes are used to develop process network models, as we will learn in Section 9.2 and Chapter 10.

For input–output networks, data for each sector and flows between sectors are compiled by government agencies and are available in public databases. As we learned in Section 8.3.2, these data are even more aggregated than those of process

Table 9.2 Typical life cycle inventory data from input–output models [2].

Flow	Wheat, corn, etc. (1111b0)	Electricity (221100)	Gasoline etc. (324110)	Couriers (492000)	Management consulting (541610)	Vehicle repair (811100)
CO_2 to air (kg/\$)	1.28E+00	6.27E+00	5.03E−01	1.82E−01	9.97E−02	1.53E−01
Cropland (m^2yr/\$)	1.19E+01	6.24E−03	6.38E−03	6.63E−03	5.77E−03	5.35E−03
Benzene to air (kg/\$)	2.20E−05	1.62E−05	6.14E−05	8.65E−06	1.75E−06	2.09E−06
Glyphosate to water (kg/\$)	2.95E−05	1.02E−08	1.17E−08	9.86E−09	1.04E−08	8.05E−09
Groundwater (m^3/\$)	3.71E−01	6.35E−03	2.82E−03	1.03E−03	7.12E−04	9.98E−04

The coefficients represent life cycle flows per dollar of final demand in terms of producer prices.

models, but they comprehensively cover an entire region. Some typical data from input–output life cycle networks are shown in Table 9.2. The coefficient is the total direct and indirect or life cycle flow per dollar of final demand, which is the monetary value of the consumption from the selected sector. The final demand is the consumption by consumers, not other economic sectors. Mathematical details about the underlying calculations are presented in Chapter 10.

Some popular sources of process and input–output inventory data are summarized in Table 9.3. Methods and tools that rely on such data will be our focus in subsequent chapters.

9.2 Calculations

With LCI data for multiple processes, it is possible to calculate various quantities for the entire life cycle network formed by connecting the process modules. The data in each process module are usually normalized to a unit output of the primary product. For example, the inventory in Table 9.1 is normalized to produce 1 kWh of the main product: electricity from natural gas. For relatively simple networks, it is easy to determine the scaling factors for each module in order to connect them and thus obtain results for the selected functional unit, as illustrated in the following examples. For more complicated networks, it may be necessary to rely on the mathematical framework described in Section 10.1 and software.

Example 9.1 A module P1 for fuel production has an input of 50 L of crude oil to produce 20 L of fuel. It emits 2 kg SO_2 and 10 kg CO_2. A module P2 for electricity generation produces 10 kWh of electricity from 2 L fuel. It emits 0.1 kg SO_2 and

Table 9.3 Sources of life cycle inventory data.

Source	Scale	Features	Reference
NREL	Process	Free database for the USA compiled by the National Renewable Energy Laboratory	[1]
GREET	Process	Free tool for assessing the life cycle of transportation fuels	[3]
Ecoinvent	Process	Commercial and comprehensive database for large number of processes with wide geographical coverage	[4]
NEI	Economic sectors	National emissions inventory of air pollutants. Compiled by the US Environmental Protection Agency	[5]
TRI	Economic sectors	Toxics release inventory. Compiled by the US Environmental Protection Agency	[6]
FAOstat	Nations	Data related to food and agriculture for many countries. Compiled by the Food and Agriculture Organization of the United Nations	[7]

1 kg CO_2. For a network constructed by connecting these modules, determine the life cycle consumption of crude oil and the emissions of CO_2 and SO_2 in generating 1000 kWh electricity.

Solution

The two modules are shown in Figure 9.1. To generate 1000 kWh of electricity, P2 needs to be scaled by a factor of $\frac{1000}{10} = 100$. This scaled-up module will require 200 L of fuel. To supply this amount of fuel, P1 needs to be scaled by a factor of $\frac{200}{20} = 10$. The resulting life cycle network is shown in Figure 9.2. As can be seen in this figure, the total consumption of crude oil is 500 L, the emission of SO_2 is 30 kg, and that of CO_2 is 200 kg.

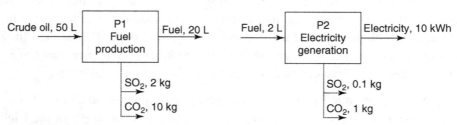

Figure 9.1 Modules for fuel production and electricity generation.

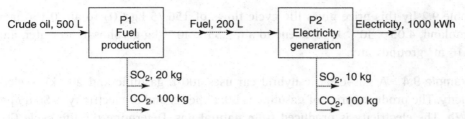

Figure 9.2 Life cycle network for producing 1000 kWh of electricity. The modules in Figure 9.1 were scaled linearly to obtain this network. Module P1 is scaled by a factor of 10 and P2 by a factor of 100.

Example 9.2 Determine the total amount of benzene emitted in the extraction and conversion of natural gas into 500 kWh of electricity.

Solution

This problem involves two processes: the extraction of natural gas, and the conversion of natural gas into electricity. The data in Table 9.1 are for the conversion process. They show that 10^{-8} kg of benzene is emitted per kWh of electricity. For the extraction of natural gas, we need similar data. Sources such as the NREL inventory database [1] contain such data. From this source, we can find the module of "natural gas, at extraction site." It shows that benzene emission is 5.31E-06 kg per cubic meter of natural gas extracted.

From the conversion module in Table 9.1, generating 500 kWh of electricity emits

$$500 \text{ kWh} \times 10^{-8} \frac{\text{kg benzene}}{\text{kWh}} = 5 \times 10^{-6} \text{ kg benzene}$$

The volume of natural gas required to produce 500 kWh electricity is

$$500 \text{ kWh} \times 0.298 \frac{\text{m}^3}{\text{kWh}} = 149 \text{ m}^3 \text{ natural gas}$$

For the extraction module, the benzene emission for 149 m^3 of natural gas is

$$5.31 \times 10^{-6} \frac{\text{kg benzene}}{\text{m}^3} \times 149 \text{ m}^3 = 7.91 \times 10^{-4} \text{ kg benzene}$$

Therefore, the total mass of benzene emitted in the extraction and conversion steps for producing 500 kWh electricity is

$$5 \times 10^{-6} + 7.91 \times 10^{-4} = 7.96 \times 10^{-4} \text{ kg benzene}$$

Example 9.3 Using the input–output (IO) data in Table 9.2, determine the life cycle emissions and resource use in the generation of 500 kWh of electricity in the USA. The producer price of electricity is 5 cents/kWh.

Solution

The producer price of 500 kWh electricity is $500 \times 0.05 = \$25$. This is the final demand of electricity. Multiplying the coefficients for the electricity sector in

Table 9.2 by this price gives life cycle flows of 156.75 kg CO_2 to air, 0.156 m^2y cropland, 4.05×10^{-4} kg benzene to air, 2.55×10^{-7} kg glyphosate to water, and 0.16 m^3 groundwater.

Example 9.4   A gas–electric hybrid car uses 100 L gasoline and 200 kWh electricity. The producer price of gasoline is $0.39 per L, and of electricity is $0.05 per kWh. The electricity is produced from natural gas. Determine the life cycle CO_2 emission for this system based on (a) an input–output model and (b) constructing a hybrid model that uses the process data in Table 9.1.

Solution
Using the input–output model for the USA in Table 9.2,

$$IO \text{ model life cycle } CO_2 \text{ emissions} = (0.503 \times 0.39 \times 100) + (6.27 \times 0.05 \times 200)$$
$$= 82.3 \text{ kg}$$

In this case, the electricity sector represents the average for all modes of generation in the USA.

Knowing that electricity is generated from natural gas, we can use detailed process data for this process and combine it with the IO data for gasoline. Thus,

$$\text{Hybrid model life cycle } CO_2 \text{ emissions} = (0.503 \times 0.39 \times 100) + (0.585 \times 200)$$
$$= 136.6 \text{ kg}$$

Note that in the hybrid model, CO_2 emissions from activities upstream of the power generation process are ignored.

Allocation. Consider a process that makes two products and emits 100 kg of CO_2. Each product is used further in other processes. A practical challenge in assessing the CO_2 emission from each product's life cycle is partitioning the 100 kg emission between the two products. Such processes with multiple coproducts are quite common. For example, as shown in Figure 8.4, corn farming produces corn and stover, the corn being used for food and the stover for maintaining soil quality or for making cellulosic ethanol. Corn fermentation produces ethanol and distillers' dried grains, the former being used as a chemical intermediate and the latter as animal feed. Similarly, power generation from coal produces electricity, heat, ash, carbon dioxide, and other byproducts, and the fractionation of crude oil produces a large number of products ranging from methane and ethane to gasoline, aviation fuel, and tar. The following illustration introduces the approach and issues in the allocation of LCI data among coproducts.

Consider a process that converts resources A and B into two products, P and Q, as shown in Figure 9.3. Since the products are likely to be used to make other items or will be sold to different buyers, the raw materials and emissions need

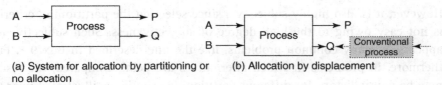

Figure 9.3 Allocation methods.

to be allocated to each product. Three approaches for addressing this allocation problem are as follows.

BOX 9.1 Allocation Problems in Daily Life

Splitting taxi fare between passengers. Three friends take a taxi to the airport. The fare is $30. How should it be split between the three passengers? The response that comes to your mind is likely to be that each passenger should pay $10. However, one passenger could argue that the fare must be split according to their bank balances, as done in Finland for speeding tickets. Another argument could be to split the fare in proportion to the mass of the passenger and her luggage, or in inverse proportionality to how long each passenger has to wait for her flight, or using some other criterion. Any of these and other such allocation methods may be used to split the fare. Thus, allocation is made by a subjective decision.

  Displacement due to leftovers. You go to a restaurant with friends and order your favorite item from the menu. However, you are unable to consume all of it and take the leftovers home, and eat them for lunch the next day. The leftover food displaces the impact of the food that you would have prepared for lunch on that day. Thus, this prevented impact can be taken as a credit to the impact of the restaurant meal that you had purchased. In this approach, we expand the system boundary to include the next day's lunch, to enable displacement.

- *Partitioning.* This approach splits or partitions the quantity of raw materials and emissions on the basis of a selected feature of the products, such as mass, energy, or monetary value. Thus, in Figure 9.3a, a quantity, say, x is partitioned between the two products as follows:

$$x_P = \frac{m_P}{m_P + m_Q} x \tag{9.1}$$

$$x_Q = \frac{m_Q}{m_P + m_Q} x \tag{9.2}$$

Here, the allocated quantities x_P and x_Q are determined in proportion to their masses, m_P and m_Q. Other allocation criteria could be monetary value, fuel value, or exergy. This is the easiest and most common-allocation approach.

However, it is also highly subjective, since selecting the partitioning criterion is not easy owing to the high degree of subjectiveness. Such subjectiveness appears in many common problems, like the one described in Box 9.1. Furthermore, in problems in which alternatives are being compared, the results may be contradictory for different partitioning criteria, as illustrated near the end of this section.

- *Displacement*. Another approach for allocation is based on displacement or system expansion. This is illustrated in Box 9.1. In this case, the byproduct is considered to substitute the same flow as that produced by a conventional process. The process being analyzed takes credit for displacing the conventional process. For example, converting sugarcane into ethanol also produces bagasse, which is commonly converted to electricity. In the displacement approach, this electricity from bagasse is considered to displace the electricity obtained from conventional sources in the region. Then the emissions from these conventional sources for producing the displaced electricity are subtracted from the sugarcane ethanol analysis. This approach is illustrated in Figure 9.3b, where the emissions from the conventional process used to produce the byproduct Q, denoted by the shaded box, are subtracted from those for the process being analyzed. A challenge that may be encountered in using this approach is that the process being displaced may itself have some coproducts that also need to be displaced and so on, resulting in an infinite displacement problem.

- *No allocation*. If a process is not fully understood, then all the products from it may not be known. In this case, partitioning or displacement methods cannot be applied. This situation is often encountered in ecological systems. For example, solar energy drives a forest ecosystem, but the products and services provided by this system may not all be known. Another reason behind this approach is that partitioning and displacement implicitly assume that the coproducts can be produced independently, but this may not be possible for many coproducts. Examples include corn and corn stover, the yellow and white portions of an egg, and goods and services from ecosystems. This approach may be illustrated with the help of Figure 9.3a. In this case, the allocated quantities x_P and x_Q may be determined as

$$x_P = x \tag{9.3}$$

$$x_Q = x \tag{9.4}$$

Comparing these equations with Equations 9.1 and 9.2, we can see that in the no-allocation approach, the allocation coefficient is unity for both products. A challenge in this approach is that when such coproducts are combined, the allocated quantities cannot be added since that will result in double counting.

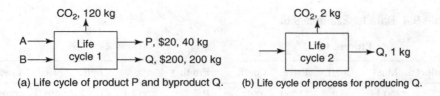

(a) Life cycle of product P and byproduct Q.   (b) Life cycle of process for producing Q.

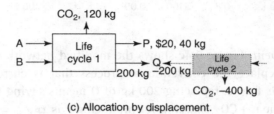

(c) Allocation by displacement.

Figure 9.4 Illustration of allocation methods.

This can be understood by considering a process that combines P and Q. With x_P and x_Q calculated by Equations 9.3 and 9.4, respectively, we have $x_P + x_Q = 2x$, which is not possible since only the amount x was used to produce P and Q. This approach is mainly used in the technique of emergy analysis, which is described and illustrated in Chapter 14 along with special rules that have been devised to avoid double counting.

Example 9.5 For the life cycle network shown in Figure 9.4, the main product is P and the byproduct is Q. The total CO_2 emitted in the supply chain is 120 tons. The masses and costs of P and Q are as shown in Figure 9.4a. Another process for producing only byproduct Q is shown in Figure 9.4b. Allocate the CO_2 emitted from life cycle 1 between the two products using partitioning, displacement, or no-allocation.

Solution
The total CO_2 emitted for life cycle 1 is $x = 120$ kg.
  Partitioning. Using monetary values and Equations 9.1 and 9.2,

$$x_{P,\$} = \frac{20}{20 + 200} \times 120$$
$$= 10.9 \text{ kg}$$
$$x_{Q,\$} = \frac{200}{20 + 200} \times 120$$
$$= 109.1 \text{ kg}$$

Similarly, using mass allocation, $x_{P,\text{mass}} = 20$ kg, and $x_{Q,\text{mass}} = 100$ kg.

Table 9.4 Data for Example 9.6.

	Process 1					Process 2			
Product	Mass	Coeff.	Price	Coeff.	Product	Mass	Coeff.	Price	Coeff.
P	200 kg	0.8	$1000	0.91	P	100 kg	0.5	$4950	0.99
Q	50 kg	0.2	$100	0.09	R	100 kg	0.5	$50	0.01

Displacement. Figure 9.4c shows the modified network with byproduct Q considered to replace the conventional process that is currently producing Q. As shown in this figure, replacing 200 kg of Q means saving the emission of 400 kg of CO_2. So the net CO_2 emission for producing P is $x_{P,disp} = 120 - 400 = -280$ kg CO_2.

No allocation. Without allocation, we use Equations 9.3 and 9.4 to get $x_{P,none} = 120$ kg CO_2 and $x_{Q,none} = 120$ kg CO_2.

Example 9.6 Two alternative processes are to be compared in terms of their water use per unit of the desired product, P. Process 1 uses 10,000 L of water to produce 200 kg of P and 50 kg of Q, which are worth $1000 and $100, respectively. Process 2 uses 12,000 L of water to produce 100 kg of P and 100 kg of R, which are worth $4950 and $50, respectively. Using mass allocation and monetary allocation, identify in each case the process that is more water-efficient per kilogram of P.

Solution

The allocation coefficients for the two processes using mass and monetary allocation are summarized in Table 9.4.

By mass allocation, the water required per kilogram of P is $0.8 \times 10,000 = 8000$ L/kg of P for Process 1, and $0.5 \times 12,000 = 6000$ L/kg for Process 2. Thus, by mass allocation, Process 2 is to be preferred for P.

By monetary allocation, the water required per kilogram of P is $0.91 \times 10,000 = 9100$ L/kg for Process 1, and $0.99 \times 12,000 = 11,880$ L/kg for Process 2. By monetary allocation, Process 1 is preferable for minimizing the water use in making P.

Example 9.6 provides conflicting results for the two allocation methods, making it difficult to choose between the two processes. Given the subjectivity involved in choosing the allocation method, the recommended approach by practitioners of life cycle assessment is to prefer the displacement approach whenever possible. If this is not possible, allocation based on monetary values is recommended. Often, multiple allocation methods are applied to a problem that involves choosing between alternatives. If the decision does not change with the allocation method, then the decision is robust to allocation.

9.3 Uncertainty

Sustainability assessment methods combine information from diverse sources, which are bound to be uncertain. Three categories of uncertain information in these methods are:

1 *Parametric uncertainty.* This is uncertainty in the measured or observed *data*. It represents the inherent variability over time of the measured variables, sensor noise, and difficulties in measurement.
2 *Scenario uncertainty.* This represents variability due to the *choices* made by the modeler. It includes the effect of the allocation method or value judgment used for combining multiple outputs.
3 *Model uncertainty.* This is uncertainty in *mathematical relationships*. It includes errors in the development of physical, input–output, or hybrid models, owing to the simplification of complicated relationships, and the extrapolation of models of similar systems.

Life cycle inventory data, like all measured data, are uncertain owing to errors in measurement, aggregation of data from multiple sources and systems, and system variability. Therefore, it is natural to account for the effect of uncertainties in the results to determine whether the difference in results is within the margin of error or is statistically significant. In this section, we will learn about some basic methods that are used in sustainability assessment.

All methods for uncertainty analysis require information about the uncertainty in the data and the models used in the analysis. This information is used to determine the uncertainty in the calculated quantities. The nature of the analysis approach changes with the availability of uncertainty information, which can be of three types.

- *Qualitative information.* Often, only qualitative information is available about uncertainty. In this case, methods such as pedigree analysis may be used.
- *Quantitative information, but without distributions.* In this case, statistical distributions are not available, but the range of variation may be assumed. In such situations, the sensitivity analysis approach is used.
- *Quantitative information with distributions.* In this case, detailed statistical distributions are available. The approach of Monte Carlo analysis may be used.

We will learn about the first two approaches.

Pedigree method. This approach is useful in the absence of quantitative information about uncertainty. It relies on selecting a numerical score representing the quality of the flows and processes in the LCI. Pedigree matrices for assigning numerical scores to LCI flows are shown in Table 9.5; smaller values indicate less uncertainty. The matrix in Table 9.5 is used to evaluate various flows in an inventory in terms of their reliability and representativeness. Assigning the scores requires insight into the nature of the inventory data, but some degree of subjectiveness may be unavoidable. If no information is available about each category, the default value of 5 is assigned, indicating the highest uncertainty.

Example 9.7   For the fuel production module in Figure 9.1, the crude oil data form an estimate from an industry document that is five years old. The data are from the same geographic region and same process type as that of the study, and represents average data over a year for the primary process being used in the region. Determine values of the pedigree matrix for this flow.

Solution
Based on Table 9.5, we see that for the category of flow reliability, the value is 4 since the data are documented. For temporal correlation, the value is 2 since the data are less than six years old. Continuing in this manner, the pedigree vector for this flow is [4, 2, 1, 4, 1].

The pedigree method results in a matrix of uncertainty indicators for each flow and process in the inventory. These data may be combined as a weighted sum of flows of the same type in each module and their uncertainties, followed by normalization by the total flow:

$$u_{ij} = \frac{\sum_k x_{ik} u_{ijk}}{\sum_k x_{ik}} \tag{9.5}$$

where i represents the impact category, j is the uncertainty type, and k is the type of inventory module.

Example 9.8   The LCI for truck transportation consists of diesel production and transportation. Information about emissions of particulate matter of sizes 2.5 and 10 microns and their pedigree matrices are given in Table 9.6. Determine indicators for the uncertainty in the life cycle emissions for each category.

Solution
Uncertainty in flow reliability for PM2.5 and PM10 are calculated by using Equation 9.5 as follows:

$$u_{PM2.5,1} = \frac{(2 \times 10^{-5} \times 5) + (6.27 \times 10^{-7} \times 1)}{2 \times 10^{-5} + 6.27 \times 10^{-7}}$$

$$= 4.9$$

$$u_{PM10,1} = \frac{(2.79 \times 10^{-5} \times 1) + (6.31 \times 10^{-7} \times 3)}{2.79 \times 10^{-5} + 6.31 \times 10^{-7}}$$

$$= 1.0$$

Table 9.5 Pedigree matrix: flow indicators [8].

Indicator		1	2	3	4	5 (default)
Flow reliability		Verified[a] data based on measurements	Verified data based on a calculation *or* non-verified data based on measurements	Non-verified data based on a calculation	Documented estimate	Undocumented estimate
Flow representa-tiveness	Temporal correlation	Less than 3 years of difference[b]	Less than 6 years of difference	Less than 10 years of difference	Less than 15 years of difference	Age of data unknown or more than 15 years
	Geographical correlation	Data from same resolution *and* same area of study	Within one level of resolution *and* a related area of study[c]	Within two levels of resolution *and* a related area of study	Outside of two levels of resolution *but* a related area of study	From a different or unknown area of study
	Technological correlation	All technology categories[d] are equivalent	*Three* of the technology categories are equivalent	*Two* of the technology categories are equivalent	*One* of the technology categories is equivalent	*None* of the technology categories are equivalent

[a] Verification may take place in several ways, e.g., by on-site checking, by recalculation, through mass balances or cross-checks with other sources.

[b] Temporal difference refers to the difference between date of data generation and the date of representativeness as defined by the goal of the project.

[c] A related area of study is defined by the user and should be documented in the geographical metadata.

[d] Technology categories are process design, operating conditions, material quality, and process scale.

Table 9.5 (cont.)

Data collection methods	Representative data from >80 percent of the relevant market[e], over an adequate period[f]	Representative data from 60–79 percent of the relevant market, over an adequate period *or representative* data from >80 percent of the market, over a shorter period of time	Representative data from 40–59 percent of the relevant market, over an adequate period *or representative* data from 60–79 percent of the market, over a shorter period of time	Representative data from <40 percent of the relevant market, over an adequate period of time *or representative* data from 40–59 percent of the market, over a shorter period of time	Unknown *or* data from a small number of sites *and* from shorter periods

[e] The relevant market should be documented in the study. The default relevant market is measured in production units.

[f] Adequate time period can be evaluated as a time period long enough to even out normal fluctuations. The default time period is 1 year, except for emerging technologies (2–6 months) or agricultural projects >3 years.

Table 9.6 Emissions and uncertainty for inventory of truck transportation [8].

Flow	Process	Amount	Unit	Flow reliability	Temporal correlation	Geographical correlation	Technological correlation	Data collection method
PM2.5/ air	Truck transport	2.00E−05	kg	5	4	3	2	1
PM2.5/ air	Diesel, at pump	6.27E−07	kg	1	2	3	4	5
PM10/ air	Truck transport	2.79E−05	kg	1	1	1	1	1
PM10/ air	Diesel, at pump	6.31E−07	kg	3	3	3	3	3
PM2.5/ air	TOTAL	2.06E−05	kg	4.9	3.9	3.0	2.1	1.1
PM10/ air	TOTAL	2.86E−05	kg	1.0	1.0	1.0	1.0	1.0

Here, the second subscript, 1, stands for the first uncertainty type of flow reliability. The calculated indicators for each uncertainty type and emission are shown in the last two rows of Table 9.6. The results convey that the data for PM10 are of higher quality since the uncertainty is lower in this case.

Sensitivity analysis. When information about the quantitative range of variation is available, this approach determines the sensitivity of the results to errors in contributing variables. This helps in identifying the variables that have the largest effect on the variability of the results. More effort may then be directed toward determining the uncertainty in these variables.

Example 9.9   The environmental impacts z_A and z_B of two product options, Product A and Product B, are calculated by combining their emissions x and y as $z_A = 2x + 5y$ and $z_B = 4x + y$. Knowing that $x = 10$ and $y = 50$, which product has a smaller impact? How robust are your results to uncertainty in the data?

Solution
Based on the specified values of x and y, we calculate $z_A = 270$ and $z_B = 90$. Thus, Product B has a smaller impact.

To perform a sensitivity analysis, since we do not have information about uncertainties in x and y, we will assume the uncertainty in each variable to be ± 10 percent. This implies that x may take values between 9 and 11, while y is between 45 and 55. The results of the sensitivity analysis are shown in Table 9.7. These results show the range of z_A to be 243–297, while that of z_B is 81–99. Thus, even after accounting for the variability, Product B has a smaller impact. The results in Table 9.7 also indicate the sensitivity of z_A and z_B to the input variables, x and y. From the first two rows we can see that for a ± 10 percent change in x, z_A varies by ± 0.8 percent while z_B varies by ± 4.7 percent. For a 10 percent change in y, z_A varies by ± 9.3 percent and z_B varies by ± 5.9 percent. Thus, for these values of x and y, z_A is more sensitive to changes in y, and z_B is almost equally sensitive to changes in both x and y, but slightly more to the latter variable. These results imply that if z_A is the variable of interest, we should pay more attention to getting more accurate measurements of y. If z_B is the variable of interest, the accuracy of both variables is equally important.

Table 9.7 Results of sensitivity analysis.

x	y	z_A	z_B
9	45	243	81
11	45	247	89
9	55	293	91
11	55	297	99

Example 9.10 For the system in Figure 9.4a, suppose that the quantity of CO_2 emitted is related to the inputs of A and B as follows: $x_{CO2} = 10x_A + 20x_B$. For the values depicted in the figure, $x_A = 2$ and $x_B = 5$. (a) Determine the sensitivity of the calculated CO_2 emissions to uncertainties in the two input variables. (b) If the error in x_A is known to be ± 20 percent and in x_B to be ± 10 percent, determine the error in x_{CO2}.

Solution
(a) Let us consider the effect of ± 10 percent variations in x_A and x_B on x_{CO2}. The range of x_A will be $[1.8, 2.2]$ and of x_B will be $[4.5, 5.5]$. The resulting range of x_{CO2} will be $[108, 132]$. The value of x_{CO2} is more sensitive to variations in x_B than in x_A.

(b) The range of x_A is $[1.6, 2.4]$, and of x_B is $[4.5, 5.5]$. Then, x_{CO2} varies between $[106, 134]$.

9.4 Summary

In this chapter we learned about the second step in sustainability assessment: inventory analysis. This step involves obtaining data for all the activities included in the life cycle model. Databases containing such data are available for individual processes and for economic sectors. For activities that produce multiple outputs, emissions and resource use need to be allocated between these outputs. Allocation is a subjective exercise, and we covered methods based on partitioning, displacement, and no allocation. Finally, since data are bound to contain errors, we learned about the pedigree method and sensitivity analysis.

Key Ideas and Concepts

- Process network
- Hybrid network
- Allocation by partitioning
- Life cycle inventory
- Input–output network
- Functional unit
- Allocation by displacement
- Uncertainty analysis

9.5 Review Questions

1. Explain the meaning of a scaling factor.
2. Define final demand.
3. Which are the three methods that can be used to allocate emissions from a process between multiple products?
4. If different allocation methods give conflicting results, what is the recommended approach?

5. Which method should be used for uncertainty analysis if information about uncertainty is qualitative?

Problems

9.1 Find the following data from the appropriate database in Table 9.3. State your assumptions, if any.
1. The fossil energy use and particulate matter emission to air from hauling a ton of goods from Los Angeles to Las Vegas.
2. The carbon dioxide emitted from the electricity used in the production of phenol formaldehyde resin.
3. Determine the top five emissions by mass from the cement sector (NAICS 3273) in the USA.
4. Which countries are the top ten emitters of nitrous oxide (N_2O) from fertilizer use in agriculture?

9.2 Process 1 produces P, Q, and R. The masses of P, Q, and R are 100 kg, 20 kg, and 5 kg, respectively, and their monetary values are $1000, $100, and $20. This process emits 500 kg of CO_2. Process 2 produces P and S with masses of 100 kg and 5 kg and monetary values of $1000 and $0. This process emits 1000 kg CO_2. Process 3 produces 1 kg of S with CO_2 emission of 0.8 kg. Choose the process for producing P that has the lowest CO_2 emission by using
1. mass-based allocation
2. monetary allocation
3. displacement and mass-based allocation
4. displacement and monetary allocation

9.3 A life cycle assessment of disposable cold cups conducted for Starbucks Coffee Company and NatureWorks LLC [9] compared cups and lids made out of polypropylene (PP), polyethylene terephthalate (PET), and polylactide acid (PLA). The report provides the following information about flow representativeness of the data for PET.
- Data are from 2002 to 2007 for the report prepared in 2009.
- Geographical coverage is for the USA.
- Technological data are based on detailed models of manufacturing.
Using this information, suggest uncertainty indicators for the four categories of flow representativeness.

9.4 Sustainability assessment of emerging technologies can be very useful for guiding their development in a way that prevents unintended harm. Since such technologies are often not yet implemented, inventory data are estimated by theoretical models. Using the pedigree matrix, determine the quality indicators for flow reliability and flow representativeness.

9.5 An integrated gasification combined cycle (IGCC) plant can convert coal and water to electricity and hydrogen. The carbon dioxide from such a process may be captured by a double-stage solvent such as Selexol. The main inputs to such a system are 5426 tons of coal and 21,787 m^3 of water to produce 560 tons of hydrogen, 30.3 MW of electricity, 11,253 tons of pure carbon dioxide, and 1293 tons of carbon dioxide in flue gas. The cost of the hydrogen is $2110 per ton, of the electricity is $1800 per MW, of the pure CO_2 is $37 per ton, and of the flue gas is zero [10]. Allocate the water use among the products of this process by the following approaches and discuss the pros and cons of each approach for this allocation problem:

1. mass-based allocation
2. energy-based allocation
3. monetary value-based allocation

What additional information will be required to do allocation by displacement in this problem?

9.6 A well produces 6000 standard cubic feet (scf) of gas for every barrel (bbl) of hydrocarbon liquid.

1. Assuming a heating value of 1000 BTU/scf for gas and six million BTU/bbl for hydrocarbon liquid, allocate the leakage (emission) from this well between the two products.
2. Considering a gas density of 25 g/scf and an oil density of 100 kg/bbl, allocate the emission in terms of mass.
3. Using prices of $3.50 per 1000 scf for gas and $90 per bbl for hydrocarbon liquid, perform a monetary allocation.

9.7 The information about some LCI modules is as follows.

- 1 kWh of electricity requires 0.64 kg of water, and the process emits 1.12 kg of CO_2, 1.1×10^{-3} of SO_2, and 1.97×10^{-3} kg of NO_x.
- 1 kg of polypropylene requires 31.9 kg of H_2O, 0.11 kWh of electricity, and the process emits 1.82 kg of CO_2 and 0.0031 kg each of SO_2 and NO_x.
- 1 kg of copper uses 4.38 kg of H_2O and 0.49 kWh of electricity, and the process emits 0.099 kg of CO_2 and 0.41 kg of SO_2
- 1 kg of steel consumes 0.22 kWh of electricity and 10.47 kg of H_2O, and the process emits 0.074 kg of CO_2.
- 1 kg of plastic extension requires 0.51 kWh of electricity, 4.7×10^{-4} kg of polypropylene, and 19.72 kg of H_2O.
- 1 paper towel dispenser requires a plastic extension that weighs 2.79 kg, 3.19 kg of polypropylene, 43.76 kWh of electricity, 0.17 kg of copper, and 0.17 kg of steel.

Construct a process-model-based life cycle network for producing a paper towel dispenser. Determine the amount of water used and the CO_2 emitted in the life cycle of one paper towel dispenser. (Provided by Tanner Anderson, Eric Bissonnette, Stephanie Gase, and Ruonan Zhao).

9.8 Answer the following questions using data from Table 9.2.

1. A company spends $10,000 per year on courier services. Determine the resulting life cycle emission of carbon dioxide and benzene. What type of network model did you use for this calculation?

2. A household's annual expenses include $1000 for gasoline, $800 for electricity, and $300 for vehicle repair. Determine the cropland needed to support these expenses. Explain which steps of the life cycle may require the cropland most.

9.9 A life cycle assessment of an office building finds the climate change impact of the building materials to be 4818 tons CO_2 equivalent, of the construction to be 822 tons CO_2 equivalent, of the heating to be 11,219 tons CO_2 equivalent, and of the electrical services to be 25,230 tons CO_2 equivalent. The pedigree vector values for data representing these activities are (2, 1, 2, 2, 2, 2), (2, 1, 2, 2, 2, 3), (2, 2, 1, 1, 1, 1), and (2, 2, 1, 1, 1, 1), respectively. Determine the pedigree vector values for the total climate change impact of these activities.

9.10 In a product life cycle, the emission of CO_2 and the use of water depend on three variables, x_i, $i = 1, 2, 3$, according to the following equations: $y_{CO_2} = 100x_1 + 11x_2 - 7x_3$; $y_{H_2O} = 22x_1 - x_2 + 50x_3$. The nominal values of x_i are 10, 50, and 5.

1. Using sensitivity analysis, determine the variable whose accuracy is most important.

2. If the range of variation of each variable is $x_1 \sim [45, 55]$, $x_2 \sim [24, 36]$, $x_3 \sim [4, 7]$, determine the range of values for CO_2 emissions and water use. Also determine the most important variables.

References

[1] National Renewable Energy Laboratory. US life cycle inventory database. www.lcacommons.gov/nrel/search, accessed November 3, 2014.

[2] Y. Yang, W. W. Ingwersen, T. R. Hawkins, M. Srocka, and D. E. Meyer. USEEIO: a new and transparent United States environmentally-extended input–output model. *Journal of Cleaner Production*, 158:308–318, 2017.

[3] Argonne National Laboratory. The greenhouse gases, regulated emissions, and energy use in transportation (GREET) model. http://greet.es.anl.gov, accessed January 31, 2012.

[4] Swiss Centre for Life Cycle Inventories. Ecoinvent life cycle inventory database. www.ecoinvent.ch, accessed January 18, 2013.

[5] Environmental Protection Agency. National emissions inventory. www.epa.gov/ttn/chief/trends, accessed August 22, 2015.

[6] Environmental Protection Agency. Toxics release inventory. www.epa.gov/triexplorer, accessed August 22, 2015.

[7] Food and Agriculture Organization. FAOSTAT. http://faostat3.fao.org, accessed July 8, 2017.

[8] A. Edelen and W. W. Ingwersen. The creation, management, and use of data quality information for life cycle assessment. *The International Journal of Life Cycle Assessment*, 23(4):759–772, 2018.

[9] PE Americas. Comparative life cycle assessment Ingeo™ biopolymer, PET, and PP drinking cups. www.natureworksllc.com/~/media/The_Ingeo_Journey/EcoProfile_LCA/LCA/PEA_Cup_Lid_LCA_FullReport_ReviewStatement_121209_pdf.pdf, 2009, accessed November 23, 2018.

[10] B. Kursun, S. Ramkumar, B. R. Bakshi, and L.-S. Fan. Coal gasification by conventional versus calcium looping process: a life cycle energy, global warming, land use and water assessment. *Industrial & Engineering Chemistry Research*, 53(49):18910–18919, 2014.

10 Mathematical Framework

> For want of a nail a shoe was lost
> For want of a shoe the horse was lost
> For want of a horse the rider was lost
> For want of a rider the message was lost
> For want of a message the battle was lost
> For want of a battle the kingdom was lost
> All for the want of a horse shoe nail.
>
> Thirteenth century proverb

In Chapter 9, we learned about sources of data for constructing a life cycle network and for calculating flows to and from the environment for this network. The calculations in Chapter 9 were for relatively simple systems. If user-friendly software is available, that level of understanding may also be adequate for doing calculations for more complicated and larger systems. However, deep understanding about the calculations requires a more rigorous mathematical framework. Such an understanding is essential for developing new methods that overcome the shortcomings of the current methods and software. It may also be needed if conventional software and readily available inventory are inadequate for the desired application. This could be the case for emerging technologies. In this chapter we will learn about the mathematical framework underlying all sustainability assessment methods that evaluate a life cycle. These methods form the foundation of the methods we will cover in the remaining chapters of this part of the book. In this chapter, most equations are presented with and without the use of matrix notation, so that readers less familiar with linear algebra can also utilize the framework.

10.1 Process Network Analysis

The most common way of constructing a life cycle network is by connecting models of individual industrial or economic activities, as illustrated in Figure 8.3a. As we learned in Section 9.1, data for individual activities are commonly available from industrial information or on simulation and are included in various life cycle inventory (LCI) databases, like those listed in Table 9.1. This section presents a systematic approach for combining multiple modules of data to build and assess a process network [1].

We start with a simple example in order to understand the approach, and then generalize it for any process network. Consider the two modules shown in Figure 9.1. In Example, 9.1, we combined these modules to determine the life cycle resource consumption and emissions for producing 1000 kWh of electricity. The resulting network is shown in Figure 9.2. Obtaining this result involves scaling the electricity generation process by a factor of 100 so that it produces the desired 1000 kWh of electricity. This process then requires 200 L of fuel. Obtaining this quantity of fuel from the fuel production process requires that process to be scaled up by a factor of 10. Connecting the two processes results in a network that produces 1000 kWh of electricity by consuming 500 liters of crude oil, while emitting 30 kg of SO_2, and 200 kg of CO_2.

10.1.1 Mathematical Framework

The example above is simple enough to permit easy combination of the two modules. However, in practice, for process inventory data like that in Table 9.1 for multiple processes, the network may become quite complicated. For example, consider the case in which the fuel production module in Figure 9.1 has an additional input of 1 kWh of electricity. For this case, since electricity is both an input and output, determining scaling factors and obtaining the network such as that in Figure 9.2 is not as straightforward as it was in Problem 9.1. Therefore, for more complicated and larger problems, a general mathematical formulation is needed for analyzing the life cycle network.

In general, life cycle modules contain two types of flows: economic and environmental.

1. Economic flows are generated by human activity and are used in other economic activities or by consumers. Such flows are usually valuable to society and are not discarded directly into the environment. Examples of such flows are the fuel and electricity in Figure 9.1. The quantity of these flows consumed by society (not by other economic sectors) is called the *final demand*.
2. The second type of flow is to or from outside the system boundary. Such flows include direct interaction with the environment, and are inputs from nature and emissions to it. In Figure 9.1, these flows include crude oil from the environment, and emissions of sulfur dioxide and carbon dioxide to the environment.

For this illustration, both fuel and electricity can have a final demand since both of these are economic flows that could have uses outside the system boundary. Let f_1 and f_2 represent the final demand of these two flows. In Figure 9.2, the final demand for fuel is zero, while that of electricity is 1000 kWh. Let s_1 and s_2 represent scaling factors or multipliers for the two modules, so that they may be connected, as shown in Figure 9.2, to produce the specified outputs. We can write the following two equations based on the conservation of each economic flow:

$$20s_1 - 2s_2 = f_1 \tag{10.1}$$

$$10s_2 = f_2 \tag{10.2}$$

Equation 10.1 is for fuel and indicates that the quantity of fuel produced is $20s_1$, while the quantity used is $2s_2$, their difference being the final demand. Similarly, Equation 10.2 is the balance for electricity. These equations may be written in matrix notation as

$$\begin{bmatrix} 20 & -2 \\ 0 & 10 \end{bmatrix} \begin{bmatrix} s_1 \\ s_2 \end{bmatrix} = \begin{bmatrix} f_1 \\ f_2 \end{bmatrix} \tag{10.3}$$

or, more generally, as

$$As = f \tag{10.4}$$

The matrix A consists of the flowrates of the economic products from each module, with inputs having a negative sign. Note that A is a matrix of products times processes, that is each row represents a product and each column represents a process. This is the *technology matrix*.

Similar balance equations may also be written for environmental flows:

$$r_{Crude} = -50s_1 \tag{10.5}$$

$$r_{SO_2} = 2s_1 + 0.1s_2 \tag{10.6}$$

$$r_{CO_2} = 10s_1 + s_2 \tag{10.7}$$

Here, r_i represents the total environmental flow of the ith component. Equations 10.5, 10.6, and 10.7 are the material balance equations for crude oil, SO_2, and CO_2, respectively. These equations may be represented in matrix notation as

$$\begin{bmatrix} -50 & 0 \\ 2 & 0.1 \\ 10 & 1 \end{bmatrix} \begin{bmatrix} s_1 \\ s_2 \end{bmatrix} = \begin{bmatrix} r_{Crude} \\ r_{SO_2} \\ r_{CO_2} \end{bmatrix} \tag{10.8}$$

Again, the negative sign indicates inputs, and the positive sign indicates outputs. More generally, we can write

$$Bs = r \tag{10.9}$$

The matrix B is called the *intervention matrix*, and r is the vector of resource consumption and emissions. For a specific consumption of products in society or final demand f the resource consumption and emissions may be calculated as

$$r = BA^{-1}f \qquad (10.10)$$

Let us now apply this framework to several examples.

Example 10.1 Consider the fuel and electricity modules shown in Figure 9.1. Calculate the life cycle flows of crude oil, sulfur dioxide, and carbon dioxide in the production of 1000 kWh of electricity. Note that this example is identical to Example 9.1.

Solution
For this example, the final demand is specified as $f_1 = 0$ and $f_2 = 1000$ kWh. Solving Equations 10.1 and 10.2 simultaneously, the scaling factors can be found to be $s_1 = 10$ and $s_2 = 100$. Substituting these scaling factors into Equations 10.5–10.7, we get $r_{crude} = -500$ L, $r_{SO_2} = 30$ kg, and $r_{CO_2} = 200$ kg.

Using matrices, define $f = \begin{bmatrix} 0 \\ 1000 \end{bmatrix}$. The resource consumption and emissions can then be expressed as $r = \begin{bmatrix} -500 \\ 30 \\ 200 \end{bmatrix}$, which matches the values in Figure 9.2.

Example 10.2 Consider the final demand to be 100 kWh of electricity and 100 liters of fuel. Calculate the life cycle flows based on the modules in Figure 9.1.

Solution
Now, $f_1 = 100$ L and $f_2 = 100$ kWh. Therefore,

$$20s_1 - 2s_2 = 100$$
$$10s_2 = 100$$

Solving both equations simultaneously yields $s_1 = 6$ and $s_2 = 10$. Then,

$$r_{Crude} = -50 \times 6$$
$$= -300 \text{ L}$$
$$r_{SO_2} = 2(6) + 0.1(10)$$
$$= 13 \text{ kg}$$
$$r_{CO_2} = 10(6) + 1(10)$$
$$= 70 \text{ kg}$$

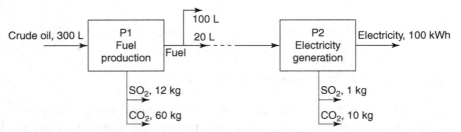

Figure 10.1 Life cycle network for producing 100 kWh of electricity and 100 liters of fuel. Note that the scaling factors for fuel production and electricity generation are 6 and 10, respectively.

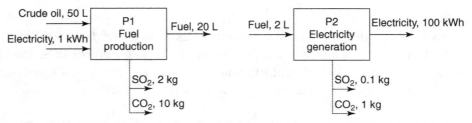

Figure 10.2 Modified module for fuel generation. This module also requires 1 kWh of electricity. The electricity generation module is unchanged.

Using matrices, the final demand vector $f = \begin{bmatrix} 100 \\ 100 \end{bmatrix}$. Using Equation (10.10), the resource consumption

$$r = \begin{bmatrix} -50 & 0 \\ 2 & 0.1 \\ 10 & 1 \end{bmatrix} \begin{bmatrix} 20 & -2 \\ 0 & 10 \end{bmatrix}^{-1} \begin{bmatrix} 100 \\ 100 \end{bmatrix}$$

$$= \begin{bmatrix} -300 \\ 13 \\ 70 \end{bmatrix}$$

The network corresponding to this problem is shown in Figure 10.1.

Example 10.3   Now consider a modified fuel generation module that has inputs of crude oil and electricity, as shown in Figure 10.2. The electricity generation module remains unchanged. Calculate the life cycle flows for producing 1000 kWh of electricity.

Solution

The final demands are $f_1 = 0$ L, $f_2 = 1000$ kWh. The balance equations are

$$20s_1 - 2s_2 = 0$$
$$-s_1 + 10s_2 = 1000$$

Solving both equations simultaneously yields $s_1 = 10.1$ and $s_2 = 101$. Then,

$$r_{Crude} = -50 \times 10.1$$
$$= -505 \text{ L}$$
$$r_{SO_2} = 2(10.1) + 0.1(101)$$
$$= 30.3 \text{ kg}$$
$$r_{CO_2} = 10(10.1) + 1(101)$$
$$= 202 \text{ kg}$$

In matrix notation, the technology matrix changes to

$$A = \begin{bmatrix} 20 & -2 \\ -1 & 10 \end{bmatrix}$$

Now, for the final demand of $f = \begin{bmatrix} 0 \\ 1000 \end{bmatrix}$, the resource consumption and emissions are found to be $r = \begin{bmatrix} -505 \\ 30.3 \\ 202 \end{bmatrix}$. Comparing this answer with that for the modules in Figure 9.1, we can see that all numbers increase slightly due to the need for electricity in fuel production.

Example 10.4   Using the inventory data provided in Table 10.1, determine the life cycle flows for generating 1 kWh of electricity. Note that we will revisit this example in all the chapters of Part III to illustrate and compare various methods for sustainable assessment.

Solution
Based on the inventory in Table 10.1, the technology matrix for this system may be written as

$$A = \begin{bmatrix} 100 & -50 & 0 \\ 0 & 20 & -2 \\ -1 & 0 & 10 \end{bmatrix} \tag{10.11}$$

The rows of this matrix represent products: oil extracted, fuel, and electricity. The columns represent processes: oil extraction, fuel production, and power generation. The intervention matrix is

$$B = \begin{bmatrix} -102 & 0 & 0 \\ 50 & 2 & 0.1 \\ 20 & 1.5 & 0.1 \\ 100 & 10 & 1 \end{bmatrix} \tag{10.12}$$

Table 10.1 Life cycle inventory for Example 10.4.

	Oil extraction	Fuel production	Power generation
	Inputs		
Oil under ground	102 L	0	0
Electricity	1 kWh	0	0
Oil extracted	0	50	0
Fuel	0	0	2 L
	Products		
Oil extracted	100 L	0	0
Fuel	0	20 L	0
Electricity	0	0	10 kWh
	Emissions		
SO_2	50 kg	2 kg	0.1 kg
NO_x	20 kg	1.5 kg	0.1 kg
CO_2	100 kg	10 kg	1 kg

The final demand vector for producing 1 kWh electricity is

$$f = \begin{bmatrix} 0 \\ 0 \\ 1 \end{bmatrix} \tag{10.13}$$

The scaling factors for these modules may be calculated as

$$s = A^{-1}f = \begin{bmatrix} 0.005 \\ 0.010 \\ 0.101 \end{bmatrix} \tag{10.14}$$

and the life cycle flows are given by

$$r = BA^{-1}f = \begin{bmatrix} -0.51 \\ 0.28 \\ 0.13 \\ 0.70 \end{bmatrix} \tag{10.15}$$

Thus, the life cycle of the production and use of 1 kWh of electricity requires 0.51 L crude oil and the emission of 0.28 kg of SO_2, 0.13 kg of NO_x, and 0.7 kg of CO_2.

Characterization Factors

It is often necessary to convert the environmental flows r into quantities that correspond to environmental impact. As we will learn in subsequent chapters, this conversion is enabled by using characterization factors to convert various physical flows into impact flows. For example, emissions of nitrogen dioxide (NO_2)

and sulfur dioxide (SO_2) may be represented in common units of sulfur dioxide equivalents (SO_2eq) to capture the acidification potential of these molecules. This involves multiplying the mass of NO_2 emission by 0.7 kg SO_2eq/kg NO_2. A general equation for such a calculation is

$$h_i = \sum_j \phi_{ij} r_j \tag{10.16}$$

where ϕ_{ij} is the characterization factor for the ith category and jth flow. In matrix notation,

$$h = \Phi r \tag{10.17}$$

where Φ is the characterization matrix, in which each row represents the factors for converting specific emissions into an impact category.

Example 10.5   Consider a life cycle with emissions of 10 kg of CO_2, 2 kg of CH_4, 5 kg of SO_2, and 3 kg of NO_2. Convert these emissions into their global warming potentials (GWPs) and acidification potentials (APs). The characterization factors are: 1 kg CO_2eq/kg CO_2, 25 kg CO_2eq/kg CH_4, 1 kg SO_2eq/kg SO_2, 0.7 kg SO_2eq/kg NO_2.

Solution
Using Equation 10.16,

$$h_{GWP} = 1(10) + 25(2) + 0(5) + 0(3)$$
$$= 60 \, \text{kg CO}_2\text{eq}$$
$$h_{AP} = 0(10) + 0(2) + 1(5) + 0.7(3)$$
$$= 7.1 \, \text{kg SO}_2\text{eq}$$

Using matrices and Equation 10.17,

$$h = \begin{bmatrix} 1 & 25 & 0 & 0 \\ 0 & 0 & 1 & 0.7 \end{bmatrix} \begin{bmatrix} 10 \\ 2 \\ 5 \\ 3 \end{bmatrix}$$
$$= \begin{bmatrix} 60 \\ 7.1 \end{bmatrix}$$

Thus, these emissions are equivalent to 60 kg CO_2eq and 7.1 kg SO_2eq. Four emission flows have been converted into two impact flows.

We will learn more about characterization factors and their use for impact assessment in Chapters 11 and 15 on carbon footprint and life cycle impact assessment, respectively.

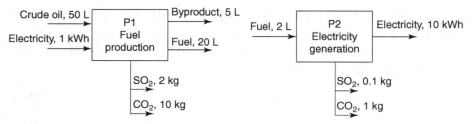

Figure 10.3 Module for fuel production with byproducts. The module for electricity generation is the same as in Figure 9.1.

10.1.2   Allocation Methods

The examples we have considered so far in this chapter produce only one product or economic flow from each module. Often, a process gives rise to multiple products simultaneously. For example, generating electricity from coal produces ash and heat along with electricity. Extending network analysis to such systems requires the use of the allocation or displacement methods that we first discussed in Section 9.2.

Let us consider a modification of the fuel module in Figure 10.2 that also produces a byproduct along with the fuel, as shown in Figure 10.3. The electricity module is unchanged. These two modules may now be represented by the following equations:

$$20s_1 - 2s_2 = f_1$$
$$5s_1 = f_2$$
$$-s_1 + 10s_2 = f_3$$

In matrix notation,

$$\begin{bmatrix} 20 & -2 \\ 5 & 0 \\ -1 & 10 \end{bmatrix} \begin{bmatrix} s_1 \\ s_2 \end{bmatrix} = \begin{bmatrix} f_1 \\ f_2 \\ f_3 \end{bmatrix} \tag{10.18}$$

These equations include a final demand f_2 of the byproduct as the second row. Since the emissions are the same as before, Equation 10.8 remains unchanged. Note that the above equations cannot be solved unambiguously after specifying the final demands since we have three equations and two unknowns. Allocation is needed to resolve this situation.

Partitioning

Allocation by partitioning requires splitting a process with multiple products into multiple processes, each with a single product. This partitioning relies on allocation weights, w_i. Thus, the fuel production process in Figure 10.3 is partitioned as shown in Figure 10.4. Here, we allocate on the basis of the volumes of the two

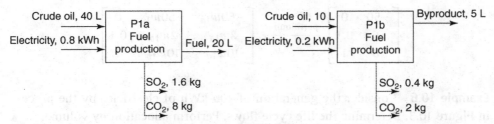

Figure 10.4 Module for fuel production with volume allocation between products.

products, resulting in allocation coefficients $w_{1a} = \frac{20}{25} = 0.8$ and $w_{1b} = \frac{5}{25} = 0.2$. Now, the balance equations for the processes P1a, P1b, and P2 may be written as

$$20s_{1a} - 2s_2 = f_1$$
$$5s_{1b} = f_2 \qquad (10.19)$$
$$-0.8s_{1a} - 0.2s_{1b} + 10s_2 = f_3$$

To solve Equation 10.18 by partitioning, we may represent the process with multiple products as multiple processes, each with a single product. As shown below, the technology matrix now has an additional column and the scaling vector has an additional row to enable partitioning between the two coproducts of process P1, f_1 and f_2.

$$\begin{bmatrix} 20 & 0 & -2 \\ 0 & 5 & 0 \\ -0.8 & -0.2 & 10 \end{bmatrix} \begin{bmatrix} s_{1a} \\ s_{1b} \\ s_2 \end{bmatrix} = \begin{bmatrix} f_1 \\ f_2 \\ f_3 \end{bmatrix} \qquad (10.20)$$

In general, partitioning involves the following transformation of the technology matrix. Columns with multiple positive values in the same column are separated into multiple columns such that each column has only one positive value or output. The negative values in these separated columns are multiplied by the weighting factors. Columns with a single positive value remain unchanged. Thus,

$$\begin{bmatrix} 20 & -2 \\ 5 & 0 \\ -1 & 10 \end{bmatrix} \xrightarrow{\text{partitioning}} \begin{bmatrix} 20 & 0 & -2 \\ 0 & 5 & 0 \\ -w_{1a} & -w_{1b} & 10 \end{bmatrix}$$

The intervention matrix is also modified by the weighting coefficients, to result in

$$\begin{bmatrix} -40 & -10 & 0 \\ 1.6 & 0.4 & 0.1 \\ 8 & 2 & 1 \end{bmatrix} \begin{bmatrix} s_{1a} \\ s_{1b} \\ s_2 \end{bmatrix} = \begin{bmatrix} r_{crude} \\ r_{SO_2} \\ r_{CO_2} \end{bmatrix} \qquad (10.21)$$

By comparing this equation with Equation 10.8, we see that the first column of the intervention matrix in Equation 10.8 is split into two columns based on the allocation factors calculated by Equations 9.1 and 9.2. That is,

$$
\begin{bmatrix}
-50 & 0 \\
2 & 0.1 \\
10 & 1
\end{bmatrix}
\xrightarrow{\text{partitioning}}
\begin{bmatrix}
-50w_{1a} & -50w_{1b} & 0 \\
2w_{1a} & 2w_{1b} & 0.1 \\
10w_{1a} & 10w_{1b} & 1
\end{bmatrix}
$$

Example 10.6 Consider the generation of 100 kWh of electricity by the process in Figure 10.3. Determine the life cycle flows. Perform allocation by volume.

Solution

To determine the emissions of the process network based on connecting the electricity generation process in Figure 10.3 with the fuel module, we utilize $f_1 = f_2 = 0$ and $f_3 = 100$. Then,

$$20s_{1a} - 2s_2 = 0$$

$$5s_{1b} = 0$$

$$-0.8s_{1a} - 0.2s_{1b} + 10s_2 = 100$$

Solving these equations, we get $s_{1a} = 1.01$, $s_{1b} = 0$, and $s_2 = 10.1$ Using Equation 10.21, the resource use is 40 L of crude oil and the emissions are 2.6 kg of CO_2 and 18 kg SO_2. Note that these emissions are for the electricity *product*, and not for the process network obtained by connecting the fuel production and electricity generation processes. Flows allocated to the byproduct from P1 will need to be added to obtain the process flows.

Displacement

For applying the displacement approach, we need data about a conventional process that produces each byproduct. Let us consider this process to be as shown in Figure 10.5. This process is displaced by the byproduct from fuel production, which may be represented by running the process in the opposite direction, as shown in Figure 9.3b and by the equations; note that the last column corresponds to the displaced process and is multiplied by −1 to indicate that the process in Figure 10.5 is run in reverse.

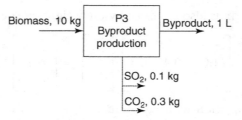

Figure 10.5 Module of byproduct production by a conventional route that would be displaced by the byproduct from the fuel production module.

$$\begin{bmatrix} 20 & -2 & 0 \\ 5 & 0 & -1 \\ 0 & 10 & 0 \end{bmatrix} \begin{bmatrix} s_1 \\ s_2 \\ s_3 \end{bmatrix} = \begin{bmatrix} f_1 \\ f_2 \\ f_3 \end{bmatrix} \qquad (10.22)$$

$$\begin{bmatrix} -50 & 0 & 0 \\ 2 & 0.1 & -0.1 \\ 10 & 1 & -0.3 \end{bmatrix} \begin{bmatrix} s_1 \\ s_2 \\ s_3 \end{bmatrix} = \begin{bmatrix} r_{crude} \\ r_{SO_2} \\ r_{CO_2} \end{bmatrix} \qquad (10.23)$$

Example 10.7 Consider the fuel module in Figure 9.1, but with an additional byproduct output of 5 L. The electricity module is unchanged. Using the displacement approach and the module in Figure 10.5, determine the resource use and emissions for producing 100 kWh of electricity.

Solution

For this problem, $f_1 = f_2 = 0$ L and $f_3 = 100$ kWh. Solving Equation 10.22 results in $s_1 = 1$, $s_2 = 10$, and $s_3 = 5$. With these scaling factors and Equation 10.23, the emissions for a final demand of 100 kWh of electricity are found to be 2.5 kg of SO_2 and 18.5 kg of CO_2, and 50 kg of crude oil are used. The corresponding network is shown in Figure 10.6. This figure also shows that biomass use will decrease by 250 kg.

No Allocation

In the absence of allocation, all allocation weights are taken as equal to 1, as we saw in Section 9.2. Thus, $w_{1a} = w_{1b} = 1$ and we can write down the following equations:

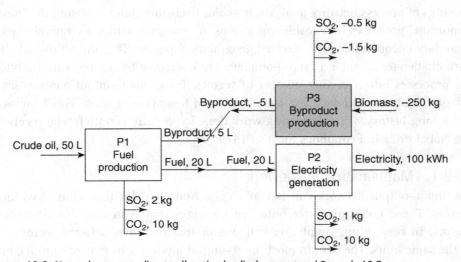

Figure 10.6 Network corresponding to allocation by displacement and Example 10.7.

$$\begin{bmatrix} 20 & 0 & -2 \\ 0 & 5 & 0 \\ -1 & -1 & 10 \end{bmatrix} \begin{bmatrix} s_{1a} \\ s_{1b} \\ s_2 \end{bmatrix} = \begin{bmatrix} f_1 \\ f_2 \\ f_3 \end{bmatrix} \qquad (10.24)$$

and

$$\begin{bmatrix} -50 & -50 & 0 \\ 2 & 2 & 0.1 \\ 10 & 10 & 1 \end{bmatrix} \begin{bmatrix} s_{1a} \\ s_{1b} \\ s_2 \end{bmatrix} = \begin{bmatrix} r_{crude} \\ r_{SO_2} \\ r_{CO_2} \end{bmatrix} \qquad (10.25)$$

Example 10.8   Solve Example 10.6 without allocation.

Solution

We are given that $f_1 = f_2 = 0$ and $f_3 = 100$ kWh. Solving Equation 10.24, we get $s_{1a} = 1.01$, $s_{1b} = 0$, and $s_2 = 10.1$. Using Equation 10.25, we find that the crude oil requirement is 50.5 L, the CO_2 emission is 3.03 kg, and the SO_2 emission is 20.2 kg for the specified final demand of 100 kWh electricity.

As we learned in Chapter 9, choosing the allocation approach leads to a subjective decision in life cycle methods. The displacement method is recommended whenever possible. Otherwise allocation in proportion to monetary value or other quantities may be used.

10.2   Input–Output Analysis

As we learned in Chapter 8, input–output analysis overcomes some of the short-comings of process network analysis. It avoids decisions about choosing the "most important" processes by considering groups of processes within a relatively large boundary encompassing the selected geographical region. The computational and data challenges of such a large boundary are overcome by aggregating individual processes into a smaller number of sectors. Thus, input–output analysis uses a larger analysis boundary, but aggregated and less accurate data. This approach has a long history, with pioneering work done by Wassily Leontief, who received the Nobel Prize in Economics for it in 1973.

10.2.1   Mathematical Framework

The input–output network consists of a collection of nodes representing various sectors. These nodes are interconnected by edges that represent flows between sectors. To keep things simple, we will assume that the flows between sectors are in the same units. The data in most input–output models is in monetary units, but the framework can also be used with flows represented in physical units.

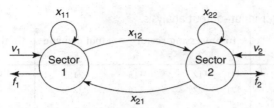

Figure 10.7 Input–output network for a two-sector system.

Consider a network consisting of n sectors, with x_{ij} representing the flow from sector i to sector j, f_i representing the final demand from sector i, and v_i representing the value added to this sector. Just as the final demand is a flow leaving the system, the value added is a flow entering the system from outside. For an economic model, these flows typically represent the interaction with society, final demand being societal consumption while value added is a societal contribution such as labor.

A network consisting of two sectors is shown in Figure 10.7. Each node is balanced, that is, the total input equals the total output. This condition implies satisfaction of the following equations for each node:

$$
\begin{array}{ccccc}
x_{11} + x_{12} + \cdots + x_{1n} + f_1 &=& x_{11} + x_{21} + \cdots + x_{n1} + v_1 &=& x_1 \\
x_{21} + x_{22} + \cdots + x_{2n} + f_2 &=& x_{12} + x_{22} + \cdots + x_{n2} + v_2 &=& x_2 \\
\vdots & & \vdots & & \vdots \\
x_{n1} + x_{n2} + \cdots + x_{nn} + f_n &=& x_{1n} + x_{2n} + \cdots + x_{nn} + v_n &=& x_n
\end{array}
\tag{10.26}
$$

The quantities $x_1, x_2, \ldots, x_n$ represent the total flow (input or output) or throughput from each sector. This network may also be represented as shown in Table 10.2. The extreme left-hand side and extreme right-hand side of Equation 10.26 may also be written as

$$
\begin{array}{ccc}
a_{11}x_1 + a_{12}x_2 + \cdots + a_{1n}x_n + f_1 &=& x_1 \\
a_{21}x_1 + a_{22}x_2 + \cdots + a_{2n}x_n + f_2 &=& x_2 \\
\vdots & & \\
a_{n1}x_1 + a_{n2}x_2 + \cdots + a_{nn}x_n + f_n &=& x_n
\end{array}
\tag{10.27}
$$

where

$$
a_{ij} = \frac{x_{ij}}{x_j}
\tag{10.28}
$$

The coefficients a_{ij} represent the flow between sectors i and j per unit of throughput (input or output) from sector j. An important assumption in input–output analysis is that these coefficients are constant for a given system. This assumption implies that if the throughput x_j of a sector changes, the flows to this sector x_{ij} change in proportion to the change in throughput. Thus, if the production from

Table 10.2 Data for input–output analysis.

		To processing sectors				Final demand	Total output
		1	2	$\cdots$	n	f	x
From	1	x_{11}	x_{12}	$\cdots$	x_{1n}	f_1	x_1
processing sectors	2	x_{21}	x_{22}	$\cdots$	x_{2n}	f_2	x_2
$\vdots$		$\vdots$	$\vdots$		$\vdots$	$\vdots$	$\vdots$
	n	x_{n1}	x_{n2}	$\cdots$	x_{nn}	f_n	x_n
Value added	v	v_1	v_2	$\cdots$	v_n		
Total input	x	x_1	x_2	$\cdots$	x_n		

the power generation sector doubles, the coal required from the coal mining sector along with inputs from other sectors to the power generation sector will also double. Similarly, all outputs, including byproducts from this sector, will also double. This assumption implies linear scaling, that is, no economy of scale for activities that are already part of the selected economy, and is identical to the assumption made in the process analysis method of Section 10.1.

In matrix notation, all the equations in Equation 10.27 may be written as follows:

$$
\begin{bmatrix}
a_{11} & a_{12} & \cdots & a_{1n} \\
a_{21} & a_{22} & \cdots & a_{2n} \\
\vdots & & \ddots & \vdots \\
a_{n1} & a_{n2} & \cdots & a_{nn}
\end{bmatrix}
\begin{bmatrix} x_1 \\ x_2 \\ \vdots \\ x_n \end{bmatrix}
+
\begin{bmatrix} f_1 \\ f_2 \\ \vdots \\ f_n \end{bmatrix}
=
\begin{bmatrix} x_1 \\ x_2 \\ \vdots \\ x_n \end{bmatrix}
\tag{10.29}
$$

that is

$$
A_{io}x + f = x \tag{10.30}
$$

The calculation of matrix A_{io} may also be written in matrix notation as

$$
A_{io} = Z\hat{x}^{-1} \tag{10.31}
$$

where Z is the matrix of x_{ij}s and $\hat{x}$ is the diagonal matrix with x_is on the diagonal. Grouping the throughputs together, the equations in Equation 10.27 may be written as

$$
\begin{aligned}
(1 - a_{11})x_1 - a_{12}x_2 - \cdots - a_{1n}x_n &= f_1 \\
-a_{21}x_1 + (1 - a_{22})x_2 - \cdots - a_{2n}x_n &= f_2 \\
\vdots \\
-a_{n1}x_1 - a_{n2}x_2 - \cdots + (1 - a_{nn})x_n &= f_n
\end{aligned}
\tag{10.32}
$$

which in matrix notation is

$$
(I - A_{io})x = f \tag{10.33}
$$

Table 10.3 Sources of economic input–output data for some countries.

Country	Source	Website
USA	Bureau of Economic Analysis	http://bea.gov
Many European countries	Organization for Economic Cooperation and Development (OECD)	http://oecd.org
Models for 40 countries and rest of the world	World Input–Output Database	http://wiod.org

Table 10.4 Data for input–output network of Figure 10.7 for Example 10.9.

	1	2	f	x
1	150	500	350	1000
2	200	100	1700	2000
v	650	1400		
x	1000	2000		

The matrix A_{io} is called the *direct requirements matrix*. Given the values of the final demands, the corresponding values of the throughputs may be found by solving Equation 10.32 simultaneously, or in matrix notation by the following equation:

$$x = (I - A_{io})^{-1} f \tag{10.34}$$

The factor $(I - A_{io})^{-1}$ is called the Leontief inverse or *total requirements matrix*. This equation provides the total or life cycle activity in the entire network that satisfies the specified final demand. It may also be written in infinite series form as

$$x = (I + A_{io} + A_{io}^2 + A_{io}^3 + \cdots) f \tag{10.35}$$

The first term of this infinite series representation is the direct contribution of the final demand to the throughput. The second term is the first-order contribution, and so on. Example 10.10 provides more insight into this interpretation of the Leontief inverse. Sources of input–output models for some countries are listed in Table 10.3.

Example 10.9 Consider the two-sector network shown in Figure 10.7 with the data in Table 10.4. Determine the throughput from both sectors for final demands

Table 10.5 Data for input–output network of Figure 10.7, with final demands $f_1 = \$600$, $f_2 = \$1500$.

	1	2	f	x
1	187.12	460.39	600	1247.46
2	249.49	92.08	1500	1841.55
v	810.85	1289.08		
x	1247.46	1841.55		

All flows are in units of $.

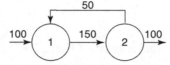

Figure 10.8 Network for Example 10.10.

of $600 and $1500 from sectors 1 and 2, respectively. Determine all the network flows.

Solution

Using the data in Table 10.4 and Equation 10.28, $a_{11} = \frac{150}{1000} = 0.15$, $a_{12} = \frac{500}{2000} = 0.25$, $a_{21} = \frac{200}{1000} = 0.2$, and $a_{22} = \frac{100}{2000} = 0.05$. The input–output equations for this network are

$$0.15x_1 + 0.25x_2 + f_1 = x_1$$
$$0.2x_1 + 0.05x_2 + f_2 = x_2$$

Thus, the direct requirements matrix for this example is

$$A_{io} = \begin{bmatrix} 0.15 & 0.25 \\ 0.2 & 0.05 \end{bmatrix} \tag{10.36}$$

Given final demands of $f_1 = \$600$ and $f_2 = \$1500$, the throughput may be calculated to be $x_1 = \$1247.46$ and $x_2 = \$1841.55$. All other flows may also be calculated from the coefficients a_{ij}. The results are shown in Table 10.5.

Example 10.10 For the network in Figure 10.8, determine the direct and total requirement matrices. Calculate each term of the infinite series expansion in Equation 10.35 and interpret its contribution to the throughput. Note that all the flows are in consistent units, which could be monetary or physical.

Solution

Based on the network in Figure 10.8, $Z = \begin{bmatrix} 0 & 150 \\ 50 & 0 \end{bmatrix}$ and $x = \begin{bmatrix} 150 \\ 150 \end{bmatrix}$. Then the

direct requirements matrix is $A_{io} = \begin{bmatrix} 0 & 1 \\ \frac{1}{3} & 0 \end{bmatrix}$ and the total requirements matrix is

$(I - A_{io})^{-1} = \begin{bmatrix} 1.5 & 1.5 \\ 0.5 & 1.5 \end{bmatrix}$.

Let us now consider each term in the infinite series expansion of Equation 10.35. With $f = \begin{bmatrix} 0 & 100 \end{bmatrix}^T$, the infinite series for this example is

$$x = \begin{bmatrix} 0 \\ 100 \end{bmatrix} + \begin{bmatrix} 100 \\ 0 \end{bmatrix} + \begin{bmatrix} 0 \\ \frac{100}{3} \end{bmatrix} + \begin{bmatrix} \frac{100}{3} \\ 0 \end{bmatrix} + \begin{bmatrix} 0 \\ \frac{100}{9} \end{bmatrix} + \cdots$$

Now consider the contribution of each term to the throughput. The first term is the final demand itself. The second term is the effect of the final demand, which is the flow from sector 1 to match the flow in sector 2. The third term is the effect of the second term through the recycle stream. Note that the fraction recycled is $\frac{1}{3}$ of the flow to sector 2. Continuing in this manner gives $x_1 = 0 + 100 + 0 + \frac{100}{3} + 0 + \frac{100}{9} + \cdots = 150$ and $x_2 = 100 + 0 + \frac{100}{3} + 0 + \frac{100}{9} + 0 + \cdots = 150$.

10.2.2 Environmentally Extended Input–Output Models

Most input–output models represent flows between economic sectors in monetary terms. For such information to be useful for assessing environmental impacts and sustainability, we will combine it with data about the interaction of each sector with the environment. The resulting environmentally extended input–output (EEIO) models can be used to quantify the wider direct and indirect implications of economic activities in the entire region covered by the input–output model. Figure 10.9 is an environmentally extended version of Figure 10.7. It shows the direct dependence of sector 1 on input 1 (m_{11}) from the environment, such as water, and of sector 2 on inputs 1 and 2 (m_{12}, m_{22}), such as water and coal from the environment. These inputs can be included in the input–output table as rows, as shown in Table 10.6. Note that for each input m_{ij}, i denotes the input type and j denotes the

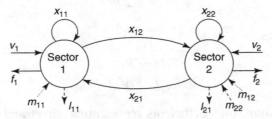

Figure 10.9 Environmentally extended input–output network for a two-sector system.

Table 10.6 Data for environmentally extended input–output analysis.

| | | To processing sectors | | | | Final demand | Total output | To ecosystems | |
		1	2	$\cdots$	n	f	x	1	2
From	1	x_{11}	x_{12}	$\cdots$	x_{1n}	f_1	x_1	l_{11}	l_{12}
processing	2	x_{21}	x_{22}	$\cdots$	x_{2n}	f_2	x_2	l_{21}	l_{22}
sectors									
	$\vdots$	$\vdots$	$\vdots$		$\vdots$	$\vdots$	$\vdots$	$\vdots$	$\vdots$
	n	x_{n1}	x_{n2}	$\cdots$	x_{nn}	f_n	x_n	l_{n1}	l_{n2}
Value	v	v_1	v_2	$\cdots$	v_n				
added									
Total input	x	x_1	x_2	$\cdots$	x_n				
From	m	m_{11}	m_{12}	$\cdots$	m_{1n}				
ecosystems		m_{21}	m_{22}	$\cdots$	m_{2n}				

sector that receives it. Similarly, the emissions from each sector to the environment may be represented as l_{ij}. Figure 10.9 shows the amounts of emission 1 from sectors 1 and 2 as l_{11} and l_{21}, respectively. Here, l_{ij} represents the emission from sector i of type j. These flows are shown as new columns at the right of Table 10.6.

In a manner analogous to defining the direct requirement coefficients a_{ij} in Equation 10.28, we define

$$r_{ij} = \frac{m_{ij}}{x_j} \text{ and } q_{ij} = \frac{l_{ij}}{x_i} \tag{10.37}$$

These new coefficients represent the direct resource use and emissions per unit of economic output from the relevant sector, respectively. In matrix notation, we have $R = M\hat{x}^{-1}$ and $Q = \hat{x}^{-1}L$. As in conventional input–output analysis, these coefficients are assumed to be constant for a given system.

Given a specific final demand f, the throughput x may be calculated by solving Equations 10.32 or 10.33. The total resource use and emissions may then be calculated by multiplying the respective coefficients and throughputs and adding the flows for all sectors:

$$m_i = \sum_j r_{ij} x_j \tag{10.38}$$

$$l_j = \sum_i q_{ij} x_i \tag{10.39}$$

In terms of matrices,

$$M = Rx \tag{10.40}$$

$$L = Q^T x \tag{10.41}$$

Using matrix notation, the coefficients representing direct and indirect resource use and emissions may now be written as

Table 10.7 Some environmentally extended input–output models.

Name	Description	References
Economic Input–Output LCA (EIOLCA)	Extends US input–output models by including data about several emissions and resources such as energy, water, and land.	[2]
Ecologically-based LCA (Eco-LCA)	Accounts for contribution of goods from ecosystems to US economic sectors, and includes some emissions. Aggregates flows in thermodynamic units.	[3]
United States Environmentally-Extended Input–Output Model (US EEIO)	EEIO model for the US economy.	[4]
Carbon footprint of nations	Accounts for life cycle greenhouse gas emissions associated with global trade.	[5]
World multi-regional input–output model (World MRIO)	Connects all countries and their energy use, greenhouse gas emissions, water use, etc.	[6]

$$R^* = R(I - A_{io})^{-1} \qquad (10.42)$$
$$Q^* = Q^T(I - A_{io})^{-1} \qquad (10.43)$$

With these coefficients, the life cycle or total resource use and the total emissions may each be calculated as a product of one or other of the coefficients and the final demand,

$$M = R^*f \qquad (10.44)$$
$$L = Q^*f \qquad (10.45)$$

These equations are used for various life cycle-oriented methods that use the input–output framework and data. Some EEIO models are listed in Table 10.7.

Example 10.11   Consider the EEIO table shown in Table 10.8. This is an extension of the model in Example 10.9, and its direct requirements matrix A_{io} is given in Equation 10.36. Determine the resource use and emissions coefficients. Calculate the life cycle interaction with the environment for a final demand vector $\begin{bmatrix} 600 & 1500 \end{bmatrix}^T$.

Solution
Using Equation 10.37, the direct resource consumption coefficients are

$$r_{11} = \frac{20}{1000} = 0.02 \text{ kg/\$}$$

Table 10.8 Data for input–output network of Figure 10.9.

	1	2	f	x	l_1	l_2
1	150	500	350	1000	100	2
2	200	100	1700	2000	50	5
v	650	1400				
x	1000	2000				
m	20	0				

Flows m, l_1, and l_2 are in kilograms. All other flows are in dollars.

$$r_{12} = \frac{0}{2000} = 0 \text{ kg/\$}$$

The direct emission coefficients are

$$q_{11} = \frac{100}{1000} = 0.1 \text{ kg/\$}$$

$$q_{12} = \frac{2}{1000} = 0.002 \text{ kg/\$}$$

$$q_{21} = \frac{50}{2000} = 0.025 \text{ kg/\$}$$

$$q_{22} = \frac{5}{2000} = 0.0025 \text{ kg/\$}$$

These coefficients may also be calculated using matrix notation as follows:

$$R = M\hat{x}^{-1}$$

$$= \begin{bmatrix} 20 & 0 \end{bmatrix} \begin{bmatrix} 1000 & 0 \\ 0 & 2000 \end{bmatrix}^{-1}$$

$$= \begin{bmatrix} 0.02 & 0 \end{bmatrix} \text{ kg/\$}$$

$$Q = \hat{x}^{-1}L$$

$$= \begin{bmatrix} 1000 & 0 \\ 0 & 2000 \end{bmatrix}^{-1} \begin{bmatrix} 100 & 2 \\ 50 & 5 \end{bmatrix}$$

$$= \begin{bmatrix} 0.1 & 0.002 \\ 0.025 & 0.0025 \end{bmatrix} \text{ kg/\$}$$

For the specified final demands $f_1 = \$600$ and $f_2 = \$1500$, the throughputs are $x_1 = \$1247.46$ and $x_2 = \$1841.55$, as calculated in Example 10.9. Using these throughputs and the coefficients calculated above, we can determine the total resource use by using Equation 10.38:

$$m_1 = (0.02 \times 1247.46) + (0 \times 1841.55)$$

$$= 24.95 \text{ kg}$$

Similarly, the total emissions are

$$l_1 = (0.1 \times 1247.46) + (0.025 \times 1841.55)$$
$$= 170.8 \text{ kg}$$
$$l_2 = (0.002 \times 1247.46) + (0.0025 \times 1841.55)$$
$$= 7.1 \text{ kg}$$

From Equations 10.42 and 10.43, the total resource and emission coefficients are

$$R^* = \begin{bmatrix} 0.0251 & 0.0066 \end{bmatrix} \text{ kg/\$}$$

$$Q^* = \begin{bmatrix} 0.1320 & 0.0611 \\ 0.0032 & 0.0035 \end{bmatrix} \text{ kg/\$}$$

For the final demand vector $f = \$ \begin{bmatrix} 350 & 1700 \end{bmatrix}^T$ it is easy to verify that we get the original resource flows. For a final demand $f = \$ \begin{bmatrix} 600 & 1500 \end{bmatrix}^T$, the total environmental flows are $m = 24.95$ kg and $l = \begin{bmatrix} 170.8 & 7.1 \end{bmatrix}^T$ kg, respectively.

Example 10.12 Consider the system in Example 10.10 but with 10 kg of water used in sector 1, and 2 kg and 5 kg of CO_2 emissions from sectors 1 and 2, respectively. Determine the life cycle flow of water and CO_2 for a final demand $f = \$ \begin{bmatrix} 0 & 200 \end{bmatrix}^T$. Using the infinite series expansion, determine the first three orders of water use.

Solution
In this example, the resource use is $M = \begin{bmatrix} 10 & 0 \end{bmatrix}$ kg and the emission flow is $N = \begin{bmatrix} 2 & 5 \end{bmatrix}^T$ kg. For the given final demand, using Equations 10.44 and 10.45,

$$M_2 = \begin{bmatrix} 10 & 0 \end{bmatrix} \begin{bmatrix} 150 & 0 \\ 0 & 150 \end{bmatrix}^{-1} \begin{bmatrix} 1.5 & 1.5 \\ 0.5 & 1.5 \end{bmatrix} \begin{bmatrix} 0 \\ 200 \end{bmatrix}$$
$$= 20 \text{ kg}$$

Using the infinite series representation in Equation 10.35, we can calculate M_2 as

$$M_2 = \begin{bmatrix} 10 & 0 \end{bmatrix} \begin{bmatrix} 150 & 0 \\ 0 & 150 \end{bmatrix}^{-1} \left(\begin{bmatrix} 1 & 0 \\ 0 & 1 \end{bmatrix} + \begin{bmatrix} 0 & 1 \\ \frac{1}{3} & 0 \end{bmatrix} + \begin{bmatrix} \frac{1}{3} & 0 \\ 0 & \frac{1}{3} \end{bmatrix} + \cdots \right) \begin{bmatrix} 0 \\ 200 \end{bmatrix}$$
$$= 0 + \frac{200}{15} + 0 + \frac{200}{45} + \cdots$$
$$= 20 \text{ kg}$$

The first term in the series is 0 because the final demand is in sector 2, which does not directly require any water. The first-order flow is due to the $200 flow from sector 1, which involves a water flow to Sector 1 of $\frac{200}{15}$, and so on. This is analogous to the reasoning in Example 10.10.

10.3   Hybrid Models

Building a life cycle network by using process models as described in Section 10.1 is attractive owing to the likely accuracy of such data obtained from actual processes. However, obtaining such detailed data for all processes in a life cycle network is difficult because the network is infinitely large. As we discussed in Section 8.3 and illustrated in Figure 8.3a, it is common to build a life cycle model by choosing a finite number of processes that are considered to be most important. The selection of these processes determines the life cycle boundary, and is often done in a subjective manner. At the other extreme is the input–output network illustrated in Figure 8.3b. It considers a large system boundary but aggregates processes into a smaller number of sectors. A hybrid model combines the features of the process and input–output models, and its mathematical framework is described here.

Let us consider the process network with two modules shown in Figure 9.1. The boundary of this process network ignores the activity of crude oil extraction, which is needed for the crude oil to be available for conversion to fuel. Rather than completely ignoring this activity, we can approximate it as the oil and gas extraction sector that commonly exists in input–output models. Notice that this sector aggregates data about the extraction of crude oil and natural gas of various types and from diverse sources. However, accounting for crude oil extraction with this aggregate model is likely to be better than completely ignoring it. From a process network calculation like those in Examples 9.1 and 10.1 we obtain the quantity of crude oil needed for meeting the specified final demands. This quantity of oil can be used as the final demand from the input–output network to approximate the crude oil extraction process. However, the physical quantity of crude oil obtained from the process network needs to be converted into its monetary value before it can be used in the input–output model. This conversion requires price information. Depending on the nature of the input–output model, the consumer (market) or producer (wholesale) price may be required.

Example 10.13   Consider the process network shown in Figure 9.1 when the final demand of electricity is 1000 kWh. From the EEIO model of the relevant region, we know that the total requirement coefficients for the oil and gas extraction sector are 0.01 kg CO_2/$ and 0.002 kg SO_2/$. Also, the cost of crude oil is $0.25/L. Determine the life cycle emissions of SO_2 and CO_2 by developing a hybrid model.

Solution

From Examples 9.1 and 10.1, we know that generating 1000 kWh electricity requires a flow $r_{crude} = 500$ L of crude oil, and it emits $r_{SO2} = 30$ kg of SO_2 and $r_{CO2} = 200$ kg of CO_2.

For building the hybrid model, we will assume that 500 L of crude oil is obtained from the oil and gas extraction sector. The monetary value of this oil is

$$f_{io,crude} = p_{crude} r_{crude}$$
$$= 0.25 \times 500$$
$$= \$125$$

Emissions from the EEIO model for this final demand of crude oil are calculated by using Equation 10.45:

$$r_{io,SO_2} = 0.002 \times 125$$
$$= 0.25 \text{ kg SO}_2$$
$$r_{io,CO_2} = 0.01 \times 125$$
$$= 1.25 \text{ kg CO}_2$$

Combining the emissions from the process network and input–output network, we get

$$r_{h,SO_2} = r_{SO_2} + r_{io,SO_2}$$
$$= 30.25 \text{ kg}$$
$$r_{h,CO_2} = r_{CO_2} + r_{io,CO_2}$$
$$= 201.25 \text{ kg}$$

Thus, the hybrid flows are obtained by adding results from the process and input–output models.

The approach used in this example results in a "tiered hybrid" life cycle model. Let us now develop the general equations for developing such a model. A typical tiered hybrid model is shown in Figure 10.10. This is an approximation of the full hybrid model shown in Figure 8.3c because it assumes that the processes and sectors have no overlap. In practice, processes are nested inside economic sectors. For example, the agricultural activity of growing corn is nested inside the grain farming sector, while the manufacture of ethylene is nested inside the economic sector of petrochemical manufacturing, which includes the production of propylene, butylene, benzene, toluene, etc. The tiered hybrid modeling approach ignores such overlaps.

Building a tiered hybrid model involves the following steps.

1. For the specified final demand f_{pr}, the process network model is developed using the approach in Section 10.1. The resulting network is depicted by the

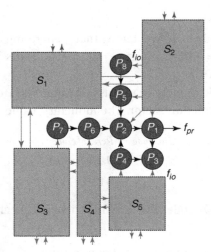

Figure 10.10 Network representation for a tiered hybrid life cycle model.

circles P_1 through P_8 in Figure 10.10, f_{pr} being the output of P_1. Equation 10.10 provides values for the flows to and from the process network:

$$r = BA^{-1}f_{pr} \qquad (10.46)$$

It will include flows associated with resources that may be obtained from other economic activities (r_{io}), and flows associated with direct exchanges with the environment (r_{env}). In Figure 10.10, flows from other economic activities include those from sector S_2 to processes P_1, P_2, P_5, and P_8, from sector S_3 to process P_7, sector S_4 to P_6, and S_5 to P_3 and P_4. These flows provide the links between the process and input–output networks.

2. The flows r_{io} are usually in physical units. Using price information, p_{io}, convert them into monetary units:

$$f_{io} = p_{io}r_{io}$$

This provides the final demand from the economic sectors, f_{io}.

3. Using the approach in Section 10.2.2, calculate the resource use and emissions for the economic activity quantified by the final demand f_{io}. Resource use is calculated by Equation 10.44 and emissions by Equation 10.45.

4. Add the results from the process network and EEIO model to calculate the total resource use and impact.

Figure 10.11 shows a tiered hybrid network for part of a corn ethanol life cycle. In this case, f_{pr} is the ethanol produced; direct environmental flows associated with processes (r_{env}) include soil and sunlight; and resource flows from the input–output model to the process model (f_{io}) are shown by the flows between the two dashed rectangles and include natural gas, fertilizer, and water.

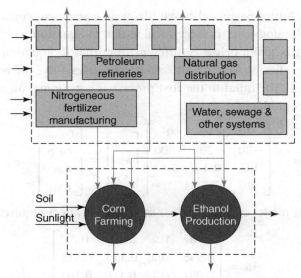

Figure 10.11 Illustration of tiered hybrid model for corn ethanol life cycle.

Example 10.14 Build a tiered hybrid life cycle model for electricity generation by combining data from the electricity module in Figure 9.1 and the input–output model in Example 10.11. You may consider sector 2 in the IO model to be a coal-mining sector that has as its output coal, which is the raw material for generating electricity. The cost of fuel is $10 per liter.

Solution

For the electricity module in the process, we will consider electricity to be the economic flow, and the fuel, SO_2, and CO_2 to be environmental flows or flows from outside the system. For this system, $A = 10$, $B = \begin{bmatrix} -2 & 0.1 & 1 \end{bmatrix}^T$. Since $f_{pr} = 1$ kWh, $r_{pr} = BA^{-1}f_{pr} = \begin{bmatrix} -0.2 & 0.01 & 0.1 \end{bmatrix}^T$. Thus, the fuel input requirement is 0.2 L, which will need to be obtained from an economic sector. This is denoted as r_{io}. The other two flows in r_{pr} are direct interactions with the environment, r_{env}. Given that the price of fuel is $10 per liter, $f_{io} = 10 \times 0.2 = \$2$. The emissions from the input–output model from Example 10.11 for flows to and from the hybrid system, namely, CO_2 and SO_2, are

$$r_h = \left(\begin{bmatrix} 0.1 \\ 1 \end{bmatrix} \times 10^{-1} \times 1 \right) + \left(\begin{bmatrix} 0.1320 & 0.0611 \\ 0.0032 & 0.0035 \end{bmatrix} \times \begin{bmatrix} 0 \\ 2 \end{bmatrix} \right)$$

$$= \begin{bmatrix} 0.1322 \\ 0.107 \end{bmatrix}$$

Here the first term is for emissions from the process model and the second for the input–output model. Thus, the emissions of SO_2 and CO_2 are 0.1322 kg and 0.107 kg, respectively.

Example 10.15 A process-model-based life cycle assessment consists of two processes whose technology matrix is identical to that in Equation 10.20. However, the intervention matrix given below is different from that in Equation 10.21 because this system also relies on water. The first process uses 100 L of water, while the second uses 500 L. The input to the first process is partitioned on the basis of the volumes of the coproducts:

$$\begin{bmatrix} -40 & -10 & 0 \\ -80 & -20 & -500 \\ 1.6 & 0.4 & 0.1 \\ 8 & 2 & 1 \end{bmatrix} \begin{bmatrix} s_{1a} \\ s_{1b} \\ s_2 \end{bmatrix} = \begin{bmatrix} r_{crude} \\ r_{H_2O} \\ r_{SO_2} \\ r_{CO_2} \end{bmatrix} \tag{10.47}$$

The input–output model consists of four sectors with direct requirements matrix

$$A_{io} = \begin{bmatrix} 0.182 & 0.2 & 0.111 & 0.25 \\ 0.545 & 0.2 & 0.333 & 0.417 \\ 0.091 & 0.1 & 0.111 & 0.167 \\ 0.091 & 0.3 & 0.222 & 0.167 \end{bmatrix} \tag{10.48}$$

The throughput in each sector for this A_{io} is given by $x = \begin{bmatrix} 550 & 1000 & 450 & 600 \end{bmatrix}$. The resource use and emissions matrices are as follows:

$$N = \begin{bmatrix} 5 & 20 \\ 1 & 5 \\ 2 & 10 \\ 0.5 & 5 \end{bmatrix} ; \; M = \begin{bmatrix} 50 & 10 & 100 & 2 \\ 10 & 0 & 0 & 0 \end{bmatrix} \tag{10.49}$$

Crude oil is bought from sector 1 for $\$10$ per liter while water is provided by sector 3 for $\$0.1$ per liter. Determine the life cycle CO_2 and SO_2 emissions and the water use per kilowatt-hour of electricity by a process-based model and also by a hybrid model.

Solution
For the process-based model, $f_1 = f_2 = 0$, $f_3 = 1$. Using Equation 10.10,

$$r = \begin{bmatrix} -40 & -10 & 0 \\ -80 & -20 & -500 \\ 1.6 & 0.4 & 0.1 \\ 8 & 2 & 1 \end{bmatrix} \begin{bmatrix} 20 & 0 & -2 \\ 0 & 5 & 0 \\ -0.8 & -0.2 & 10 \end{bmatrix}^{-1} \begin{bmatrix} 0 \\ 0 \\ 1 \end{bmatrix} \tag{10.50}$$

$$= \begin{bmatrix} -0.40 \\ -51.21 \\ 0.026 \\ 0.18 \end{bmatrix} \tag{10.51}$$

Thus, 51.21 L of H_2O and 0.40 L of crude oil are consumed while 0.026 kg of SO_2 and 0.18 kg of CO_2 are emitted in the life cycle.

For the hybrid model we use the EEIO model to account for the life cycle of the inputs to the processes, crude oil and water. Knowing that the price of fuel is $10 per liter and of water is $0.1 per liter, the final demand from the input–output model is $0.4 \times 10 = \$4$ for fuel and $51.21 \times 0.1 = \$5.12$ for water. Thus, $f_{io,1} = 4$, $f_{io,3} = 5.12$ and $f_{io,2} = f_{io,4} = 0$. The coefficients representing total resource use and emissions are calculated as follows:

$$R^* = M\hat{x}^{-1}(I - A_{io})^{-1}$$

$$= \begin{bmatrix} 0.4572 & 0.3563 & 0.5478 & 0.4294 \\ 0.0458 & 0.0262 & 0.023 & 0.0316 \end{bmatrix}$$

$$Q^* = N^T\hat{x}^{-1}(I - A_{io})^{-1}$$

$$= \begin{bmatrix} 0.0309 & 0.0216 & 0.0234 & 0.0257 \\ 0.1383 & 0.1014 & 0.1114 & 0.1245 \end{bmatrix}$$

The result for the hybrid model is found by adding the results of the process and input–output models for $f_{io} = \begin{bmatrix} 4 & 0 & 5.12 & 0 \end{bmatrix}^T$:

$$r_{th} = \begin{bmatrix} -0.40 \\ -51.21 \\ 0.026 \\ 0.18 \end{bmatrix} + \begin{bmatrix} -4.633 \\ -0.303 \\ 0.2437 \\ 1.1242 \end{bmatrix}$$

$$= \begin{bmatrix} -5.031 \\ -51.513 \\ 0.2697 \\ 1.3042 \end{bmatrix}$$

Thus, the result from the tiered hybrid model is a life cycle consumption of 5.031 L of crude oil and 51.513 L of water, and a life cycle emission of 0.2697 kg of SO_2 and 1.3042 kg of CO_2.

10.4 Summary

This chapter introduced the mathematical framework for developing process model networks, input–output networks, and their combination. These methods form the underlying basis of many techniques used for evaluating the broader implications of human activities. These techniques include various types of footprint analysis, life cycle assessment, energy analysis, and cumulative exergy consumption analysis. We will learn about these in the chapters that follow.

Key Ideas and Concepts

- Technology matrix
- Final demand
- Leontief inverse
- Hybrid model

- Intervention matrix
- Value added
- EEIO model

10.5 Review Questions

1. What is a technology matrix?
2. What is the intervention matrix?
3. Explain allocation by displacement with an example.
4. Explain the direct and total requirement matrices.
5. What is a tiered hybrid model? Explain with an example and figure.

Problems

10.1 Process A uses 10 m^3 of oil and 10 kWh of electricity to produce 8 m^3 of fuel. It emits 1.5 m^3 of SO$_2$. Process B uses 1 m^3 of fuel to produce 5 kWh of electricity and emits 0.2 m^3 of SO$_2$. Answer the following questions for this system.

　1. If the system produces 1000 kWh of electricity to sell to consumers, what is its total (direct and indirect) SO$_2$ emission?
　2. If it sells 0.1 m^3 of fuel and 10 kWh of electricity, determine the total quantity of crude oil used and the SO$_2$ emitted.

10.2 The life cycle network for manufacturing widgets is shown in Figure 10.12. (Developed by Dr. Rebecca Hanes.)

　1. Write down the technology and intervention matrices for this network.
　2. Calculate the life cycle flows for producing one unit of the widget.
　3. Calculate the life cycle flows for producing 10 kWh of electricity along with one widget.

10.3 For the widget manufacturing system in Figure 10.12, consider a modified case in which the plastics production process produces 1.5 kg of expanded (exp.) plastic and 0.85 kg of thermoformed plastic. All the other flows remain unchanged. (Developed by Dr. Rebecca Hanes.)

　1. Write down the technology and intervention matrices for this network.
　2. Using mass allocation, determine the life cycle flows for manufacturing 100 widgets along with 1000 kWh of electricity.
　3. A process for making thermoformed (thm.) plastic is shown in Figure 10.13. Calculate the life cycle flows if this process is displaced by thermoformed plastic produced as a byproduct.

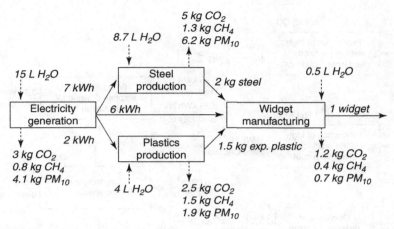

Figure 10.12 Life cycle network for widget example.

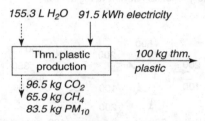

Figure 10.13 Process for making thermoformed plastic that may be displaced.

10.4 Five life cycle modules involved in washing a cup in a standard size residential dishwasher are shown in Figure 10.14. For simplicity, many operations, emissions, and resources are not included, and flows are reported with only one or two significant figures. The utility mix (75 percent coal, 25 percent natural gas) is somewhat arbitrary, while a 10 percent loss in transmission and distribution is typical. (Developed by Dr. Laura Wood.)

1. Write down the technology and intervention matrices for the life cycle of a glass cup.
2. Assuming that a single glass cup is used 800 times, determine the corresponding final demand, and calculate the life cycle emissions. Determine the scaling vector and draw the corresponding life cycle network diagram. You may assume that one dishwasher load is 80 cups.

10.5 Inventory data for a paper towel dispenser are given in Problem 9.7. Using this information, determine the technology and environmental intervention matrices for this product.

10.6 Consider the network shown in Figure 10.15. It consists of three sectors (nodes) and shows the monetary exchange between the sectors. The final

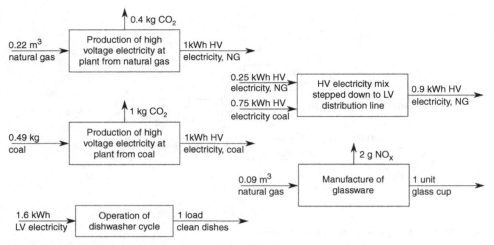

Figure 10.14 Life cycle inventory modules relevant to a glass cup.

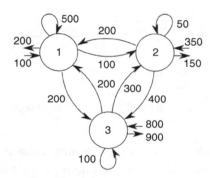

Figure 10.15 Input–output network. All flows are in dollars.

demand and value added flows are shown as flows to and from human resources. Assuming this diagram to represent a static snapshot of the network, answer the following questions.

1. Represent the network as an input–output table and find the total throughput x_i for each sector.
2. Determine the input–output model relating the total throughput from each sector x_i to the final demand f_i. This will require you to find the appropriate coefficients, $a_{ij} = x_{ij}/x_j$.
3. Use the model derived above to determine the throughput from each sector for final demands $f_1 = \$100, f_2 = \$300, f_3 = \$0$.
4. From the results obtained in (3), prepare the corresponding network diagram or input–output table. You will need to show the monetary flows between all the sectors and human resources.

10.7 An input–output model consists of the following flows in dollars: $x_{11} = 10$, $x_{12} = 20$, $x_{21} = 30$, $x_{22} = 10$. The throughputs are $x_1 = 50$ and $x_2 = 40$, and the emissions of CO_2 from sectors 1 and 2 are 50 kg and 100 kg, respectively. Answer the following questions for this system.

   1. Determine the final demand and value added for each sector.

   2. Determine the direct requirements matrix.

   3. For final demands $f_1 = 10$ and $f_2 = 10$, determine the throughputs.

10.8 The flows in an economic input–output model are as follows: $x_{11} = 50$, $x_{12} = 50$, $x_{21} = 30$, $x_{22} = 0$, $f_1 = 10$, $f_2 = 25$.

   1. Determine the input–output model for this network. You need to derive two equations in terms of the throughput and final demand from each sector.

   2. For the given flows, sector 1 extracts crude oil and emits 5 m^3 of SO_2. Sector 2 transports crude oil and emits 8 m^3 of SO_2. Determine the life cycle emission of SO_2 for delivering $10,000 worth of crude oil.

10.9 Consider the economic input–output model represented by the following equations:

$$0.833x_1 - 0.182x_2 = f_1$$
$$-0.667x_1 + 0.818x_2 = f_2$$

The energy intensity of sectors 1 and 2 is 10 J/$ and 20 J/$, respectively. Calculate the total life cycle energy use for final demands $f_1 = \$15$ and $f_2 = \$25$, respectively.

10.10 This problem utilizes the process model in Problem 10.1 and the EEIO model in Problem 10.8. Develop a hybrid model to determine the life cycle SO_2 emission for producing 1000 kWh of electricity.

10.11 For the system in Example 10.1, determine the sensitivity of the flows of crude oil, SO_2, and CO_2 for producing 100 kWh of electricity. You may consider a ± 10 percent variation in the values given in Figure 9.1. Which flows in this figure need to be the most accurate?

References

[1] R. Heijungs and S. Suh. *The Computational Structure of Life Cycle Assessment.* Kluwer Academic Publishers, 2002.

[2] L. B. Lave, E. Cobas-Flores, C. T. Hendrickson, and F. C. McMichael. Using input–output analysis to estimate economy-wide discharges. *Environmental Science & Technology*, 29(9):420–426, 1995.

[3] Y. Zhang, A. Baral, and B. R. Bakshi. Accounting for ecosystem services in life cycle assessment, part II: toward an ecologically based LCA. *Environmental Science & Technology*, 44(7):2624–2631, 2010.

[4] Y. Yang, W. W. Ingwersen, T. R. Hawkins, M. Srocka, and D. E. Meyer. USEEIO: a new and transparent United States environmentally-extended input–output model. *Journal of Cleaner Production*, 158:308–318, 2017.

[5] E. G. Hertwich and G. P. Peters. Carbon footprint of nations: a global, trade-linked analysis. *Environmental Science & Technology*, 43(16):6414–6420, 2009.

[6] M. Lenzen, D. Moran, K. Kanemoto, and A Geschke. Building EORA: a global multi-region input–output database at high country and sector resolution. *Economic Systems Research*, 25(1):20–49, 2013.

11 Footprint Assessment

> We forget that the water cycle and the life cycle are one.
>
> Jacques Cousteau

In sustainability assessment, the term *footprint* refers to the direct and indirect environmental impact of an activity represented in terms of a single unit. This approach has been popular owing to its relative simplicity and to focus on important environmental challenges such as climate change, freshwater scarcity, and exceeding planetary biocapacity. In this chapter we will learn about two commonly used methods, based on either the carbon footprint or the water footprint. We will also understand the shortcomings of the footprint approach and the nexus between multiple footprints.

11.1 Carbon Footprint

Interest in the carbon footprint has been driven by concerns about anthropogenic climate change due to emission of greenhouse gases and their increasing atmospheric concentration. The global trend of some greenhouse gases over the last few hundred thousand years is summarized in Figure 2.18. In addition to CO_2, other major greenhouse gases (GHGs) include methane, nitrous oxide, and various chlorofluorocarbon (CFC) and hydrochlorofluorocarbon (HCFC) compounds. Their trends in recent years are shown in Figure 11.1. Concentrations of CO_2 and N_2O have continually increased, the former mainly due to the use of fossil fuels, while the latter is mainly from application of artificial nitrogen fertilizers. Oscillations in these graphs are at the scale of about a year, and are due to seasonal variation. The concentration of CFCs has started decreasing, thanks to their being banned by the Montreal Protocol. However, HCFC concentrations continue to increase. Methane emissions are mainly from cows, rice paddies, and the natural gas infrastructure. Its concentration also continues to increase, and is beyond roughly three times the highest concentration in the last several hundred thousand years.

Since the contribution of each GHG to radiative forcing is different, to determine the overall impact all GHG emissions are represented in terms of carbon dioxide equivalents (CO_2eq). The contributions of various GHGs to CO_2eq are enabled by factors such as those summarized in Table 11.1. The global warming potentials

Table 11.1 Global warming potentials relative to CO_2 [2].

Common name	Chemical formula	GWP for given time horizon		
		20-yr	100-yr	500-yr
Carbon dioxide	CO_2	1	1	1
Methane	CH_4	72	25	7.6
Nitrous oxide	N_2O	289	298	153
CFC-11	CCl_3F	6730	4750	1620
Carbon tetrachloride	CCl_4	2700	1400	435
HCFC-22	$CHClF_2$	5160	1810	549
HFC-134a	CH_2FCF_3	3830	1430	435
Sulfur hexafluoride	SF_6	16,300	22,800	32,600
Methyl chloride	CH_3Cl	45	13	4

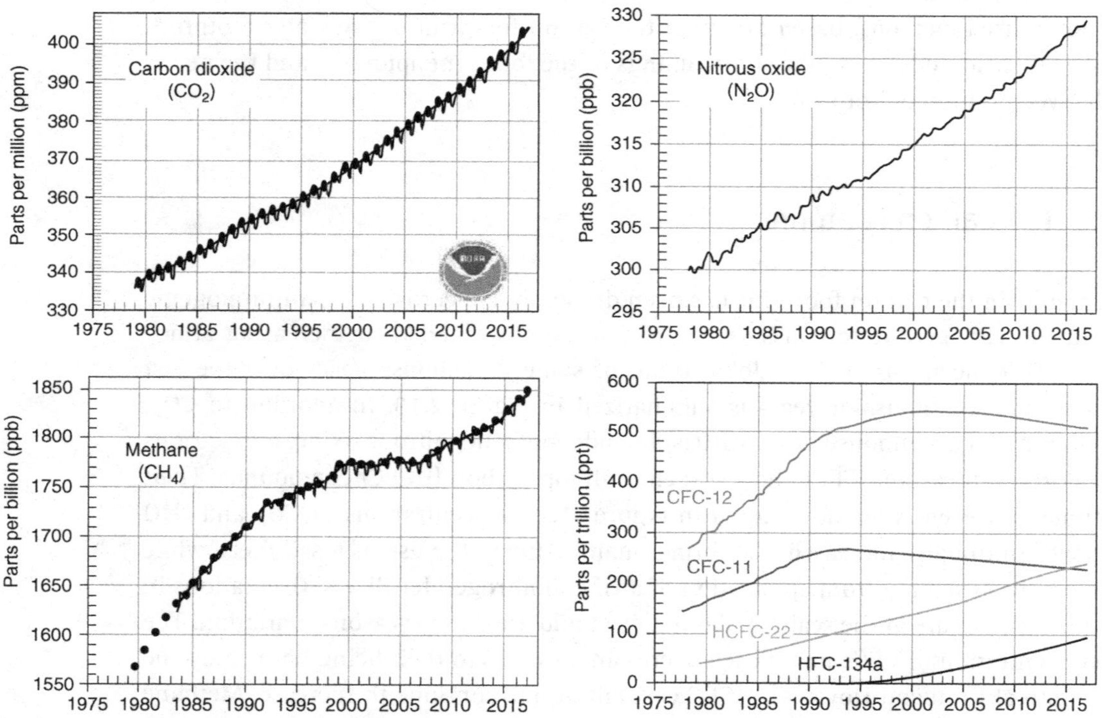

Figure 11.1 Atmospheric concentration of major greenhouse gases [1].

(GWP) in this table show that if we consider a 100-year period then, based on their mass, methane is 25 times more potent than CO_2, while N_2O is 298 times more potent. Their potency changes with time according to how long each gas lasts in the atmosphere and how it changes. In 2016, in terms of CO_2 equivalents, the atmosphere contained 489 ppm CO_2eq, an increase of almost 40 percent since 1990. [1] Aggregating GHGs into the single unit of CO_2eq is done using

$$GWP = \sum_j m_j GWP_j$$

where GWP_j is the GWP of the jth GHG and m_j is its mass. Note that this equation is equivalent to Equation 10.16.

Example 11.1   A life cycle inventory consists of the following emissions: 20 kg of CO_2, 0.001 kg of CFC-11, 0.5 kg of N_2O, and 2 kg of NO_2. Determine the GWP using a 100-year time horizon.

Solution

Using Table 11.1, the GWP of these emissions is

$$GWP = (20 \times 1) + (0.001 \times 4750) + (0.5 \times 298)$$
$$= 173.8 \text{ kg } CO_2\text{eq}$$

Note that NO_2 does not contribute to global warming and is excluded from this calculation.

Data about GHG emissions from various human activities are compiled in public and private databases. The US Environmental Protection Agency compiles data about GHG flows in the USA and the world [3]. The US inventory contains sources and sinks, and is summarized in Figure 11.2. As can be seen in this figure, total emissions in 2016 were about 6.5 billion tons of CO_2eq, while sequestration was about 1 billion tons CO_2eq. This indicates an overshoot of about 5.5 billion tons CO_2eq in 2016 above nature's capacity to sequester CO_2 in the USA. This quantity is sequestered by ecosystems in other parts of the world or it accumulates in the atmosphere.

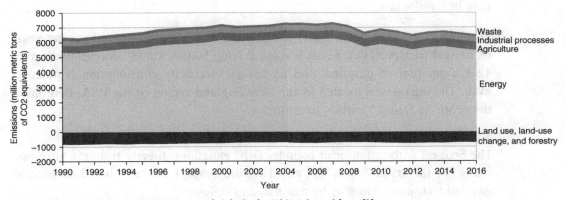

Figure 11.2 Greenhouse gas emissions and sinks in the USA. Adapted from [3].

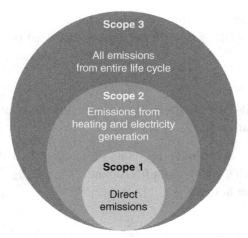

Figure 11.3 Scope of carbon footprint calculations.

The boundary of carbon footprint calculations is often classified as Scope 1, 2, or 3. As depicted in Figure 11.3, Scope 1 is a narrow boundary that considers direct GHG emissions from only selected processes; Scope 2 includes GHG emissions in generating the electricity and those caused by the heating that is used in the processes included in Scope 1; and Scope 3 is the full life cycle GHG emissions.

The overall approach of carbon footprint calculations is the same as that covered in Chapters 8, 9, and 10, but with only GHG flows included in the model and characterization by GWP values like those in Table 11.1. GHG emission factors for some common materials and activities are given in Table 11.2, while carbon footprints of some common products are shown in Table 11.3. The carbon footprints of common food items normalized in different ways are shown in Figure 11.4 [4]. The carbon footprint of global trade is described in Box 11.1. This demonstrates the importance of considering GHG emissions at the life cycle scale (Scope 3), since otherwise the shifting of emissions to other geographic regions due to global trade may be overlooked.

Example 11.2   Consider a business that burns 10,000 m³ of natural gas per year; its annual transportation needs involve driving 50,000 km in vehicles that travel 12 km per liter of gasoline, and its annual electricity consumption is 100,000 kWh. The business is located in the New England region of the USA. Determine the Scope 1, 2, and 3 carbon footprints.

Solution

The Scope 1 carbon footprint includes GHG emissions directly from the business. Using the factors in Tables 11.1 and 11.2, the Scope 1 footprint based on natural gas and transportation may be calculated as follows:

Table 11.2 Greenhouse gas emissions factors for selected products and activities [5].

Item	CO_2 factor, kg/unit	CH_4 factor, kg/unit	N_2O factor, kg/unit	Unit
Bituminous coal	2.563	0.302	0.044	kg
Natural gas	1.922	0.036	0.004	m^3
Kerosene	366.9	41.22	6.077	L
Motor gasoline	2.32			L
Diesel	2.697			L
Ethanol	1.519			L
CNG	1.924			m^3
Emissions due to the generation of electricity for selected US regions				
New England	327.8	0.033	0.006	MWh
Northwest	382.5	0.007	0.006	MWh
Rockies	861.1	0.01	0.013	MWh
Midwest	822.1	0.009	0.013	MWh
South	614.8	0.01	0.009	MWh
California	277.3	0.013	0.003	MWh

$$GWP_{S1} = 10{,}000 \times [(1.922 \times 1) + (0.036 \times 25) + (0.004 \times 298)] + \frac{50{,}000}{12} \times 2.32$$
$$= 49{,}807 \text{ kg } CO_2\text{eq}$$

The Scope 2 footprint includes the emissions due to electricity and heating, which for this example are calculated as follows:

$$GWP_{S2} = GWP_{S1} + \frac{100{,}000}{1000} \times [(327.8 \times 1) + (0.033 \times 25) + (0.006 \times 298)]$$
$$= 82{,}848 \text{ kg } CO_2\text{eq}$$

For the Scope 3 calculations, we may use a process life cycle assessment (LCA) or the input–output LCA model. This example uses the latter with the 2002 EIO (economic input–output) model. The cost of natural gas may be found to be 10.42 cents/m^3, gasoline is \$0.359 per liter, and electricity is 4.88 cents/kWh. Thus, the business buys \$1042 of natural gas, \$1496 of gasoline, and \$4880 of electricity. The total CO_2eq emission intensities for sectors producing a dollar's worth of these items may be found from environmentally extended input–output tables to be 2.43 kg/\$, 2.79 kg/\$, and 9.37 kg/\$, respectively. Then, the Scope 3 carbon footprint is

$$GWP_{S3} = GWP_{S2} + (2.43 \times 1042) + (2.79 \times 1496) + (9.37 \times 4880)$$
$$= 135{,}280 \text{ kg } CO_2\text{eq}$$

Table 11.3 Carbon footprint in gCO2eq/kWh of some pathways for generating electricity in the USA [6].

Item	C footprint
Coal – fluidized bed	1144
Coal – integrated gasification combustion cycle	903
Coal – pulverized	989
Coal – supercritical	768
Natural gas – natural gas combined cycle	449
Natural gas – natural gas combustion turbine	588
Nuclear – boiling water reactor	13
Nuclear – light water reactor	9
Nuclear – pressurized water reactor	12
Wind – offshore and onshore	11
Hydropower	7
Geothermal – enhanced geothermal system	25
Geothermal – flash steam	56
Concentrated solar power – dish	22
Concentrated solar power – tower	33
Concentrated solar power – trough	27
Photovoltaic – monocrystalline Si, ground-mounted	40
Photovoltaic – monocrystalline Si, rooftop	40
Photovoltaic – polycrystalline Si, ground-mounted	69
Photovoltaic – polycrystalline Si, rooftop	46
Biopower – agricultural residues	37
Biopower – animal wastes	40
Biopower – forest residues	34
Biopower – herbaceous crops	50
Biopower – mill wastes	15
Biopower – other wastes	51
Biopower – urban wastes	37
Biopower – woody crops	43
Biopower – cofiring	48
Biopower – direct combustion	35
Biopower – gasification	47

Example 11.3   For the problem in Example 10.4, determine the carbon footprint of 1 kWh electricity.

Solution

From the result for r, the carbon footprint of this system is 0.7 kg CO_2eq/kWh, since among the emissions, only CO_2 has a GWP.

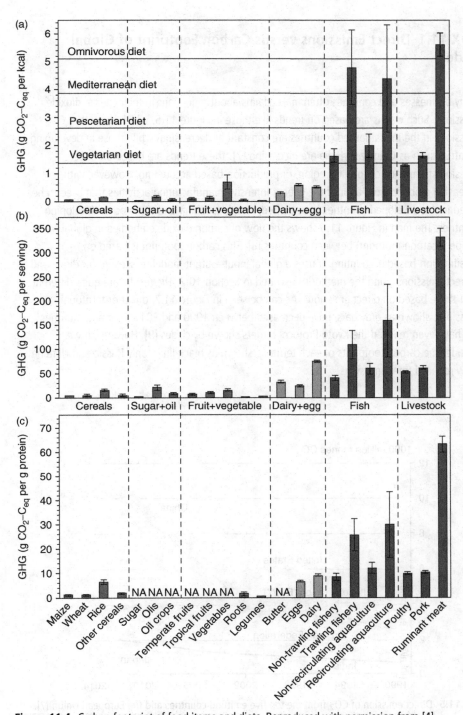

Figure 11.4 Carbon footprint of food items and diets. Reproduced with permission from [4].

BOX 11.1 Direct Emissions versus Carbon Footprint of Global Trade

Many businesses and countries often make claims about reductions in their carbon dioxide emissions. Such claims are based on trends like those in Figure 11.5, which show that CO_2 emissions in the industrialized countries are constant or decreasing, while those in developing countries such as China and India are increasing [7]. These trends are based on direct emissions from each country or region, or production-based accounting. However, with increasing globalization, a lot of the energy-intensive manufacturing activities that used to be based in the developed countries have shifted, mainly to China and some other developing countries. The map in Figure 11.6 shows the flow of carbon dioxide embodied in global trade (Scope 3 carbon footprint) between countries [8]. This carbon footprint is based on consumption-based accounting. It uses a global input–output model to account for direct and indirect emissions using the methods described in Section 10.2. The results are quite different from those based on direct emissions. As can be seen in Figure 11.7, direct or territorial emissions, shown as gray bars, have decreased between 1990 and 2011 for many countries and have even reached the Kyoto Protocol targets shown by circles [9]. However, if we consider the carbon footprints of each territory, shown as black lines, only Russia and Ukraine show reduced total emissions.

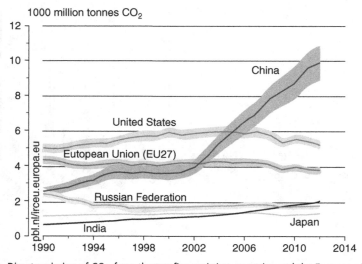

Figure 11.5 Direct emission of CO_2 from the top five emitting countries and the European Union [7].

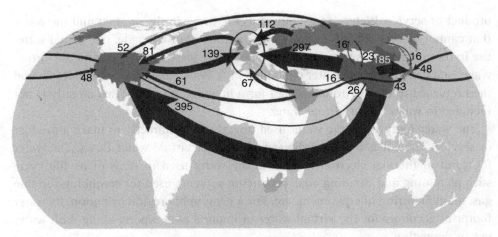

Figure 11.6 CO_2 emissions embodied in trade in Mt CO_2/yr from net exporting countries to net importing countries. Reproduced with permission from [8].

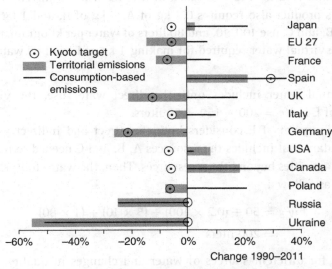

Figure 11.7 Direct (territorial) and total (life cycle or consumption-based) CO_2 emissions from selected territories. [9]

11.2 Water Footprint

The water footprint approach [10] provides insight into the direct and indirect dependence of products, processes, and other systems on water. Such insight is increasingly important, particularly for water-intensive activities in water-stressed regions of the world. We learned about such regions in Section 2.3. The water footprint relies on the concept of *virtual water*, which is the water used to make a product or service. Water use is defined as the water in the product and the water that cannot be used directly for any other purpose. For example, the virtual water use in thermal power is mainly due to evaporative losses and leaks, and the virtual water in a beverage is the water contained in it. Water that may be withdrawn and then returned to the environment in a form in which it can be used directly is not included in water use or virtual water.

The water footprint is the water used directly and indirectly to make a product or service. Thus, it is a sum of direct and indirect virtual water flows. The water footprint of thermal electricity includes the water used in steps of its life cycle such as mining and cleaning coal, producing solvents used for scrubbing the flue gas, manufacturing of equipment, etc. For a geographic region or nation, its water footprint accounts for the virtual water in imports and exports along with water use in the region.

Example 11.4   The production of 1 kg of product E withdraws 200 liters of water from a river, and returns 150 liters at a downstream location. The manufacture of 1 kg of this product also requires 0.2 kg of A, 5 kg of B, and 1 kg of C. These resources, A, B, and C, use 100, 10, and 80 liters of water per kilogram, respectively. Determine the virtual water required for making 1 kg of E, and its water footprint.

Solution

Since the virtual water includes only the direct water use, the virtual water requirement of E is $V_E = 200 - 150 = 50$ liters.

The water footprint of E considers both the direct and indirect use. We may define a boundary that includes the resources A, B, and C needed to make product E but ignores activities beyond these resources. Then, the water footprint of E may be calculated as follows:

$$F_{W,E} = 50 + (0.2 \times 100) + (5 \times 10) + (1 \times 80)$$
$$= 200 \text{ liters}$$

To account for different sources of water and changes in quality, most water footprint calculations consider the following three categories of water.

- *Blue water.* This category includes fresh water available at the surface in rivers, lakes, and canals, and below the surface as groundwater.
- *Green water.* This consists of precipitation that falls on land and does not runoff or recharge the aquifer.
- *Gray water.* This category focuses on the quality of water. It is the water needed to dilute pollutants to acceptable levels.

The water footprint of many common products is shown in Table 11.4.

Determining the quantity of gray water often requires a material balance calculation with information about the current and desired concentrations of contaminants in the water. Let the pollutant loading be P (mass/time), the maximum

Table 11.4 Water footprint in liters of selected products.

Item	Quantity	Water footprint (L)
Apple	1 (150 g)	125
Banana	1 (200 g)	160
Beef	1 kg	15,415
Beer	1 glass (250 ml)	74
Corn ethanol	1 liter	2854
Cheese	1 kg	3178
Chicken meat	1 kg	4325
Cotton	1 shirt (250 g)	2495
Leather (bovine)	1 kg	17,093
Pizza	1 pizza	1259
Pork	1 kg	5988
Rice	1 kg	2497

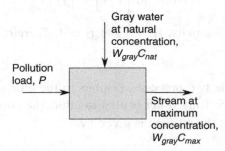

Figure 11.8 Pollutant balance for determining gray water.

allowed concentration of contaminant be C_{max} (mass/volume), and the natural concentration of this contaminant in the environment be C_{nat}. Then, the gray water footprint for this flow may be calculated as

$$W_{gray} = \frac{P}{C_{max} - C_{nat}}$$ (11.1)

From a material balance point of view, this equation describes the pollution load being combined with a volume of gray water at its natural concentration to result in a stream with the maximum allowable concentration. This is illustrated in Figure 11.8.

Example 11.5 The fuel and electricity modules shown in Figure 11.9 are identical to those in Figure 9.1, except that the fuel production module uses 100 L of water from a local river, while the electricity production module uses 10 L from a reservoir. Fuel production results in a wastewater stream of 10 L leaving the facility with a 10 percent concentration of contaminant by volume. The maximum allowed contaminant concentration is 5 percent, and the natural concentration is 2 percent. Calculate the water footprint of 1 kWh of electricity.

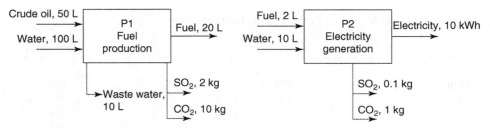

Figure 11.9 Modules for Example 11.5.

Solution

The modules for this problem are shown in Figure 11.9. As shown in Example 9.1, the technology matrix for this example is given in Equation 10.3 as

$$\begin{bmatrix} 20 & -2 \\ 0 & 10 \end{bmatrix} \begin{bmatrix} s_1 \\ s_2 \end{bmatrix} = \begin{bmatrix} f_1 \\ f_2 \end{bmatrix}$$

To produce 1 kWh of electricity, $f_1 = 0$ and $f_2 = 1$. Therefore, the scaling factors are $s = \begin{bmatrix} 0.01 \\ 0.1 \end{bmatrix}$.

The water used in the two processes belongs to the blue water category. There is no use of green water. Gray water is used to dilute the effluent.

The pollutant load in the effluent is given by

$$P = C_{eff} V_{eff}$$
$$= 0.1 \times 10$$
$$= 1 \text{ L}$$

Given that $C_{max} = 0.05$ and $C_{nat} = 0.02$, using Equation 11.1,

$$W_{gray} = \frac{1}{0.05 - 0.02}$$
$$= 33 \text{ L}$$

The intervention matrix for the three water categories may be written as

$$\begin{bmatrix} 0 & 0 \\ -100 & -10 \\ -33 & 0 \end{bmatrix} \begin{bmatrix} s_1 \\ s_2 \end{bmatrix} = \begin{bmatrix} r_{green} \\ r_{blue} \\ r_{gray} \end{bmatrix}$$

For 1 kWh of electricity, $f_1 = 0, f_2 = 1$. Solving the above equations results in $r = \begin{bmatrix} 0 \\ -2 \\ -0.3 \end{bmatrix}$. Adding the elements in r provides the water footprint as 2.3 L.

Table 11.5 Carbon and water footprinting and requirements for sustainability assessment methods.

Demand for ecosystem services	The carbon footprint considers the demand for the carbon sequestration ecosystem service in terms of CO_2 emissions, while the water footprint focuses mainly on the water provisioning ecosystem services in terms of the volume of water consumed.
Supply of ecosystem services	Neither method accounts for the capacity of ecosystems to supply the relevant ecosystem service.
Spatial scales	Methods for carbon and water footprinting account for multiple spatial scales, using the network analysis we covered in Chapter 10.
Temporal interactions	Neither method considers dynamic effects.
Cross-disciplinary interactions	Neither method considers the effect on other disciplines of decisions to reduce these footprints.
Multiple flows	Footprint methods aim to provide univariate indicators, and therefore do not account for flows other than a selected few.

11.3   Characteristics of Footprint Methods

Metrics based on carbon and water footprints are popular owing to their simplicity and ease of interpretation. Let us now consider how well these methods satisfy the requirements for sustainability assessment methods that were summarized in Box 3.3. As described in Table 11.5, footprint methods are best used for considering the demand for some ecosystem goods and services and for considering multiple spatial scales. Advancing these methods to satisfy other requirements is part of ongoing research.

One important shortcoming of footprint methods is that they do not take into account the interaction between different flows such as GHGs, food, energy, and water. These flows are often dependent on each other, as illustrated in Table 11.6 for the water–energy nexus. The top part of this table shows how energy needs water, while the bottom part shows how water needs energy. Thus, efforts to reduce water or energy use must consider this nexus to prevent the shifting of impacts from one type of resource flow to another. The use of univariate metrics like the carbon and water footprints can miss such interactions. Owing to this water–energy nexus, efforts to increase the availability of fresh water by treatment methods such as reverse osmosis are likely to increase energy use, just as efforts to use biofuels are likely to increase water use.

Table 11.6 Water–energy nexus.

Energy depends on water [11]		
Item	Water	Unit
Coal	160	L/GJ
Crude oil	1060	L/GJ
Uranium	90	L/GJ
Solar thermal	270	L/GJ
Hydroelectric	22,000	L/GJ
Biofuel	70,000	L/GJ
Water depends on Energy [12]		
Item	Energy	Unit
Disinfection	360	J/L
Reverse osmosis	2844	J/L
Desalination	14,400	J/L
Bottled water	8,000,000	J/L

In addition to the carbon and water footprints, other footprints have also been developed. The nitrogen footprint quantifies flows associated with introducing reactive nitrogen in the environment. The major impacts of these flows are acidification and eutrophication. The ecological footprint is another approach that represents multiple flows and represents them in the common unit of land area. We will learn about the ecological footprint in Chapter 16.

Key Ideas and Concepts

- Global warming potential
- Blue water
- Gray water
- Energy–water nexus
- Carbon dioxide equivalents
- Green water
- Virtual water
- Ecological footprint

11.4 Review Questions

1. Define the carbon footprint and water footprint.
2. What are Scope 1, 2, and 3 footprint calculations?
3. What is the net overshoot of CO_2 emission and sequestration in the USA?
4. What is the meaning of the term "embodied carbon in trade"?
5. What is virtual water? How is it related to the water footprint?

Table 11.7 Greenhouse gas concentration in atmosphere before 1750 and in 2013 [13].

Gas	Pre-1750 tropospheric concentration	Recent tropospheric concentration	Units
Carbon dioxide (CO_2)	280	395.4	ppm
Methane (CH_4)	722	1827.5	ppb
Nitrous oxide (N_2O)	270	325	ppb
Tropospheric ozone (O_3)	237	337	ppb
CFC-11 (CCl_3F)	Zero	235	ppt
CFC-12 (CCl_2F_2)	Zero	527	ppt
CFC-113 (CCl_2CClF_2)	Zero	74	ppt
HCFC-22 ($CHClF_2$)	Zero	220.5	ppt
HCFC-141b (CH_3CCl_2F)	Zero	22.5	ppt
HCFC-142b (CH_3CClF_2)	Zero	22	ppt
Halon 1211 ($CBrClF_2$)	Zero	4.05	ppt
Halon 1301 ($CBrF_3$)	Zero	3.3	ppt
HFC-134a (CH_2FCF_3)	Zero	69.5	ppt
Carbon tetrachloride (CCl_4)	Zero	84	ppt
Sulfur hexafluoride (SF_6)	Zero	7.59	ppt
Other halocarbons	Zero	Varies by substance	

Problems

11.1 Calculate the CO_2 equivalents for the atmosphere before 1750 and in 2013 using the data in Table 11.7. Use a 100-year time horizon. Identify the largest contributors to the increase in GWP, and discuss their primary sources.

11.2 Many universities across the world have committed to becoming carbon neutral within the next few decades. This is encouraging many efforts to reduce waste such as the "My Cup" program, where the university provides a reusable cup to students for hot and cold beverages, and a small discount when it is used. The carbon footprint of a conventional paper coffee cup for a selected American campus is determined to be 1.5011 kg CO_2eq per cup. The reusable cup has a footprint of 3.055 kg CO_2eq. A campus survey found that during the academic year, a typical student consumes a hot beverage 255 times. For an appropriate functional unit, what will be the annual reduction in the campus carbon footprint per student? Setting up this reusable cup program, providing the cups, and operating the program costs the university $5 per cup. If the reduced CO_2 emissions are worth $30 per ton CO_2eq, is the cost of this program justified? (Based on a problem developed by Timothy Becker, Brian Shawd, Garrett Greco, and Daniel Meeks.)

11.3 An economic input–output model is represented by the following equations:

$$0.833x_1 - 0.182x_2 = f_1$$
$$-0.667x_1 + 0.818x_2 = f_2$$

For throughputs of $x_1 = \$200$ and $x_2 = \$400$, sector 1 uses 50 L rainwater and 20 L of river water, and emits 100 kg of CO_2 and 1 kg of N_2O, while Sector 2 uses 200 L of lake water and emits 200 kg of CO_2. Determine the carbon and water footprints of a dollar's worth of product from each sector.

11.4 Why does the GWP of CO_2 remain constant over time, that of CH_4 decreases, and that of SF_6 increases?

11.5 A typical data center, housing servers for cloud computing, consumes 7 MWh of electricity per year. The company plans to develop five new data centers, with one each in New England, the Northwest, and the Midwest, and two in California. Calculate the increase in the company's carbon footprint due to this expansion. If all the electricity was obtained from concentrated solar power with a parabolic trough, what would be the reduction in the carbon footprint of the new data centers? If all the electricity came from nuclear light water reactors, what would be the carbon footprint? Discuss the possible side-effects and unintended harm of the solar and nuclear options.

11.6 The quantity of natural gas (CH_4) produced in the USA in 2015 was 28 trillion cubic feet. Net emissions of methane to the environment during this production were 6.4 million metric tons. The emission factors for bituminous coal and natural gas are 93.3 and 53.07 kg of CO_2 per million Btu.

1. For generating one million Btu of heat from natural gas versus bituminous coal, calculate the difference in CO_2 emissions. Ignore the emissions of methane.
2. Do the same calculation as (1) with emissions of methane included in equivalents of CO_2.

11.7 Electric cars are considered by many to be environmentally friendlier than cars with a conventional internal combustion engine. An electric sedan can travel 100 miles using 34 kWh of electricity. A similar gasoline sedan travels 100 miles with four gallons of gasoline. Electricity costs $0.12/kWh, while gasoline costs $2.49 per gallon. Calculate the costs of the two cars traveling 300 miles. Using data in Tables 11.2 and 11.3, determine the sources of electricity that must be used if the electric car is to have a smaller carbon footprint than the gasoline car.

11.8 A small town withdraws water from the local river and disinfects it before delivering it to its residents. Most residents own reverse osmosis water purification units, which are used for producing drinking water. The town treats 10,000 L of water per month, of which the residents use 5 percent for drinking after reverse osmosis. The effluent from the town has an average biological oxygen demand (BOD) of 10 mg/L. The acceptable BOD is 1 mg/L, and the natural BOD is 0.1 mg/L.

1. Calculate the direct green, blue, and gray water flows in this town.
2. The town produces electricity by means of a coal integrated gasification combustion cycle process. Estimate the water and carbon footprints of this electricity. You may assume that both disinfection and reverse osmosis use this electricity.

11.9 A farmer uses 10,000 L of water to produce 100 kg of biomass and 25 kg of food product. The farming process also uses 2 kg of CH_4 as fuel and emits 15 kg of CO_2 and 0.5 kg of N_2O. The harvested biomass is stored, during which time some of it rots and emits 0.1 kg of CH_4 per kilogram of stored biomass. When the stored biomass is converted into fuel it produces 12.5 kg of CH_4 and 11 kg of CO_2 per 50 kg of biomass. This conversion process needs 5 L of water as input. Using this information, prepare diagrams representing farming, storage, and fuel conversion modules. Calculate the carbon and water footprints per kilogram of CH_4 produced.

11.10 Many utilities have programs to encourage energy-efficient use of electricity. These include incentives to consumers to use more efficient light bulbs and home appliances. In the state of California, between July 2015 and June 2016, such programs resulted in a saving of 1651 GWh of electricity. During this period, the state also implemented various water conservation programs to address an ongoing drought. The electricity saved due to the water conservation measures was estimated to be 1830 GWh. Explain how water conservation in households can result in less electricity use across the state. Consider the direct and indirect effects.

References

[1] J. H. Butler and S. A. Montzka. The NOAA annual greenhouse gas index (AGGI). www.esrl.noaa.gov/gmd/aggi/aggi.html, 2017, accessed November 23, 2018.

[2] P. Forster, V. Ramaswamy, P. Artaxo, et al. Changes in atmospheric constituents and in radiative forcing. In S. Solomon, D. Qin, M. Manning, et al., editors, *Climate*

Change 2007: The Physical Science Basis. Contribution of Working Group I to the Fourth Assessment Report of the Intergovernmental Panel on Climate Change. Cambridge University Press, 2007.

[3] Environmental Protection Agency. Inventory of U.S. greenhouse gas emissions and sinks: 1990–2016. www.epa.gov/ghgemissions/inventory-us -greenhouse-gas-emissions-and-sinks-1990-2016, 2018, accessed November 17, 2018.

[4] D. Tilman and M. Clark. Global diets link environmental sustainability and human health. *Nature*, 515(7528):518–522, 2014.

[5] US EPA. Emissions factors for greenhouse gas inventories. www.epa.gov/sites /production/files/2018-03/documents/emission-factors_mar _2018_0.pdf, April 4, 2014, accessed November 23, 2018.

[6] Life Cycle Assessment Harmonization Project. LCA harmonization. https:// openei.org/apps/LCA/. accessed July 9, 2018.

[7] J. G. J. Olivier, G. Janssens-Maenhout, M. Muntean., and J. A. H. W. Peters. *Trends in Global CO2 Emissions: 2014 Report*. Netherlands Environmental Assessment Agency, 2014.

[8] S. J. Davis and K. Caldeira. Consumption-based accounting of CO2 emissions. *Proceedings of the National Academy of Sciences*, 107(12):5687–5692, 2010.

[9] K. Kanemoto, D. Moran, M. Lenzen, and A. Geschke. International trade undermines national emission reduction targets: new evidence from air pollution. *Global Environmental Change*, 24:52–59, 2014.

[10] A. Y. Hoekstra, A. K. Chapagain, M. M. Aldaya, and M. M. Mekonnen. *The Water Footprint Assessment Manual*. Earthscan, 2011.

[11] Water Footprint Network. P. W. Gerbens-Leenes, A. Y. Hoekstra, Th. H. van der Meer. Water footprint of bioenergy and other primary energy carriers. Report. UNESCO-IHE Institute for Water Education, March 2008.

[12] P. H. Gleick and H. S. Cooley. Energy implications of bottled water. *Environmental Research Letters*, 4(1):014009, 2009.

[13] T. J. Blasing. Recent greenhouse gas concentrations. http://dx.doi.org/10 .3334/CDIAC/atg.032, 2014, accessed November 23, 2018.

12 Energy and Material Flow Analysis

Every society is molded by energies it consumes and embodies.

Vaclav Smil [1]

It is easy to appreciate the importance of materials and energy for human activities and well-being. Human activities invariably involve their utilization from diverse sources, including minerals, fossil resources, and renewables such as wood, biomass, and solar energy. We learned about trends in global materials and energy use in Chapter 2. The importance of energy use for human development can be seen in Figure 12.1, which plots the Human Development Index (HDI) versus per capita energy use. It shows the strong correlation between energy use per capita and HDI for an energy use per person of less than about 2500 kg of oil equivalent, and little correlation beyond this value. It implies that energy use is essential for human development, but that beyond a certain point additional energy use per person need not enhance human development. In addition to its role in human activities, energy is also essential for all ecological activities, making it a universal currency. However, in nature, the type of energy that is commonly used is renewable. The role of materials is also similar to that of energy. In fact, material resources are essential for the generation of energy resources such as fuels and electricity.

Given the importance of materials and energy, many techniques have been developed for understanding their roles and for identifying improvement opportunities. In this chapter we will learn how information about energy and material flow can help in understanding the impact and sustainability of large systems. We will become familiar with concepts such as energy return on investment, and methods to analyze systems at small scales, such as industrial processes, and at large scales, such as national economies.

12.1 Energy Analysis

The dictionary definition of energy is "the capacity for doing work." However, from a thermodynamic point of view, this definition is incorrect. As we will learn in Chapter 13, this is the definition of exergy, not energy. In this chapter, the term

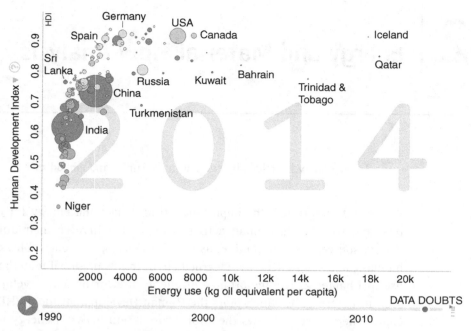

Figure 12.1 Human Development Index versus energy use per capita (data from the United Nations Development Program and the World Bank).

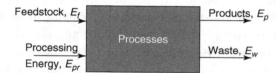

Figure 12.2 Energy flows in a typical process network.

energy represents fuel value or enthalpy. Commonly encountered types of energy include the following:

- *The fuel value* is the amount of heat that can be generated by combusting the fuel. The heat of combustion or heating value for various fuels is determined experimentally and is often reported for a fuel as a low heating value (LHV) and a high heating value (HHV).
- *Enthalpy* is the thermal energy in a substance and is calculated as

$$\Delta H = mC_p\Delta T \tag{12.1}$$

Here, m is the mass, C_p is the specific heat, and ΔT represents the temperature difference with respect to a reference temperature.

Flows relevant to energy analysis are shown in Figure 12.2. Here, the feedstock energy E_f is the energy content of the resource being extracted or converted into the product, which has energy content E_p. The processing energy E_{pr} is the fuel

Table 12.1 Examples of different types of energy conveyed in Figure 12.2.			
Activity	Feedstock energy E_f	Product energy E_p	Processing energy E_{pr}
Coal mining	Fuel value of coal under ground	Fuel value of coal above ground	Energy for bringing coal above ground from the mine
Refining crude oil	Energy in crude oil	Energy in refined products	Energy for running refining process
Cheetah hunting a gazelle	Energy in live gazelle	Energy in captured gazelle	Energy expended by the cheetah in catching the gazelle

required to convert the feedstock into the product. The examples in Table 12.1 should help in understanding the meaning of these terms for different types of systems.

These flows are related by the first law of thermodynamics as

$$E_f + E_{pr} = E_p + E_w \qquad (12.2)$$

Energy efficiency is commonly defined as the ratio of the energy content of the useful product(s) to the total energy entering the process. Thus, for the system in Figure 12.2,

$$\eta = \frac{E_p}{E_f + E_{pr}} \qquad (12.3)$$

This quantity is always less than or equal to 1.

The energy return on investment (EROI) is defined as the ratio of the energy obtained from the process (usually in the form of useful products) to the processing energy:

$$\text{EROI} = \frac{E_p}{E_{pr}} \qquad (12.4)$$

This quantity may be greater or less than 1. The net energy is defined as the difference between the product and processing energies:

$$E_{net} = E_p - E_{pr} \qquad (12.5)$$

As we will learn in Section 12.3, for a life cycle to be "energetically profitable," the energy contained in the product must be more than the energy needed to make the product, $E_p > E_{pr}$. That is, EROI must be larger than 1, which also implies that the net energy must be positive.

Example 12.1 A process for mining coal requires 0.5 MJ per kilogram of mined coal. The fuel value of the coal is 27 MJ/kg. Determine the energy efficiency, EROI, and net energy of this process.

Solution

Consider a feedstock of 1 kg of coal. Then the feedstock fuel value is

$$E_f = 27 \text{ MJ}$$

The processing energy for extracting this quantity of coal is

$$E_{pr} = 0.5 \text{ MJ}$$

Assuming no loss in moving coal from inside the mine to the surface, the product is also 1 kg of coal, so its fuel value is

$$E_p = 27 \text{ MJ}$$

Using Equations 12.3–12.5,

$$\eta = \frac{27}{27 + 0.5}$$
$$= 98.18 \text{ percent}$$
$$\text{EROI} = \frac{27}{0.5}$$
$$= 54$$
$$E_{net} = 27 - 0.5$$
$$= 26.5 \text{ MJ}$$

12.2   Energy Analysis of Processes

The energy analysis of individual processes requires the fuel value and enthalpy of streams entering and leaving a system. This information can help in understanding the system and in identifying opportunities for its improvement. In this section we will see this illustrated by two examples at different levels of detail. The narrow scope of these examples makes them useful for improving energy efficiency, but of limited use for sustainability assessment. However, the underlying approach may be extended to consider systems at larger scales, including the life cycle, as we will learn in Section 12.3.

Consider the simple mixing process shown in Figure 12.3, where water at two different temperatures is mixed. The feed streams are at 100 °C and 50 °C. The enthalpy of the two feed streams is calculated by Equation 12.1 as

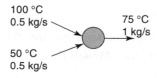

Figure 12.3 Simple mixing process.

$$H_{hot} = 0.5 \times 1 \times (100 - 25)$$
$$= 37.5 \text{ kcal/s}$$

and

$$H_{cold} = 0.5 \times 1 \times (50 - 25)$$
$$= 12.5 \text{ kcal/s}$$

Under ideal conditions, the enthalpy of the hot stream is the sum of the two input streams, which is 50 kcal/s. Its temperature is 75 °C. This system may be portrayed as a *Sankey* diagram, shown in Figure 12.4a. In this diagram, the width of each stream corresponds to the enthalpy content of the corresponding stream. This diagram captures the law of energy conservation and this case does not have any energy loss. A more practical situation, where there is a 24 percent loss of heat in the mixing process, may be depicted by the modified diagram shown in Figure 12.4b, with the dark gray triangle depicting the loss.

Sankey diagrams are popular for the energy analysis of all kinds of systems. Figures 12.5 and 12.6 show respectively the simplified flowsheet and Sankey diagram for a coal-burning power plant [2]. The Sankey diagram shows that the highest energy loss is due to the heat rejected in the condensor, where low-pressure steam is condensed before the water is recirculated. This diagram indicates that if we wish to improve the energy efficiency of this system, we should reduce this loss of heat in the condensor. We will revisit this process in the next chapter and analyze it using exergy analysis.

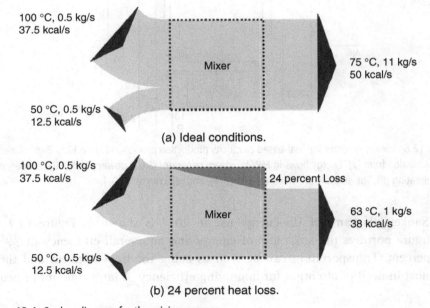

(a) Ideal conditions.

(b) 24 percent heat loss.

Figure 12.4 Sankey diagram for the mixing process.

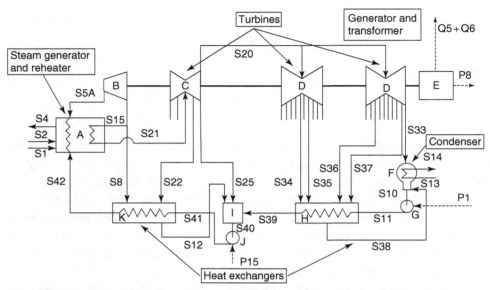

Figure 12.5 Coal-based electricity production process. Reproduced with permission from [2].

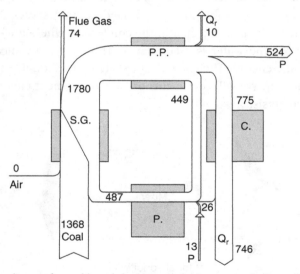

Figure 12.6 Sankey diagram for coal-based electricity production process in Figure 12.5. Reproduced with permission from [2]. Energy flows in MW for steam generator (S.G.), power plant (P.P.), condensor (C), preheating (P). The streams are electrical power (P), and rejected heat (Q_r).

A Sankey diagram for US energy use in 2017 is shown in Figure 12.7 [3]. This figure portrays the large loss of energy and an overall efficiency of $\frac{31.1}{97.7} =$ 31.8 percent. Transportation can be seen as being the least efficient and there-fore most in need of attention for improving efficiency. A large amount of heat is

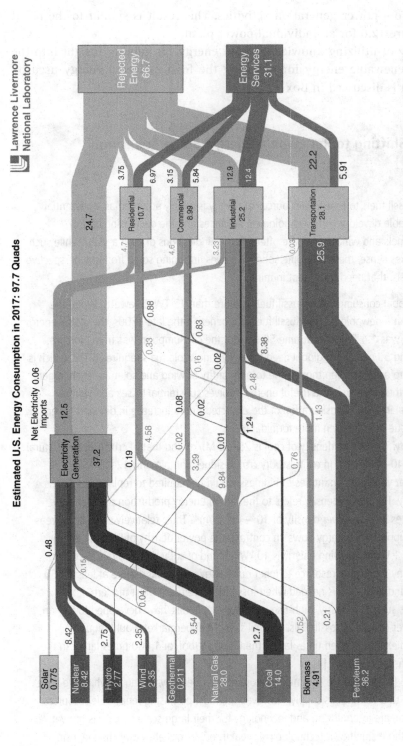

Figure 12.7 Sankey diagram for the United States in the year 2017 [3]. (1 Quad = 10^15 BTU = 1.055 × 10^18 joules = 1.055 exajoules.)

Estimated U.S. Energy Consumption in 2017: 97.7 Quads

Lawrence Livermore
National Laboratory

Net Electricity 0.06
Imports

Solar 0.775
Nuclear 8.42
Hydro 2.77
Wind 2.35
Geothermal 0.211
Natural Gas 28.0
Coal 14.0
Biomass 4.91
Petroleum 36.2

Electricity Generation 37.2

Rejected Energy 66.7
Energy Services 31.1

Residential 10.7
Commercial 8.99
Industrial 25.2
Transportation 28.1

Source: LLNL April, 2018. Data is based on DOE/EIA MER (2017). If this information or a reproduction of it is used, credit must be given to the Lawrence Livermore National Laboratory and the Department of Energy, under whose auspices the work was performed. This chart was revised in 2017 to reflect changes made in mid–2016 to the Energy Information Administration's analysis methodology and reporting. The efficiency of electricity production is calculated as the total retail electricity delivered divided by the primary energy input into electricity generation. End use efficiency is estimated as 65% for the residential sector, 65% for the commercial sector, 21% for the transportation sector, and 49% for the industrial sector which was updated in 2017 to reflect DOE's analysis of manufacturing. Totals may not equal sum of components due to independent rounding. LLNL-MI-410527

also rejected from power-generation activities. This result is similar to the result we saw in Figure 12.6 for an individual power plant.

Another way of utilizing knowledge about energy for gaining insight into the feasibility of renewable energy for replacing the fossil-powered energy used by our civilization is discussed in Box 12.1.

BOX 12.1 Shifting to Renewable Energy: Some Sobering Thoughts [4]

Shifting from fossil fuels to renewable sources of energy is widely accepted as an essential step for sustainable development. Technological advances and the decreasing cost of photovoltaic panels and wind energy are often cited as indications of such a shift. While such a transition makes sense, there are some sobering issues, including some from the perspective of energy analysis, that need to be kept in mind.

1 *Scale.* The global consumption of fossil fuels is more than 12 TW (terawatt). When the transition from renewable fuels to fossil fuels happened in the late 1800s, the global energy consumption was 1.4 TW. At that time 50 percent the consumption was from biomass. Even satisfying 50 percent of global need today by renewable fuels requires 6 TW, which is very difficult to achieve. Even though the total quantity of wind and solar are each larger than 12 TW, their conversion to useful energy involves substantial losses and various environmental impacts. Consideration of the expected future increase in demand for energy makes the issue of scale even more formidable.

2 *Energy density.* The energy density of coal is 25–30 MJ/kg and that of crude oil and natural gas is about 40–45 MJ/kg. In contrast, dry crop residues contain only 12–15 MJ/kg, indicating that very large quantities of biomass would be required to replace fossil fuels.

3 *Power density.* The power density refers to the rate of energy production per unit area. Fossil resources have a power density of 10^2–10^3 W/m². Thus, relatively small land areas are able to supply large energy flows. In contrast, the power density of biomass energy production is 1 W/m², of wind energy is 10 W/m², and of solar photovoltaic is 20 W/m². Thus, using any renewable resource to meet our energy needs will require at least 5–50 times more land area than is needed when fossil fuels are used. Most modern infrastructure, such as industry, high-rise buildings, and malls, have developed to rely on the high power density of fossil fuels. As Smil says, "A supermarket would require a photovoltaic field roughly ten times larger than its own roof, or 1,000 times larger in the case of a high-rise building" [4].

4 *Intermittency.* Fossil fuels are an amazing store house of energy, owing to which we have become used to a continuous supply. However, the most promising renewable resources, wind and solar, are intermittent, and technology for their large-scale storage is not yet available. Without significant technological breakthroughs or behavioral change, using

fossil resources may be necessary to provide a base load for times when the wind is not blowing or the sun is not shining.

5 *Geographical distribution.* Modern transportation infrastructure has made fossil fuels available across the world. Similar or larger infrastructure will be needed to make renewable energy available in all corners of the world since the geographical distribution of solar energy, wind, and geothermal is highly uneven.

These factors indicate that without significant breakthroughs in technology and changes in human behavior, it will be difficult for renewable resources to become the main source of energy.

12.3 Net Energy Analysis

Net energy analysis was developed in the 1970s to consider the direct and indirect roles of energy resources in supporting economic activities [5, 6]. This approach accounts for the use of fossil energy sources in economic activities – crude oil, coal, and natural gas – to calculate their life cycle fuel consumption. Other names for this approach are full fuel cycle analysis, life cycle energy analysis, or energy footprint analysis. It relies on the steps for sustainability assessment we learned in Chapters 8, 9, and 10. Results from such calculations are used to calculate metrics such as the energy return on investment and the net energy, which we defined in Equations 12.4 and 12.5. These calculations may be done for any boundary, ranging from an individual process to the full life cycle based on process, input–output, or hybrid network models. Efforts toward sustainable engineering usually focus on the life cycle boundary to prevent the shifting of impacts. At this scale, the EROI may be interpreted as the ratio of the energy (fuel value) returned to society to the energy (fuel value) required to get that energy.

Net energy analysis is best suited for analyzing systems that produce products with fuel value. The energy return on investment values of some common fuels in various countries and their variations over time are listed in Table 12.2. These data show that the EROI of domestic oil extraction decreased from about 100 in 1950 to about 20 in 1970, with the increase in production. In 2011, the EROI of domestic oil was around 16. The average EROI of oil across the world has also been decreasing with increasing production. Many researchers have argued about oil production reaching a peak, as we discussed in Chapter 2. However, the development of new technologies and discovery of new deposits seem to have prevented a peak so far. Nevertheless, the EROI continues to decrease. Coal has also seen its EROI decrease but not as drastically as that of crude oil.

Table 12.2 EROI of some common fuels [7, 8].

Resource	Year	Region	EROI
Oil and gas production	1950	USA	100
Oil and gas production	1970	USA	20
Oil and gas production	1999	Global	35
Oil and gas production	2006	Global	18
Oil and gas production	2010	China	10
Natural gas	1993	Canada	38
Natural gas	2000	Canada	26
Natural gas	2009	Canada	20
Coal	1950	USA	100
Coal	2000	USA	80
Coal	2007	USA	60
Shale oil	2008	USA	5
Shale gas	2013	USA	12
Nuclear	n/a	USA	5–15
Hydropower	n/a	n/a	> 100
Wind	n/a	n/a	18
Photovoltaic	n/a	n/a	6–12
Gasoline	n/a	n/a	8
Sugarcane ethanol	n/a	n/a	0.8–10
Corn ethanol	n/a	n/a	0.8–1.6

The reason for the decreasing EROI is that in the early years of extraction, the resource that is easiest to extract is usually accessed. The effort or processing energy required to extract this resource is then usually quite small. For example, before its commercial exploitation, crude oil was available directly on the earth's surface in some places, thus requiring virtually no effort in mining it. As the easy-to-access resources are depleted or as demand increases, other resources are accessed, but they usually require more processing energy, so have a smaller EROI. Thus, for nonrenewable resources, it is common for the EROI to decline over time. If the EROI falls below 1, it is no longer energetically profitable to process the resource. Even though decisions are made based on monetary return, a reduction in EROI often correlates well with a decrease in monetary return, especially in the long run.

The EROIs of alternatives for power generation are shown in Figure 12.8. This shows that hydropower has the highest return, but it is important to keep in mind that the large impact of hydropower on land use is not captured by this approach. The EROI of wind energy is also a healthy 20. Electricity from nuclear resources, coal, and natural gas vary between 10 and 20. Solar and geothermal EROI values

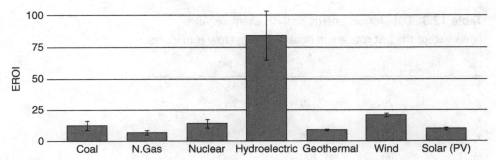

Figure 12.8 EROIs of power generation from various sources. Reproduced with permission from [8].

are around 10. In contrast, most biomass-based fuels in temperate regions have EROIs of no more than 3 or 4, while tropical biofuels can have returns of as high as 8 for sugarcane ethanol. The EROI of corn ethanol is only around 1.

Example 12.2 Consider the network in Table 12.3, with the monetary flow represented by the following equations:

$$0.833x_1 - 0.182x_2 = f_1 \qquad (12.6)$$
$$-0.667x_1 + 0.818x_2 = f_2 \qquad (12.7)$$

Determine the direct and total energy intensities of the two sectors.

Solution
The m vector in this table represents the fossil energy used by each sector for its activities. Using Equation 10.37, the direct energy intensities of the two sectors are

$$r_1 = \frac{m_{11}}{x_1}$$
$$= 10 \text{ J/\$}$$
$$r_2 = \frac{m_{12}}{x_2}$$
$$= 1.82 \text{ J/\$}$$

Using Equation 10.42, the total or life cycle energy intensity is

$$R^* = \begin{bmatrix} 10 & 1.82 \end{bmatrix} \begin{bmatrix} 0.833 & -0.182 \\ -0.667 & 0.818 \end{bmatrix}^{-1}$$
$$= \begin{bmatrix} 16.77 & 5.96 \end{bmatrix} \text{ J/\$}$$

These coefficients may be used to calculate the life cycle energy use or net energy for any final demand, as shown in Equation 10.44 and illustrated in the next example.

Table 12.3 Data for net energy analysis example. All flows except the last row are in dollars. The last row is in joules.

	1	2	f	x
1	5	10	15	30
2	20	10	25	55
v	5	35		
x	30	55		
m(J)	300	100		

Example 12.3 Suppose sector 2 in Example 12.2 produces 100 J/\$ of fuel. Calculate the EROI of this fuel.

Solution

Since we are determining the EROI of the output from sector 2, let the final demand from this sector be \$1. That is, $f_1 = 0$ and $f_2 = 1$. We can now calculate the life cycle energy needed to produce this final demand as follows.

With this final demand, we can simultaneously solve Equations 12.6 and 12.7 to get $x_1 = \$0.33, x_2 = \$1.49$. Using Equation 10.38,

$$m_1 = (10 \times 0.33) + (1.82 \times 1.49)$$
$$= 5.95 \text{ J}$$

Alternatively, we can use the total energy intensity coefficients and Equation 10.44:

$$M = R^*f$$
$$= \begin{bmatrix} 16.77 & 5.95 \end{bmatrix} \begin{bmatrix} 0 \\ 1 \end{bmatrix}$$
$$= 5.95 \text{ J}$$

Thus, 5.95 J is the total energy use to produce \$1 of final demand from sector 2, which has a fuel value of 100 J. Therefore,

$$\text{EROI} = \frac{100}{5.95}$$
$$= 16.81$$

Example 12.4 Continuing the problem in Example 10.4, determine the energy return on an investment of electricity. Of the energy used for oil extraction, 2 L of oil and 1 kWh of electricity are used as processing energy. No processing energy is needed in the fuel production or power generation steps.

Solution

The processing energy is 2 L of oil and 1 kWh of electricity in the oil extraction step. These inputs are needed to obtain 100 L of oil on the ground. Since the scaling factor for the oil extraction module is 0.005 (see Equation 10.14), the processing energy to produce 100×0.005 L of oil is $2 \times 0.005 = 0.01$ L of oil and $1 \times 0.005 = 0.005$ kWh of electricity. The total processing energy to produce 1 kWh or 3.6×10^6 J of electricity is

$$E_{pr} = \left(0.005 \text{ kWh} \times \frac{3.6 \times 10^6 \text{ J}}{1 \text{ kWh}} \right) + (0.01 \text{ L oil} \times 3.8 \times 10^7 \text{ J/L})$$

$$= 3.98 \times 10^5 \text{ J}$$

Here, 3.8×10^7 J/L is the fuel value of crude oil. Thus,

$$\text{EROI} = \frac{1 \text{ kWh} \times 3.6 \times 10^6 \text{ J/kWh}}{3.98 \times 10^5 \text{ J}}$$

$$= 9.04$$

12.4 Material Flow Analysis

Material flow analysis (MFA) considers the mass of materials flowing in and out of a selected system, and the accumulating or depleting stock in the system. The goal is to gain insight into the relative flows of feedstock, products, stock, recycling, and waste. The MFA approach may be applied to systems ranging from a single piece of equipment to the entire economy. If the focus is on the flow of a specific substance such as chlorine or copper, it is often called substance flow analysis. Such analysis is useful for identifying opportunities for improving the selected system in terms of its environmental impact and economic aspects.

Like other sustainability assessment methods, MFA also requires defining the goal and scope of the study, obtaining relevant data, assessing impact, and identifying opportunities for improvement. Quantifying material flows and metrics often relies on mass balance calculations, and the results are commonly represented in graphical format, as illustrated by the examples in this section.

Example 12.5 Cooking 1 cup of dry rice requires 2 cups of water. In the cooking process 40 percent of the water evaporates, but with a covered pot most of it is recycled in such a way that the net loss of water is only 10 percent. Draw a Sankey diagram for this process. What is the material efficiency of this process? One cup of dry rice weighs 184 g and 2 cups of water weigh 480 g.

Solution

The water evaporated during this process is $0.4 \times 480 = 192$ g, and the net loss is 10 percent of 480 g, which is 48 g. Thus, water recycled is $192 - 48 = 144$ g. Using an overall mass balance, the cooked rice will weigh $480 + 184 - 48 = 616$ g. The Sankey diagram for these mass flows is shown in Figure 12.9. Its material efficiency is

$$
\begin{aligned}
\eta &= \frac{\text{Useful mass}}{\text{Total mass}} \\
&= \frac{616}{616 + 48} \\
&= 92.8 \text{ percent}
\end{aligned}
$$

A general diagram representing MFA at the scale of an economy is shown in Figure 12.10. Inputs include imports and resources extracted from the domestic environment, while outputs include exports and domestic process output (DPO). Wastes may be exported or become part of the DPO. In the economy, materials may accumulate (or deplete) as stocks due to new infrastructure or other economic activities. End-of-life activities may recycle some materials as well.

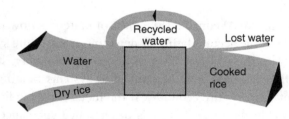

Figure 12.9 Sankey diagram for MFA of cooking rice.

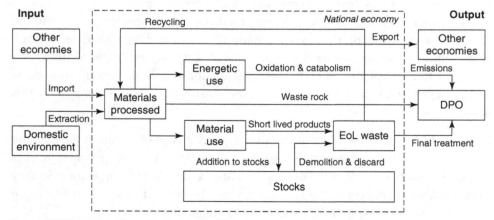

Figure 12.10 Material flows in an economy [9].

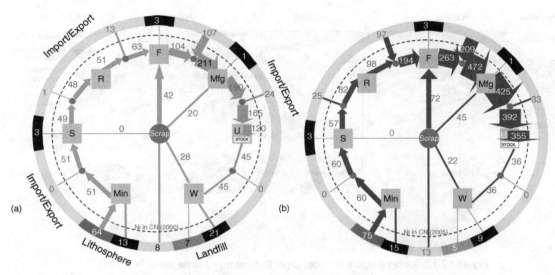

Figure 12.11 Nickel flows (Gg/y) in China in (a) 2000 and (b) 2005. Reproduced with permission from [10]. Min, mining; S, smelting; R, refining; F, fabrication; Mfg, manufacturing; U, use; W, waste.

Let us now consider MFA studies for some common materials to get an idea of the type of insight that may be obtained from such analysis.

- *Nickel use in China in 2000 and 2005.* Figure 12.11 shows the mass flow of nickel in China through various stages of the life cycle, including imports and exports [10]. By comparing the two figures it is evident how China tripled its in-use stock of nickel by increasing its own domestic extraction, but still relying mainly on imports of fabricated and pre-fabricated nickel. Flow in waste to landfills did not change much. This increase in nickel use was mainly due to a large increase in the use of stainless steel for its domestic infrastructure.
- *Global paper flows in 2012.* Figure 12.12 shows the flow of paper in multiple steps from harvest to end of life. The large reliance on wood and the large recycling and energy recovery flows are apparent.
- *Global material flows in 2005.* Figure 12.13 shows the flow of materials in the world and in the European Union.

Material flow analysis is appealing due to its simplicity and ability to provide unique insight into materials use. However, just as energy analysis does not have the ability to account for material resources, MFA is not able to account for energy resources except in terms of their mass. Such studies are finding a use in efforts toward developing a circular economy, which we will learn about in Chapter 19.

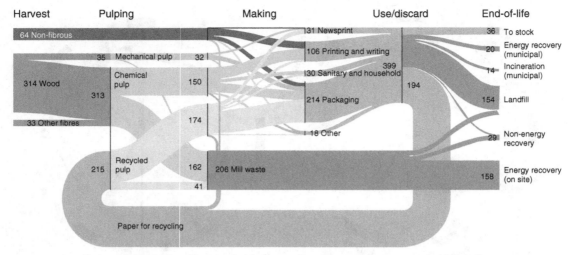

Figure 12.12 Sankey diagram of global paper flow measured in megatons in 2012 [11].

Example 12.6 Determine the material efficiency of each step of the global paper flow system shown in Figure 12.12 and the efficiency of the overall system.

Solution
The material efficiency is the ratio of the useful material output to the total material input. For the pulping step, the material efficiency is

$$\eta_{pulping} = \frac{150 + 32}{35 + 313 + 33}$$
$$= 0.48$$

For the manufacturing step,

$$\eta_{manufacturing} = \frac{31 + 106 + 30 + 214 + 18}{64 + 32 + 150 + 174}$$
$$= 0.95$$

For the use/discard step,

$$\eta_{use/discard} = \frac{194 + 36}{399}$$
$$= 0.58$$

For the overall material efficiency, the useful material output is 36 Mt of stock. Therefore,

$$\eta_{paper} = \frac{36}{64 + 35 + 314 + 33}$$
$$= 0.08$$

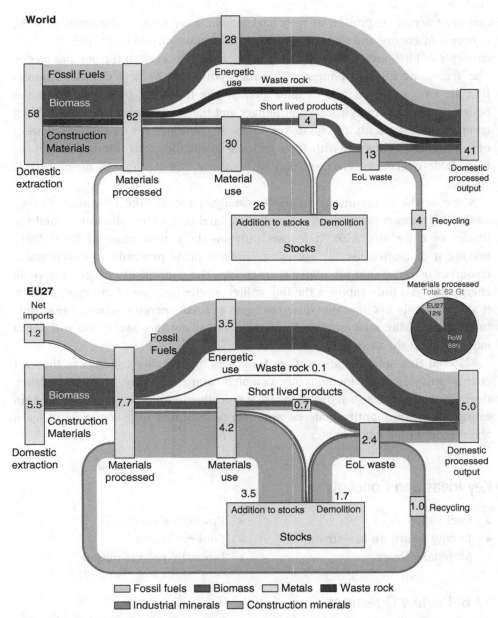

Figure 12.13 Materials flow in the world and the European Union in 2005 in gigatons per year (Gt/yr) [9]. RoW, rest of world; EoL, end of life.

12.5   Summary

Energy analysis and MFA are appealing approaches for providing holistic insights because all activities involve the flow and transformation of energy and materials, and most people, regardless of their background, have some understanding of the concepts of energy and materials, and of their importance in daily life. Therefore, concepts of EROI and net energy have become popular even in the popular media. The decreasing EROI of nonrenewable fuels points to the depletion of resources that are easy to access and of high quality. The low EROI of many renewables conveys their larger environmental impact and the small likelihood of their being complete replacements for fossil resources. In addition, the low power density of renewables, combined with their regional availability and intermittency, pose formidable challenges for transition to a society based primarily on renewable energy.

Some of the shortcomings of energy analysis include the following. Energy analysis ignores resources such as materials, land use, water withdrawal, and the impact of emissions. Also, it focuses only on the consumption of fossil fuels, making it of limited use for assessing products made primarily from renewable resources or those that are material-intensive. The concept of energy as used in energy analysis only captures the fuel value and the first law of thermodynamics. It also implicitly assumes that different types of fossil energy resources are substitutable and of the same quality. In practice, this is not the case, as we will see in more detail in the next few chapters.

Material flow analysis also suffers from very similar shortcomings to those of energy analysis. Both methods are complementary and may be used together. However, there are also more thermodynamically rigorous approaches based on exergy that can quantify material and energy resources together. We will learn about them in the next chapter.

Key Ideas and Concepts

- Fuel value
- Energy return on investment
- Material efficiency
- Processing energy
- Sankey diagram
- Domestic process output

12.6 Review Questions

1. Does a Sankey diagram take into account the first and second laws of thermodynamics?
2. Why does the EROI have to be larger than 1 for a system to be energetically feasible?
3. Why has the EROI of crude oil decreased over time?
4. From the US Sankey diagram, identify the least efficient and most efficient economic activities.

5. Compare the material efficiency of the economies of the world and the European Union.

Problems

....

12.1 On the plains of Africa, the Thomson's gazelle is a favorite prey of cheetahs. Consider a gazelle that, if caught, will yield a calorific value of 1000 kcal to the hunting cheetah. Determine how far the gazelle has to run to escape from a cheetah, knowing that cheetahs desire an energy return on investment of 22. A typical cheetah weighs 50 kg and can reach a speed of 60 km/h in two seconds.

12.2 Generating a joule of electricity requires 1.2 joules of biomass and 0.3 joules of additional energy. Determine the energy return on investment and energy efficiency.

12.3 For the US Sankey diagram shown in Figure 12.7, calculate the energy efficiency of power generation and the four sectors. Obtain the Sankey diagrams for the USA from the cited source for earlier years and compare how the overall energy efficiency and the efficiency of the five activities have changed since 2010.

12.4 Sankey diagrams for various countries are available at http:// flowcharts.llnl.gov. Obtain the flowcharts for China, Germany, Saudi Arabia, and USA for 2011. Compare the overall efficiency of each country and the efficiencies of electricity and heat generation in the residential, commercial, industrial, and transportation sectors.

12.5 A home heating system maintains the indoor temperature at 20 °C. The outside temperature is 0 °C. A furnace burns natural gas to heat a 1 : 4 mixture of outside and inside air. The resulting air is at a temperature of 40 °C. Draw a Sankey diagram of this system for a furnace that loses 20 percent of heat in its exhaust. You may ignore changes in the density of air with temperature. The fuel value of natural gas is 46 MJ/kg.

12.6 Consider the economic input–output system shown in Problem 10.6. Suppose the energy consumptions of the three sectors are $t_1 = 100$ MJ/\$, $t_2 = 500$ MJ/\$, and $t_3 = 200$ MJ/\$.

1. Calculate the life cycle energy consumption per dollar of final demand for each sector. (Hint: For the life cycle energy consumption of sector 1, use $f_1 = 1$ and $f_2 = f_3 = 0$.)
2. Calculate the energy return on investment for sector 1, given that the price of energy produced from this sector is \$2 per joule.

12.7 An economic input–output model consists of the following flows: $x_{11} = 50$, $x_{12} = 50$, $x_{21} = 30$, $x_{22} = 0$, $f_1 = 10$, $f_2 = 25$.

Table 12.4 Data for net energy analysis of some transportation fuels.

| | Energy input (in MJ) to generate 1 MJ from fuel | | | | | |
	Petroleum	Natural gas	Coal	Other	TOTAL	Greenhouse gases
Gasoline	1.10	0.03	0.05	0.01	1.19	94
Corn ethanol	0.04	0.28	0.41	0.04	0.77	77
Cellulosic ethanol	0.08	0.02	−0.02	0.02	0.10	11

1. Determine the input–output model for this network. You need to derive two equations in terms of the throughput and final demand for each sector.
2. If the direct energy inputs to each sector are $m_{11} = 1000$ J and $m_{12} = 500$ J, calculate the energy return on investment for the product from sector 2 whose fuel value is 200 J/$.

12.8 An input–output model is given as

$$0.7x_1 - 0.1x_2 = f_1$$
$$-0.6x_1 + 0.6x_2 = f_2$$

This system uses 300 J of energy for $50 of throughput in sector 1, and 100 J of energy for $100 of throughput from sector 2. All of this energy is used for processing. Calculate the energy return on investment for sector 1 if it produces 100 J of fuel per dollar of final demand.

12.9 Professors David Pimentel and Bruce Dale debated the use of net energy analysis for determining "The Cost of Biofuels" [12]. Answer the following questions based on this article. Table 12.4 contains data that the authors use for their arguments.

1. Calculate the energy ROI for gasoline and ethanol today. Make reasonable assumptions.
2. Dr. Dale argues: "Net energy analysis is misleading because ethanol's net energy was never compared with gasoline's net energy. The comparison is easily done. [From the table above,] ethanol's net energy is 1.0 minus the sum of (0.04 + 0.28 + 0.41) times 100, which equals +27%, while gasoline's net energy is 1.0 minus the sum of (1.1 + 0.03 + 0.05) times 100, which equals −18%. Gasoline's net energy is therefore significantly lower than ethanol's." Check the validity of this calculation. You may need to use your calculation in (1). Why does gasoline look so much worse than ethanol in this calculation?

12.10 Consider the electricity generation system in the USA as shown in the Sankey diagram of Figure 12.7. Assuming that 70 percent of the total feedstock entering this system is converted into electricity, calculate the energy return on investment of the system.

12.11 For the rice preparation problem in Example 12.5, if three cups of water are used to wash the rice before cooking, draw the Sankey diagram and determine the material efficiency.

12.12 For nickel flows in China, how has the material efficiency changed between 2000 and 2005? Perform this calculation with and without accounting for imports and exports.

References

[1] V. Smil. *Energy in Nature and Society*. MIT Press, 2008.

[2] M. A. Rosen. Energy- and exergy-based comparison of coal-fired and nuclear steam power plants. *Exergy: An International Journal*, 1(3):180–192, 2001.

[3] Lawrence Livermore National Laboratory, United States Department of Energy. Estimated energy use in 2016: 97.7 quads. https://flowcharts.llnl.gov/. accessed April 20, 2018.

[4] V. Smil. 21st century energy: some sobering thoughts. *OECD Observer*, 258/59:22–23, 2006.

[5] C. W. Bullard and R. A. Herendeen. The energy cost of goods and services. *Energy Policy*, 3:268–278, 1975.

[6] D. T. Spreng. *Net-Energy Analysis*. Praeger, 1988.

[7] C. J. Cleveland. Net energy from the extraction of oil and gas in the United States. *Energy*, 30(5):769–782, 2005.

[8] C. A. S. Hall, J. G. Lambert, and S. B. Balogh. EROI of different fuels and the implications for society. *Energy Policy*, 64(0):141–152, 2014.

[9] W. Haas, F. Krausmann, D. Wiedenhofer, and M. Heinz. How circular is the global economy? An assessment of material flows, waste production, and recycling in the European Union and the world in 2005. *Journal of Industrial Ecology*, 19(5):765–777, 2015.

[10] B. K. Reck and V. S. Rotter. Comparing growth rates of nickel and stainless steel use in the early 2000s. *Journal of Industrial Ecology*, 16(4):518–528, 2012.

[11] S. Van Ewijk, J. A. Stegemann, and P. Ekins. Global life cycle paper flows, recycling metrics, and material efficiency. *Journal of Industrial Ecology*, 22(4):686–693, 2017.

[12] W. Schulz. The costs of biofuels. *Chemical and Engineering News*, 85(51):12–16, 2007.

13 Exergy Analysis

> The device by which an organism maintains itself stationary at a fairly high level of orderliness (= fairly low level of entropy) really consists in continually sucking orderliness from its environment.
>
> Erwin Schrödinger [1]

The value and usefulness of resources is related to their ability to do useful work. Determining this ability requires consideration of the second law of thermodynamics, since this law accounts for the fact that, even under ideal conditions, heat cannot be converted completely into work. In this chapter we will learn about the concept of available energy or exergy, and its use for assessing and improving the efficiency of individual processes such as manufacturing systems and economies. The use of exergy for assessing life cycles will be the focus of the next chapter.

13.1 Concept of Exergy

For any system, the ability to do work requires a difference between the system and its surroundings. For example, a stationary object has potential energy with respect to the ground only when it is elevated from it, and the ability of a battery to generate electricity depends on the concentration gradient between the electrodes. The concept of exergy represents the maximum ability of a system to do work with respect to a reference state. In other words, exergy is the maximum work that can be done as a system is brought to equilibrium with the reference state.

Consider a heat source at temperature T and an ideal reversible engine that runs between this heat source and sink, as shown in Figure 13.1a. The source and sink are assumed to be large enough (infinite) that heat transfer through the engine does not affect their temperatures. The maximum work that can be done by this engine is proportional to the temperature difference between the source and the sink. Since the lowest possible temperature of the sink is 0 K, according to the third law of thermodynamics, the maximum possible work W_{max} is proportional to the gradient, $T - 0$. Thus,

$$W_{max} \propto T \qquad (13.1)$$

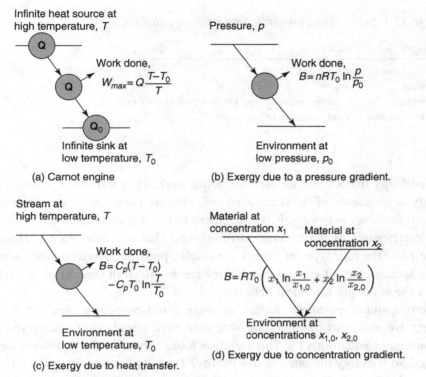

Figure 13.1 Exergy due to different types of gradients.

If the sink temperature is T_0, then the maximum possible work given this sink temperature is

$$W \propto (T - T_0) \tag{13.2}$$

Thus, the maximum efficiency of the engine is

$$\psi_{max} = \frac{W}{W_{max}} = \frac{T - T_0}{T} \tag{13.3}$$

This implies that if heat Q is transferred between the source at T and the sink at T_0, the maximum useful work that can be done is

$$B = Q\left(\frac{T - T_0}{T}\right) \tag{13.4}$$

Thus, the maximum available energy or exergy of such a heat transfer is given by Equation 13.4. Equation 13.3 is called the Carnot efficiency.

In general, exergy is defined as follows:

$$B = (H - H_0) - T_0(S - S_0) \tag{13.5}$$

Here, the subscript 0 indicates the reference state. The first term represents the total quantity of energy or enthalpy, while the second term represents the part of

Table 13.1 Standardized reference state for exergy analysis.

Property	Value
Temperature	25 °C
Pressure	1 atm
Minerals	Concentration of first kilometer of Earth's crust
Other materials	Concentration of sea water

this enthalpy that cannot be used for doing work. Thus, exergy is "entropy free" energy or a measure of the thermodynamic distance from the surroundings. Since the surroundings may vary, it is convenient to use a standardized reference state that corresponds to the average surroundings. This reference state is shown in Table 13.1. The four types of exergy are kinetic, potential, physical, and chemical. Some special cases of exergy calculations are described in the rest of this section under the assumption of ideal behavior.

Exergy and free energy. Before we learn about calculating specific types of exergy, we will clear one question that may have already occurred to you after encountering Equation 13.5. This equation looks similar to the equations for free energy. So is exergy the same as free energy? The answer depends on the reference state and the path taken to get there.

- *Exergy.* If the reference state is the environment, like that in Table 13.1, then Equation 13.5 calculates exergy. This is what we will use in this chapter.
- *Gibbs free energy.* If the reference state is the final state of a chemical reaction, and pressure and volume are held constant along the path toward this state, then Equation 13.5 becomes the Gibbs free energy.
- *Helmholtz free energy.* If the reference state is the final state of a chemical reaction, and volume and temperature are held constant along the path toward this state, then Equation 13.5 becomes the Helmholtz free energy.

Pressure gradient. The exergy due to a pressure gradient is depicted in Figure 13.1b, and may be calculated as

$$
\begin{aligned}
B &= \int_{p}^{p_0} p\, dV \\
&= -nRT_0 \int_{p}^{p_0} \frac{dp}{p} \\
&= nRT_0 \ln \frac{p}{p_0}
\end{aligned}
\tag{13.6}
$$

Here, n is the number of moles of the material. Ideal gas behavior is assumed.

Temperature gradient. The exergy change as a system decreases in temperature from temperature T to T_0 without phase change and with constant specific heat is shown in Figure 13.1c and may be calculated as

$$B = (H - H_0) - T_0(S - S_0)$$

$$= \int_{T_0}^{T} mC_p dT - T_0 mC_p \int_{T_0}^{T} \frac{dT}{T}$$

$$= mC_p(T - T_0) - mC_p T_0 \ln \frac{T}{T_0} \tag{13.7}$$

Here, m is the mass of the material.

Concentration gradient. The chemical exergy of a substance is calculated with respect to its composition in the reference state [2]. The exergy of mixing for an ideal gas with composition $x_{i,0}$ to form a mixture in which its partial fraction becomes x_i is shown in Figure 13.1d and may be calculated as

$$B = \sum_i nRT_0 x_i \ln \frac{x_i}{x_{i,0}} \tag{13.8}$$

For a single compound that is present in the same form in the reference environment, its exergy may be calculated as

$$B = -nRT_0 \ln x_0 \tag{13.9}$$

where x_0 is the mole fraction of the substance in the reference environment.

Example 13.1 Heat is available from a large source at a temperature of 150 °C. Determine the maximum work that can be done if 100 J of this heat is transferred to a large sink at 25 °C.

Solution
Since the source and sink are large, we may assume that transfer of heat has a negligible effect on the source temperature. Then, using Equation 13.4, the maximum work from this transfer is

$$B = 100 \times \frac{423 - 298}{423}$$

$$= 29.55 \text{ J}$$

Example 13.2 A water stream with enthalpy 100 J at 150 °C cools down to 25 °C. What is the maximum amount of work that can be done during this change? The specific heat of water is 1 cal/g °C.

Solution
The following energy balance provides the mass of water,

$$100 \text{ J} \times \frac{1 \text{ cal}}{4.184 \text{ J}} = m \times \left(1\frac{\text{cal}}{\text{g }°\text{C}}\right) \times (150 - 25) \text{ }°\text{C}$$

Solving this equation, we get

$$m = 0.19 \text{ g}$$

The maximum work that can be done during the change from 150 °C to 25 °C is the difference between the exergies at these two temperatures. This is also equal to the change in exergy due to cooling from 150 °C to 25 °C. Using Equation 13.7 we obtain this quantity as

$$B = (0.19 \times 1 \times (423 - 298)) - \left(0.19 \times 1 \times 298 \times \ln\frac{423}{298}\right)$$

$$= 3.92 \text{ cal}$$

Thus, cooling this stream can do maximum work of 3.92 cal or 16.4 J. This is less than the 29.55 J of work that can be done if the source and sink are infinitely large, as calculated in Example 13.1.

Example 13.3   Determine the chemical exergy of pure oxygen.

Solution
The reference state for oxygen is the atmospheric concentration of 20.34 percent by volume. Then the chemical exergy of pure oxygen may be calculated by Equation 13.9 as

$$B_{O_2} = -RT_0 \ln x_{O_2} \tag{13.10}$$

$$= -8.314\frac{\text{J}}{\text{mol K}} \times 298.2 \text{ K} \times \ln(0.2034) \tag{13.11}$$

$$= 3948 \text{ J/mol} \tag{13.12}$$

Example 13.4   Calculate the exergy of an oxygen-rich stream with 30 percent O_2 and 70 percent N_2. Consider two cases: the gas stream is at 25 °C and at 100 °C. The specific heat at 100 °C is 1.009 J/(g K).

Solution
At 25 °C there is no contribution from thermal exergy since the stream is at the reference temperature. Using Equation 13.8,

$$B = RT_0 \left(0.30 \ln\frac{0.30}{0.21} + 0.70 \ln\frac{0.70}{0.79}\right)$$

$$= 55.34 \text{ J/mol}$$

At 100 °C, the thermal exergy of the gas mixture is

$$B = (1.009 \times (373 - 298)) - \left(1.009 \times 298 \ln \frac{373}{298}\right)$$

$$= 8.18 \text{ J/g}$$

The molecular weight of the mixture is $0.3 \times 32 + 0.7 \times 28 = 29.2$ g/mol. Then the total exergy (thermal + chemical) at 100 °C is

$$8.18 \frac{\text{J}}{\text{g}} \times 29.2 \frac{\text{g}}{\text{mol}} + 55.34 \frac{\text{J}}{\text{mol}} = 294.20 \frac{\text{J}}{\text{mol}}$$

Example 13.5   Air is stored at a pressure of 10 atm. Determine the maximum work that can be done by releasing this air to atmospheric pressure.

Solution
By using Equation 13.6, we get

$$B = 8.314 \times 298 \ln \frac{10}{1}$$

$$= 5704.8 \text{ J/mol}$$

Exergy of materials and reactions. If a substance is present as a different compound in the reference state then the chemistry of converting it to its reference compound needs to be considered. Thus, to determine the exergy of a compound P that is present in the reference state as R, we consider the reaction that relates P and R, $pP + qQ \rightarrow rR + sS$. Here, p, q, r, s are stoichiometric coefficients. The exergies of P, Q, R, and S are related as

$$\Delta G_{rxn} = rB_R + sB_S - pB_P - qB_Q \tag{13.13}$$

Here, ΔG_{rxn} is the Gibbs free energy of this reaction. If the total exergy of the products is larger than that of the reactants, the Gibbs free energy represents the minimum work that must be done for the reaction to occur. Conversely, if the total exergy of the products is less than that of the reactants, then the Gibbs free energy is the maximum work that can be obtained from that reaction.

Such an approach can also be used to calculate the efficiency of a reaction. For example, if R is the useful product then the reaction efficiency may be calculated as the ratio of B_R to the total exergy input if the reaction is endothermic, or the total exergy output if it is exothermic.

Example 13.6   Calculate the exergy of pure iron metal.

Solution
We consider the state of iron in the reference state, which is as Fe_2O_3. The relevant reaction is

$$2Fe_{(solid)} + \frac{3}{2}O_{2,(gas)} \rightarrow Fe_2O_{3,(solid)}$$

For this reaction, using Equation 13.13, the Gibbs free energy of reaction is related to the exergy of the relevant compounds as

$$B_{Fe_2O_3} - \frac{3}{2}B_{O_2} - 2B_{Fe} = \Delta G_{rxn}$$

The Gibbs free energy of the reaction is $\Delta G_{rxn} = -742.6$ kJ/mol, the exergy of Fe_2O_3 is zero, and that of oxygen is given in Equation 13.12. Thus,

$$B_{Fe} = \frac{1}{2}\left(-\frac{3}{2} \times 3.948 + 742.6\right)$$

$$= 368.3 \text{ kJ/mol}$$

Example 13.7   In the reaction $2NaCl + 2H_2O \rightarrow Cl_2 + 2NaOH + H_2$, the useful products are chlorine and sodium hydroxide. Calculate the ideal second law efficiency of this reaction. If the actual reaction requires an exergy input of 712 kJ/mol, calculate the actual second law efficiency.

Solution
Using standard chemical exergies of relevant molecules and Equation 13.13, the exergy balance for this reaction is

$$(2 \times 14.75) + (2 \times 0.95) + \Delta G_{rxn} = 123.6 + (2 \times 74.9) + 236.1$$

$$31.4 + \Delta G_{rxn} = 509.5$$

$$\Delta G_{rxn} = 478.1 \text{ kJ/mol}$$

This indicates that an additional 478.1 kJ/kmol is required as input for this reaction to occur. This is an endothermic reaction. The ideal second law efficiency is calculated as follows:

$$\psi_{ideal} = \frac{\text{Useful exergy output}}{\text{Total exergy input}}$$

$$= \frac{(123.6 + 149.8)}{509.5}$$

$$= 53.7 \text{ percent}$$

In the non-ideal case, the second law efficiency is

$$\psi_{actual} = \frac{273.4}{712}$$

$$= 38.4 \text{ percent}$$

Thus, there is an opportunity for improving the efficiency by 15.3 percent.

13.2   Exergy Flow in Systems

Consider the general system shown in Figure 13.2. This system consists of material flowing in and out at a rate $\dot{m}$, heat flow rate $\dot{Q}$ at a specific temperature, heat loss rate to the surroundings $\dot{Q}_0$, and rate of work done on and by the system, $\dot{W}$. The first law of thermodynamics requires conservation of energy. Thus,

$$0 = \left\{ \sum_{i+} \dot{m}H_{in} + \sum_{i+} \dot{Q}_{in} + \sum_{i+} \dot{W}_{in} \right\} - \left\{ \sum_{i-} \dot{m}H_{out} + \sum_{i-} \dot{Q}_{out} + \sum_{i-} \dot{W}_{out} \right\}$$

(13.14)

Here, the energy into the system is equal to that leaving the system under steady-state conditions.

Since exergy does not have to be conserved, the exergy balance for this system may be written as follows:

$$LW = \left\{ \sum_{i+} \dot{m}B_{in} + \sum_{i+} \dot{Q}_{in} \left(1 - \frac{T_0}{T} \right) + \sum_{i+} \dot{W}_{in} \right\}$$
$$- \left\{ \sum_{i-} \dot{m}B_{out} + \sum_{i-} \dot{Q}_{out} \left(1 - \frac{T_0}{T} \right) + \sum_{i-} \dot{W}_{out} \right\}$$

(13.15)

Here, LW is lost work and B is the exergy of the material stream. The lost work is related to entropy generation by

$$LW = T_0 S_{gen}$$

(13.16)

This relationship is known as the Gouy–Stodola theorem. In an ideal reversible process there is no lost work: $LW = 0$ since $S_{gen} = 0$. Of course, in practice,

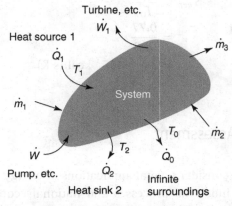

Figure 13.2 General system for energy and exergy balance equations.

such processes are not feasible, so there is always some lost work. Equation 13.15 is useful for understanding exergy flow through a system and can be used to identify opportunities for enhancing efficiency, as we will discuss and illustrate in the next section.

Example 13.8 Calculate the lost work for the mixing process in Figure 12.3 and determine the entropy generation.

Solution
Considering a reference temperature of 25 °C and specific heat of water of $C_p = 1$ kcal/kg °C, the exergy of the input streams using Equation 13.7 is

$$B_{hot} = 1 \times (373 - 298) - 1 \times 298 \times \ln\frac{373}{298}$$
$$= 8.10 \text{ kcal/kg}$$

$$B_{cold} = 1 \times (323 - 298) - 1 \times 298 \times \ln\frac{323}{298}$$
$$= 0.99 \text{ kcal/kg}$$

$$B_{mix} = 1 \times (348 - 298) - 1 \times 298 \times \ln\frac{348}{298}$$
$$= 3.78 \text{ kcal/kg}$$

Using Equation 13.15 and the mass flowrates from Figure 12.3, the lost work is

$$LW = (0.5)(8.10) + (0.5)(0.99) - (1)(3.78)$$
$$= 0.77 \text{ kcal/s}$$

Using Equation 13.16, the entropy generated is

$$S_{gen} = \frac{LW}{T_0}$$
$$= \frac{0.77}{298}$$
$$= 2.6 \times 10^{-3} \text{ kcal/(s K)}$$

13.3   Exergetic Assessment

In this section we will consider several applications of exergy analysis for improving the efficiency of industrial processes and national economies. We will also learn about the insight that exergy assessment can provide about emerging technologies.

13.3.1   Improving Efficiency

Among solutions to address the challenge of depleting resources and environmental pollution, the most common suggestion is to improve technological efficiency. This is based on the understanding that enhancing technological efficiency will result in less resource use and reduced pollution per unit of output. Exergy analysis can play an important role in identifying opportunities for improving efficiency at scales ranging from individual processes to the economy. It can also help by determining the feasibility of emerging technologies and identifying limits to efficiency improvement. Let us revisit the mixing example from Sections 12.2 and 13.2 with the following extension.

Example 13.9   Calculate the exergy efficiencies of the ideal and non-ideal (24 percent loss) mixing systems whose Sankey diagrams are given in Figure 12.4.

Solution

Ideal mixing. We calculated the exergy values of each stream in Example 13.8. Using these values, the overall exergy efficiency of the ideal mixing system is

$$\psi_1 = \frac{3.78}{0.5(8.10 + 0.99)}$$

$$= 83.05 \text{ percent}$$

Non-ideal mixing. For this case, the exergy of the mixed stream is

$$B_{mix} = 1 \times (336 - 298) - 1 \times 298 \times \ln\frac{336}{298}$$

$$= 2.23 \text{ kcal/kg}$$

Now the exergy efficiency is

$$\psi_2 = \frac{2.23}{0.5(8.10 + 0.99)}$$

$$= 49.13 \text{ percent}$$

Notice that the exergy efficiencies are less than the corresponding energy efficiencies of 100 percent and 86 percent shown in Figure 12.4.

This example demonstrates that even under the ideal condition of zero enthalpy loss, this mixing process has a loss of exergy due to irreversibilities. The exergy flow diagram for this system is shown in Figure 13.3. In Figure 13.3a, the light stream is the enthalpy value and the thin dark stream in it is the corresponding exergy. The ratio of exergy to enthalpy (B/H) is shown in this figure for each stream. It is an indicator of thermodynamic quality. Notice that the value of this ratio is larger for warmer or higher-quality streams. The exergy flows by

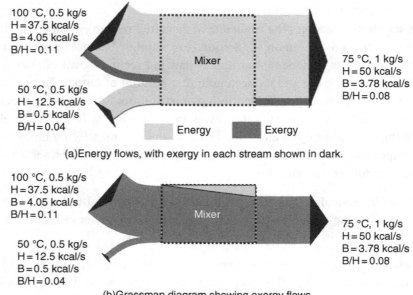

100 °C, 0.5 kg/s
H = 37.5 kcal/s
B = 4.05 kcal/s
B/H = 0.11

Mixer

75 °C, 1 kg/s
H = 50 kcal/s
B = 3.78 kcal/s
B/H = 0.08

50 °C, 0.5 kg/s
H = 12.5 kcal/s
B = 0.5 kcal/s
B/H = 0.04

Energy   Exergy

(a) Energy flows, with exergy in each stream shown in dark.

100 °C, 0.5 kg/s
H = 37.5 kcal/s
B = 4.05 kcal/s
B/H = 0.11

Mixer

75 °C, 1 kg/s
H = 50 kcal/s
B = 3.78 kcal/s
B/H = 0.08

50 °C, 0.5 kg/s
H = 12.5 kcal/s
B = 0.5 kcal/s
B/H = 0.04

(b) Grassman diagram showing exergy flows.

Figure 13.3 Energy and exergy flows in mixing process.

themselves are drawn in Figure 13.3b, which is called a Grassman diagram. This diagram shows the lost work as a triangle.

An alternative definition of exergy efficiency is based on the ratio of the actual exergy of the system to the ideal exergy,

$$\nu = \frac{\text{exergy in product}}{\text{maximum possible exergy in product}} \tag{13.17}$$

Using this definition, the efficiency of non-ideal mixing in Example 13.9 is 2.235/3.778 = 59 percent. This implies that the potential for improving efficiency in this system is 41 percent. The use of exergy analysis for improving the efficiency of a coal-burning power plant is demonstrated in the next subsection.

Example 13.10 Consider a conventional home water heater that stores hot water in a tank at a specified temperature. With such heaters, it is common for users to combine hot and cold water to get water at the desired temperature. Consider a case in which cold water is available at 25 °C and is stored in the tank at 80 °C. You desire the water to be at 50 °C when it comes out of the faucet, so you mix appropriate amounts of hot and cold streams. Determine the lost work and efficiencies for this system. Methane is used as the heating fuel, and its fuel value is 11,907.3 cal/g. Owing to heat losses from the storage tank, 25 percent more methane is used than the minimum. More modern systems do not have a

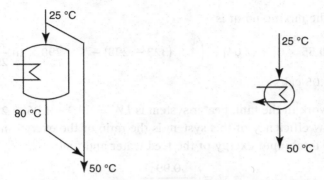

(a) Heating water with a storage tank.   (b) Heating water without a storage tank.

Figure 13.4 Heating systems for producing hot water at the desired temperature.

storage tank, and they heat the water directly to the desired temperature. Use exergy analysis to compare both systems. They are shown in Figure 13.4.

Solution

Let us consider 1 g/s as the desired flow rate of water at 50 °C as the basis for this analysis. Thus, $m_w = 1$ g/s. The specific heat of water is $C_{pw} = 1$ cal/g °C.

Heater with a tank. The quantity of cold water mixed with water from the heater to produce 1 g of water at 50 °C is calculated by an energy balance at the mixing point,

$$m_{w,c} \times 1 \times (25 - 25) + (1 - m_{w,c}) \times 1 \times (80 - 25) = 1 \times 1 \times (50 - 25)$$

Solving this equation, the cold water required is $m_{w,c} = 0.55$ g/s and $m_{w,h} = 0.45$ g/s.

Energy balance around the tank helps determine the fuel input. Note the 25 percent excess fuel requirement.

$$1.25 \times m_{w,h} C_{pw} \Delta T_w = m_{fuel} H_{fuel}$$

$$m_{fuel} = \frac{1.25 \times 0.45 \times 1 \times (80 - 25)}{11,907.3}$$

$$= 2.6 \times 10^{-3} \text{ g methane/s}$$

Exergy balance around the tank heater provides lost work as

$$LW_{tank} = \left(0.45 \times (298 - 298) - 0.45 \times 298 \ln \frac{298}{298} \right) + (2.6 \times 10^{-3} \times 11,907.3)$$

$$- \left(0.45 \times (353 - 298) - 0.45 \times 298 \ln \frac{353}{298} \right)$$

$$= 0 + 30.94 - 2.04$$

$$= 28.9 \text{ cal/s}$$

Lost work at the mixing point is

$$LW_{mix} = (0.55 \times 0) + (2.04) - \left(1 \times (323 - 298) - 1 \times 298 \times \ln \frac{323}{298}\right)$$

$$= 1.05 \text{ cal/s}$$

Thus, total lost work in the tank heater system is $LW = 28.9 + 1.05 = 29.95$ cal/s.

The second-law efficiency of this system is the ratio of the exergy in the water at 50 °C and the total input exergy of the feed water and fuel:

$$\psi = \frac{0.99}{0 + 30.94}$$

$$= 3.2 \text{ percent}$$

Heater without a tank. For the tankless heater, the methane required to heat 1 g/s water from 25 °C to 50 °C is

$$m_{fuel} = \frac{1 \times 1 \times 25}{11,907.3}$$

$$= 2.1 \times 10^{-3} \text{ g methane/s}$$

The lost work is

$$LW = \left(1 \times (298 - 298) - 1 \times 298 \ln \frac{298}{298}\right) + (2.1 \times 10^{-3} \times 11,907.3)$$

$$- \left(1 \times (323 - 298) - 1 \times 298 \ln \frac{323}{298}\right)$$

$$= 0 + 25 - 0.99$$

$$= 24.01 \text{ cal/s}$$

Thus, the tankless heater saves $29.95 - 24.01 = 5.94$ cal/s of exergy per g/s of water as compared to the conventional heater. Also, the second-law efficiency of the tankless heater is

$$\psi = \frac{0.99}{0 + 25}$$

$$= 3.96 \text{ percent}$$

From understanding the concepts of exergy and exergy balance, we know that the efficiency may be improved by addressing the phenomena that cause a loss of gradient. Some common reasons for exergy loss are listed in Table 13.2, along with potential solutions.

13.3.2   Exergy Analysis of Technologies

A unique insight from the second law and exergy analysis is about the thermo-dynamic limits to efficiency improvement. A simple example of this limit is the

Table 13.2 Some reasons for exergy loss.

Description	Mechanism	Solution
Mixing	Mixing of streams at different temperatures, compositions, or pressures	Avoid unnecessary mixing such as of hot and cold water in a faucet
Reactions far from equilibrium	Chemical interactions between reactants with a large concentration gradient	Avoid reactions such as combustion by using alternatives such as fuel cells
Heat exchange	Finite driving forces produces losses	Cascade heat to reduce losses, avoid use of high-temperature heat for low-quality applications such as space heating.
Electric resistance	Resistivity, eddy currents, and hysteresis losses in devices	Use superior materials for electrical components, reduce length of electrical wires through miniaturization.

Carnot efficiency, which indicates the highest possible efficiency in converting heat to work. Similarly, the difference between the exergy of the products and the reactants indicates the maximum work that can be obtained from the reaction or the minimum work needed to enable the reaction. Such information can be used to identify limits to efficiency improvement, as illustrated in Example 13.7, and to determine the validity of claims about emerging technologies, as illustrated in the following examples.

Example 13.11 Given the high demand for fresh water in many regions of the world that are faced with dwindling supplies, many researchers are focusing on new desalination technologies. A major obstacle facing these technologies is their need for energy. Claims about the potential of new technologies to significantly reduce the energy required for desalination are of the type quoted below [3, 4].

> It is 500 times thinner than the best filter on the market today and a thousand times stronger. The energy that's required and the pressure that's required to filter salt [are] approximately 100 times less.

This statement is about a new graphene membrane. Use exergy analysis to determine the validity of this statement, knowing that the best membranes available today require 1.6 kWh/m^3 water.

Solution

The energy requirement of the best membranes is 1.6 kWh/m³ (= 5.76 kJ/kg). According to the claim above, a 100-fold reduction in energy requirement means that desalination should be possible with an energy input of 0.0576 kJ/kg. Exergy analysis can help us evaluate this claim without knowing any details of the new graphene membrane or any other technology.

The minimum work that must be done to desalinate a given concentration of water may be calculated by exergy analysis as follows. The average concentration of seawater is known to be 3.5 percent or 0.035 kg/kg. Thus, the mole fraction of pure water in seawater can be calculated to be

$$x_w = \frac{\frac{(1-0.035)}{18}}{\frac{0.035}{58.5} + \frac{(1-0.035)}{18}}$$

$$= 0.9889$$

Using Equation 13.9,

$$B_{seawater} = -8.314 \times 298.2 \ln 0.9889$$

$$= 27.67 \text{ J/mol}$$

Knowing that the molecular weight of seawater is $0.9889 \times 18 + (1 - 0.9889) \times 58.5 = 18.45$, the exergy of mixing of seawater is $27.67 \times 18.45 = 1.5$ kJ/kg.

This indicates that, regardless of the technology used, obtaining freshwater from seawater requires exergy of at least 1.5 kJ/kg. Clearly, the claim that the graphene membrane is able to reduce the energy requirement 100-fold to 0.0576 kJ/kg is impossible, since it would violate the thermodynamic limit.

Example 13.12   The Grassman diagram for the coal-burning power plant we considered in Chapter 12 is shown in Figure 13.5. How is the insight from the Grassman diagram different from the insight provided by the Sankey diagram?

Solution

The Grassman diagram shows that the largest exergy loss is in the furnace, while the rejected heat has a very small exergy loss. This insight is different from that provided by energy analysis. As shown by the Sankey diagram in Figure 12.6, energy analysis identified rejected heat as the largest loss, while the furnace has no loss and operates with 100 percent efficiency. This perfect energy efficiency is due to assuming an ideal furnace with no losses and considering only the first law of thermodynamics, that of energy conservation. Exergy analysis accounts for the difference in quality of the resources as measured by their ability to do work. The rejected heat, being at a relatively low temperature, has little ability to do work, while the irreversibility in the furnace, being at a high temperature, has the largest exergy loss. The irreversibility in the furnace is due to the combustion

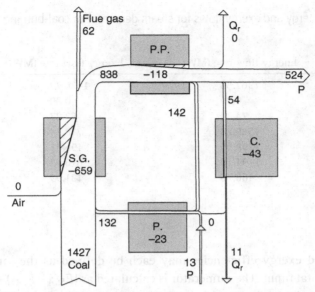

Figure 13.5 Grassman diagram for coal-burning power plant. All flows are in MW. See Figure 12.6 for the abbreviations.

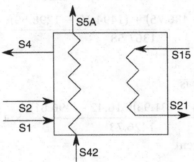

Figure 13.6 Steam generator from coal-burning power plant in Figure 12.5.

reaction, and presents the largest opportunity for efficiency improvement. This loss is present even under ideal conditions of 100 percent energy efficiency. On the basis of this insight, technologies that avoid combustion and permit the utilization of heat at high temperatures have been developed since they have a higher exergy efficiency than conventional combustion. Fuel cells and chemical looping are examples of such technologies.

Example 13.13 The steam generator in the power plant of Figure 12.5 is reproduced in Figure 13.6. Based on the energy and exergy information in Table 13.3, determine the energy and exergy efficiencies of this system. Stream S1 is 100 percent carbon, S2 is 79 percent N_2 and 21 percent O_2, and S4 is 79 percent N_2, 6 percent O_2, and 15 percent CO_2. All the other streams are 100 percent H_2O.

Table 13.3 Energy and exergy flows for steam generator of coal-burning power plant.

Stream	Energy flow rate (MW)	Exergy flow rate (MW)
S1	1367.58	1426.73
S2	0.00	0.00
S4	74.39	62.27
S5A	1585.28	718.74
S15	1298.59	496.81
S21	1494.16	616.42
S42	486.75	131.93

Solution

The energy and exergy efficiencies may each be defined as the ratio of the net output to the total input. The numerator is calculated as $(E_{S5A}-E_{S42})+(E_{S21}-E_{S15})$, while the denominator is E_{S1}, where E_i is the energy content of the ith stream. Thus, the energy efficiency of the steam generator is

$$\eta = \frac{(1585.28 - 486.75) + (1494.16 - 1298.59)}{1367.58} \times (100 \text{ percent})$$
$$= 95 \text{ percent}$$

and its exergy efficiency is

$$\psi = \frac{(718.74 - 131.93)(616.42 - 496.81)}{1426.73} \times (100 \text{ percent})$$
$$= 49 \text{ percent}$$

This large difference between the energy and exergy efficiencies occurs because even though most of the input energy is captured in the steam enthalpy, most of the input exergy is not. This exergy loss is due to the large difference in concentration between the inputs and outputs of combustion.

The energy and exergy efficiencies of many energy conversion devices are listed in Table 13.4. We note that for many devices, the energy and exergy efficiencies are close, while for others they can be vastly different. This difference depends on the types of products and the differences in their enthalpy and exergy values. Devices that produce much heat have the largest differences between their energy and exergy efficiencies because heat, particularly at lower temperatures, has little exergy. Their relationship is given by Equation 13.4. Boxes 13.1 and 13.2 describe the use of exergy analysis for determining opportunities for improving the efficiency of manufacturing processes, and how exergy analysis can provide insight into the evolution of manufacturing technologies.

Table 13.4 Energy and exergy efficiencies of selected conversion devices [5].

Device	Description	Energy efficiency	Exergy efficiency
Electricity generation from:			
Gas	Natural gas and gas works	40	38
Coal	Hard coal, lignite and derived fuels	34	32
Nuclear	Nuclear fission	33	33
Renewable	Hydro, geothermal, solar, wind, tide, and wave energy	80	80
Diesel engine	Compression ignition diesel engine: truck, car, train, ship, generator	22	21
Biomass burner	Wood/biomass combustion device: open fire/stove, boiler	34	7
Electric heater	Electric resistance heater, electric arc furnace	80	24
Light device	Lighting: tungsten, halogen, fluorescent	13	12

BOX 13.1 Opportunities for Improving Efficiency of the Top Ten Chemicals in the USA [6]

The United States Department of Energy has used exergy analysis to identify opportunities for improving efficiency in chemical manufacturing. This effort considers various chemical manufacturing sectors of the US economy and relies on data collected from industry, practical considerations about improving efficiency, and theoretical analysis based on the Gibbs free energy of relevant reactions. Some of the findings for the top ten energy-consuming chemicals are summarized in Figure 13.7. It shows that many ammonia manufacturing processes are much less efficient than the state-of-the-art technologies, indicating an opportunity for improvement simply by replacing older technologies. Sodium hydroxide production is next after ammonia. However, the state-of-the-art technologies are quite close to the practical minimum energy use, indicating the limited benefit of further research and development. However, products such as ethanol, ethylene, soda ash, and benzene have much room for improvement beyond the current state-of-the-art toward the practical minimum. The theoretical minimum is based on the second law using methods like those in Section 13.1. This thermodynamic minimum shows that products like sodium hydroxide and soda ash could be manufactured with energy production, but this is not possible in practice.

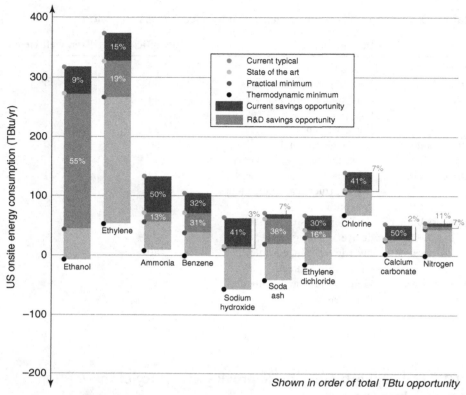

Figure 13.7 Improvement opportunities for top ten energy-consuming chemicals in the USA [6].

BOX 13.2 Insight from Exergy Analysis about the Evolution of Technologies [7]

Over the years, exergy analysis has been used to analyze a large number of manufacturing processes. Traditional manufacturing activities like those related to iron and steel involve processes such as melting, machining, and injection molding. These tasks can be carried out at a high rate, and as shown in Figure 13.8 they use roughly the same amount of electrical exergy per kilogram processed. The exergy consumed is approximated as electricity consumption, since that is the major input to these processes. Newer manufacturing technologies such as grinding, finish machining, and abrasive waterjet have a smaller processing rate and larger exergy intensity. The exergy input into these processes is mainly for energy tasks such as heating. As we consider more modern industries such as those in the electronics industry and those related to nanotechnology, the use of exergy for materials processing increases, making these technologies much more exergy intensive. At the same time, the process rate for these technologies decreases. Thus, as shown in Figure 13.8, carbon nanofiber production is about an order of magnitude more exergy intensive, with a process rate that is an order of magnitude smaller than abrasive waterjet. Other emerging technologies such as chemical vapor deposition, the production of carbon nanotubes, and laser deposition use exponentially more electrical exergy per kilogram of production.

The reason for the exponential increase in exergy requirement may be understood by realizing that newer technologies are increasingly capable of manipulating matter at smaller scales, going all the way to atomic (nano) scales. Such manipulation requires a high degree of control for the large decrease in entropy of the finished product as compared to the raw materials. Using our intuition about the second law, we can surmise that the larger the decrease in entropy in processing, the greater the increase in entropy of the surroundings, which appears as a large exergy use.

13.3.3 Exergy Analysis of Economies

The Sankey diagram of an economy like that shown in Figure 12.7 is commonly used to indicate where the highest energy losses are present. However, this can be highly misleading since it does not consider the usefulness of the lost energy. Exergy can provide more useful insight, as illustrated by the Grassman diagram in Figure 13.9 for the 2010 economy of the UK [8]. This diagram shows that the

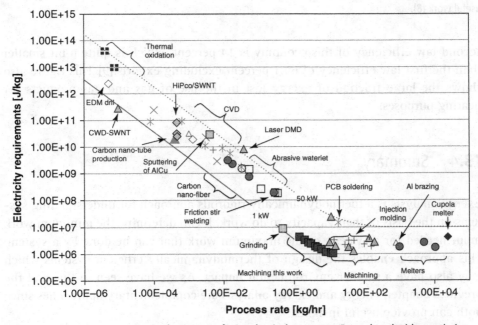

Figure 13.8 Exergy use versus production rate for mechanical processes. Reproduced with permission from [7].

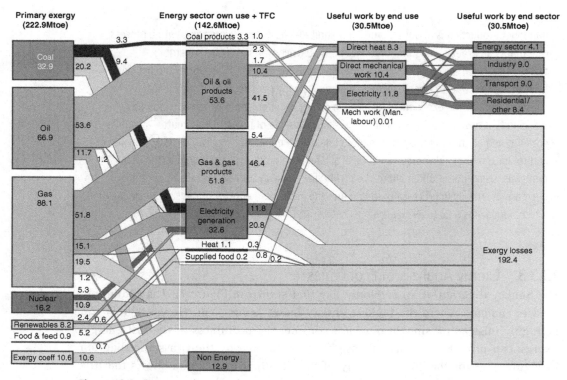

Figure 13.9 Grassman diagram of UK economy in 2010, depicting the stages from primary exergy to useful work [8].

second law efficiency of this economy is 14 percent, which is quite a bit smaller than the first law efficiency of 69.1 percent excluding exports.[9] This figure also shows the large fraction of exergy lost in the use of gas and gas products for heating purposes.

13.4 Summary

Exergy analysis is a thermodynamically rigorous approach for understanding systems on the basis of their capacity to do work. It can determine the minimum work input needed for a system, or the maximum work that can be done by a system. This approach enables the design of thermodynamically efficient systems which may also have a smaller environmental impact. As we have seen in this and the previous chapter, energy and exergy analysis are complementary approaches since both can provide useful insights.

From the requirements for sustainability assessment methods developed in Chapter 3, we can see that exergy analysis satisfies only the requirement of accounting for the input of ecosystem goods and services, since it can quantify these inputs in terms of the work done by or on nature. In the next chapter, we will learn more about this and about extensions of exergy analysis that are more capable of satisfying the requirements for sustainability assessment.

KEY IDEAS AND CONCEPTS

- Exergy
- Lost work
- Carnot efficiency
- Grassmann diagram

13.5 Review Questions

1. Define exergy.
2. What is the relationship between exergy and entropy?
3. Explain why the exergetic efficiency of a process is less than 100 percent even when its energy efficiency is 100 percent.
4. What is the lost work in a perfectly reversible process?
5. Explain the large difference in the energy and exergy efficiencies of an electric heater.

Problems

13.1 Liquid iron at 1000 °C cools to 750 °C. Determine the lost work and generated entropy. Where does this entropy go? The specific heat of liquid iron is 0.197 cal/g °C.

13.2 Consider a water heater meant to increase the temperature from 25 °C to 100 °C. A manufacturer advertises the heater as 80 percent efficient. This is the first law efficiency, which means that 20 percent of the energy input is lost [10].
   1. Calculate the energy input needed to heat a kilogram of water.
   2. Determine the least amount of energy required for heating this water, and the second law efficiency of the current device.

13.3 A domestic boiler heating system is heating the outside air at 5 °C up to 20 °C with 80 percent first law efficiency. Determine the second law efficiency of this system. Draw the corresponding Sankey and Grassmann diagrams.

13.4 A compressed air system stores the energy generated by solar panels on the company buildings' rooftops during the weekend for use during the week. The generated electricity over the weekend is used to compress air from atmospheric pressure to 10 atm; the air is then brought back to atmospheric pressure as needed during the week. What is the lost work per joule of electricity produced by the solar panel when the efficiencies of compression and decompression are both 100 percent and when they are both 80 percent?

13.5 Calculate the second law efficiency of the reaction $CH_4 + 2H_2O \rightarrow CO_2 + 4H_2$. Here, hydrogen is the useful product.

13.6 This problem considers the possibility of generating power from the concentration gradient between CO_2 in industrial emissions and the ambient atmosphere.

1. Flue gas from a coal-fired power plant has 12.7 percent CO_2. What is the maximum amount of work that can be done from the mixing of this flue gas with air? The flue gas is at 25 °C and ambient air contains 0.04 percent CO_2.

2. The annual production of flue gas from coal-fired power plants across the world is 9 Gt. In addition, 2 Gt per year of flue gas is produced from plants that burn natural gas, and 11 Gt per year is produced from mobile sources. The CO_2 concentration in these emissions may be assumed to be 7.5 percent. Calculate the maximum work that can be obtained from the global emission of flue gas. Knowing that the Hoover Dam in the USA produces 4 TW of power, is the potential to generate power from the flue gas CO_2 gradient substantial?

13.7 Large volumes of fresh water and sea water mix where rivers meet the sea. This concentration gradient could be used to do work and generate electricity. Calculate the maximum work that can be done due to the mixing of 1 m^3 of seawater of concentration 0.5 mol NaCl/L with 1 m^3 of river water with 0.01 mol NaCl/L.

13.8 A hot stream provides 2 kg/s at 200 °C, with specific heat 3 kcal/kg °C. A cold stream is available at 30 °C and with specific heat of 2 kcal/kg °C.

1. Determine the flowrate of the cold stream required to cool the hot stream to 100 °C. Determine the exergy efficiency of this heat exchange system.

2. Repeat step 1 for a cold stream final temperature of 50 °C.

3. Explain the difference in the two exergy efficiencies.

13.9 A large number of commuters walk in and out of train stations every day. The changing pressure on the floor could be utilized to generate electricity. Each weekday, 3.64 million passengers walk through Shinjuku station in Tokyo. Such a power generation device is being considered for installation in a section of the station where an average person walks 1000 steps while applying 50 kg mass on a 150 cm^2 footstep. What is the maximum electricity that could be generated per year? How many 10 W bulbs could be lit with this electricity?

13.10 Post-combustion carbon capture is increasingly being considered in power plants that burn fossil fuels such as coal or natural gas. Typical flue gas is at 40 °C and has 13 percent CO_2; the remainder by mole is inert gases. Calculate the minimum work that must be done to separate flue gas into a stream of pure CO_2. On average, US coal-fired power plants emit 9187 t CO_2 per MW y of electricity generated. If the energy for separation of CO_2 from flue gas is obtained from the generated electricity, calculate the fraction of the generated electricity that will have to be used for CO_2 separation.

References

[1] E. Schrödinger. *What is Life? With Mind and Matter and Autobiographical Sketches.* Cambridge University Press, 1945.

[2] N. Sato. *Chemical Energy and Exergy: An Introduction to Chemical Thermodynamics for Engineers.* Elsevier, 2004.

[3] D. Alexander. RPT-Pentagon weapons-maker finds method for cheap, clean water. Reuters. www.reuters.com/article/2013/03/13/usa-desalination-idUSL1N0C0DG520130313, March 13, 2013, accessed November 23, 2018.

[4] W. F. Banholzer and M. E. Jones. Chemical engineers must focus on practical solutions. *AIChE Journal*, 59(8):2708–2720, 2013.

[5] J. M. Cullen and J. M. Allwood. Theoretical efficiency limits for energy conversion devices. *Energy*, 35(5):2059–2069, 2010.

[6] S. Brueske, C. Kramer, and A. Fisher. *Bandwidth Study on Energy Use and Potential Energy Savings in U.S. Chemical Manufacturing.* US Department of Energy, 2015.

[7] T. G. Gutowski, M. S. Branham, et al. Thermodynamic analysis of resources used in manufacturing processes. *Environmental Science & Technology*, 43(5):1584–1590, 2009.

[8] P. E. Brockway, J. R. Barrett, T. J. Foxon, and J. K. Steinberger. Divergence of trends in US and UK aggregate exergy efficiencies 1960–2010. *Environmental Science & Technology*, 48(16):9874–9881, 2014.

[9] Department of Energy & Climate Change. Energy flowchart 2011. www.gov.uk/government/statistics/energy-flow-chart-2011, 2012, accessed November 23, 2018.

[10] R. U. Ayres, L. T. Peiró, and G. V. Méndez. Exergy efficiency in industry: where do we stand? *Environmental Science & Technology*, 45(24):10634–10641, 2011.

14 Cumulative Exergy Consumption and Emergy Analysis

> Energy from the sun and from the earth is running the landscape and its links to humanity. The quantity of useful energy determines the amount of structure that can exist and the speed at which processes can function. The small areas of nature, the large panoramas that include civilization, and the whole biosphere of Earth and the miniature worlds of ecological microcosms are similar. All use energy resources to produce, consume, recycle, and sustain.
>
> Howard T. Odum [1]

In the previous chapter we learned about exergy and its use for understanding and improving the performance of systems ranging from individual equipment to manufacturing processes and entire economies. In this chapter we will expand the use of exergy analysis to larger spatial scales by determining the cumulative exergy consumption, thus making exergy analysis more relevant to sustainability assessment. We will learn about two ways of expanding the system boundary. The first will be analogous to the life cycle and footprint methods covered in previous chapters. Then we will also include ecosystems in our boundary, which will allow us to account for the work done by nature. This will result in the approach of emergy analysis, which originated in systems ecology and attempts to account for the role of ecosystems in sustaining human activities.

14.1 Cumulative Exergy

The cumulative exergy consumption (CEC) is the total exergy consumed directly and indirectly in a process or product. Thus, the CEC is equivalent to "life cycle exergy consumption" or "exergy footprint," and is able to aggregate flows of diverse resources such as fossil fuels, minerals, and water in terms of their capacity to do work. Like the approaches of footprint analysis, discussed in Chapter 11, and energy analysis, discussed in Chapter 12, the cumulative exergy also accounts for the life cycle. The carbon footprint considers only greenhouse gases, while the water footprint considers only water. The cumulative exergy is able to capture a wider variety of physical flows. Thus, in terms of the requirements for sustainability methods listed in Chapter 3, cumulative exergy is able to consider the life cycle and many types of physical flows.

The exergy efficiency of a life cycle may be defined as the exergy of the useful products divided by the cumulative exergy consumption. This is called the cumulative degree of perfection (CDP):

$$\psi_C = \frac{B}{C} \tag{14.1}$$

Here, B is the exergy content of the products and C is the cumulative exergy consumed to make the products, which may be calculated as

$$C = \sum_{k_n} B_{k_n} \tag{14.2}$$

with B_{k_n} the exergies of the natural resources that are directly obtained from nature and used in industrial processes. If inputs to the industrial processes are not directly obtained from nature and involve intermediate processing steps such as mining or manufacturing, the CEC of a system may be determined by adding the CEC of each input:

$$C = \sum_{k} C_k \tag{14.3}$$

These calculations may be understood with the help of Figure 14.1. Sometimes, the term degree of perfection is used to indicate the exergy efficiency of a partial life cycle.

Example 14.1   Determine the degrees of perfection of the furnace and coal extraction processes, and the cumulative degree of perfection of the coal-based electricity life cycle shown in Figure 14.2a.

Solution
This process has an exergy input rate of 141.95 kW of coal. Extracting that much coal requires 7.05 kW of fuel oil. The final product is 34.51 kW of electricity. The

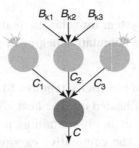

Figure 14.1 Calculating cumulative exergy consumption.

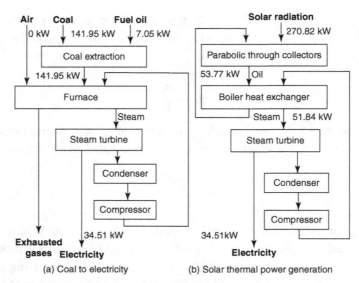

Figure 14.2 Illustrative examples for cumulative exergy consumption.

degree of perfection of the power generation is

$$\psi_1 = \frac{34.51}{141.95}$$
$$= 24.3 \text{ percent}$$

Similarly, for the coal extraction,

$$\psi_2 = \frac{141.95}{141.95 + 7.05}$$
$$= 95.3 \text{ percent}$$

The cumulative degree of perfection for this process is

$$\psi_C = \frac{34.51}{141.95 + 7.05}$$
$$= 23.2 \text{ percent}$$

Example 14.2   Consider the system for solar thermal electricity generation shown in Figure 14.2b. Determine the cumulative degree of perfection.

Solution
The generation of 34.51 kW of electricity requires 51.84 kW of steam, which may be produced from 53.77 kW of heated oil. The heat is obtained by parabolic reflectors that concentrate sunlight on an oil-containing pipe. The solar energy required for this system is 270.82 kW. The cumulative exergy consumed in this system is 270.82 kW, and the cumulative degree of perfection for this system is

$$\psi_C = \frac{34.51}{270.82}$$
$$= 12.7 \text{ percent}$$

This smaller CDP of the solar process as compared to the coal process is due to the dilute nature of sunlight as compared to coal.

Example 14.3 An input–output network is represented by the following equations:

$$0.833x_1 - 0.182x_2 = f_1$$
$$-0.667x_1 + 0.818x_2 = f_2$$

Sector 1 has direct inputs of 10 tons of iron ore (hematite) and 1000 L of water per dollar of throughput. Sector 2 has a direct input of 100 L of water per dollar of throughput. Calculate the cumulative exergy consumed per dollar of final demand in each sector. The exergy of hematite is 12.4 kJ/mol with molecular weight of 159.692 g/mol. For water these values are 0.9 kJ/mol and 18 g/mol.

Solution
Using the notation in Section 10.2.2 and considering iron ore to be input 1 and water to be input 2, we have that $m_{11} = 10$ tons of iron ore, $m_{12} = 0$ tons of iron ore, $m_{21} = 1000$ L of water, and $m_{22} = 100$ L of water.

In terms of exergy,

$$m_{11} = \frac{10^7 \text{ g}}{159.692 \text{ g/mol}} \times 12.4 \text{ kJ/mol}$$
$$= 7.77 \times 10^5 \text{ kJ}$$

Similarly, for water, $m_{21} = 5 \times 10^4$ kJ and $m_{22} = 5 \times 10^3$ kJ. Thus, the total direct exergy consumption in sector 1 is 8.27×10^5 kJ/\$ and in sector 2 is 5×10^3 kJ/\$. Using Equation 10.42, the total resource use per dollar of final demand may be calculated as

$$\begin{bmatrix} 8.27 \times 10^5 & 5 \times 10^3 \end{bmatrix} \begin{bmatrix} 0.833 & -0.182 \\ -0.667 & 0.818 \end{bmatrix}^{-1} = \begin{bmatrix} 1.21 \times 10^6 & 2.76 \times 10^5 \end{bmatrix}$$

Thus, the cumulative exergy consumption per dollar of final demand in sectors 1 and 2 is $C_1 = 1.21 \times 10^6$ kJ and $C_2 = 2.76 \times 10^5$ kJ, respectively.

Example 14.4 Continuing Example 10.4, determine the industrial cumulative exergy consumption and cumulative degree of perfection of this system for 1 kWh electricity. The density of crude oil is 0.8 kg/L and its exergy is 50 MJ/kg.

Solution
The total resource consumption for this life cycle is 0.51 L of crude oil. The CEC of this process for producing 1 kWh (3.6×10^6 J) electricity is

Table 14.1 Cumulative exergy consumption and cumulative degree of perfection of some industrial products [3].

Name	CEC (MJ/kg)	CDP %
Pig iron	28.6	30.6
Steel pipes	58.7	16.0
Cold rolled steel	47.6	14.8
Copper	147.4	1.4
Zinc	125.8	4.1
Aluminum	250.2	13.2
Sulfur	30.2	62.9
Ammonia	48.2	41.1
Sulfuric acid	11.1	15.0
Methanol	73.1	30.7
Glass	33.4	0.5
Cellulose	60.0	27.4

$$C = 0.51 \text{ L} \times 0.8 \text{ kg/L} \times 50 \text{ MJ/kg}$$
$$= 20.4 \text{ MJ}$$

Then, using Equation 14.1, the industrial CDP is

$$\psi_C = \frac{3.6 \times 10^6}{20.4 \times 10^6}$$
$$= 0.1765$$

The industrial CDP is 17.65 percent.

The CEC and CDP values for several industrial products are shown in Table 14.1. The CDP values for various energy conversion devices are shown in Table 14.2. In this table, CDP is calculated using the degrees of perfection of three common categories of activities: fuel transformation (f), electricity generation (e), and end-use device conversion (d). For devices such as diesel engines that do not rely on electricity, $\psi_e = 100$ percent. This table shows that most of the efficiency loss occurs in end-use conversion devices such as the light device or the automotive engine. Such information may be used to identify opportunities for improving energy efficiency and reducing impact throughout the life cycle [2].

This concept of cumulative exergy was developed by engineers [4], and includes only industrial processes and nonrenewable resources in the analysis boundary. However, the sustenance of industrial and human activities also requires renewable resources such as sunlight, wind, and biomass. In fact, as we learned in Chapter 2, sustaining human activities requires goods and services from nature, including

Table 14.2 CDP as a percentage in of common energy conversion devices. Reproduced with permission from [2].

Energy chain	ψ_f	ψ_e	ψ_d	ψ_C
Aircraft engine	93	100	27	25
Diesel engine	93	100	21	20
Electric motor	93	32	56	17
Motion average	93	77	24	17
Coal burner	90	100	19	17
Gas burner	91	100	13	12
Electric heater	93	32	24	7
Heat average	93	76	14	10
Light device	93	34	12	4
Cooler	93	33	7	2
Electronic	93	32	6	2
Other	93	33	8	2
Overall average	93	70	18	11

The subscript f is for fuel transformation, e is electricity generation, d is end-use device conversion. $\psi_C = \psi_f \psi_e \psi_d$.

renewable and nonrenewable resources. With this understanding, methods have been developed to include the role of nature in supporting human activities. In the rest of this chapter, we will learn about some such methods that are based on exergy. We will also appreciate the challenges faced in accounting for large systems, including ecosystems.

14.2 Aggregation and Resource Quality

As illustrated in Section 13.1, the concept of exergy is able to represent a variety of materials and fuels using a common unit, in a thermodynamically rigorous manner. This representation in terms of a common unit is appealing since it permits comparison between alternatives and calculation of aggregate metrics. We saw some such benefits in Chapter 12, where we represented all fuels in the common unit of fuel value, resulting in metrics such as the energy return on investment. The most common system for representing diverse flows and activities with a common unit is that followed in economics, where all flows are represented in terms of monetary value. This is the basis of many aggregated metrics such as gross domestic product and monetary return on investment. Other examples are given in Box 14.1 from the viewpoints of nutrition, comparison of automotive efficiency, and energy analysis of biofuels.

(a) Twinkies   (b) Banana

Figure 14.3 Despite the equal calorific values of a Twinkie and two bananas, most of us do not consider them to be substitutable food items. (a) Photo by Larry D. Moore / Wikimedia Commons CC-BY-SA-3.0. (b) Photo by Evan-Amos / Wikimedia Commons CC-BY-SA-3.0.

BOX 14.1 Comparing Resource Quality

Is a Twinkie equivalent to a banana? A Twinkie, shown in Figure 14.3a, is an American processed food item. In terms of calorie content, one Twinkie (270 cal) is equivalent to about two large bananas. Thus, from a calorie-counting point of view, having a Twinkie for breakfast instead of bananas is equivalent. Unfortunately, for the Twinkie lovers of the world and its manufacturer, most people would not consider it acceptable to have a Twinkie for breakfast instead of two bananas. That is, these two products are not considered to be substitutes. The reason is that the quality of the calories in the two products is very different. More than one-third of the calories in a Twinkie are from fat, while in a banana this is close to zero. In addition, bananas contain many other essential nutrients that a Twinkie does not.

Miles per gallon equivalent. For comparing the fuel efficiency of conventional vehicles with gas–electric hybrid and electric vehicles, the US Environment Protection Agency (EPA) determines the mile per gallon equivalent of various types of vehicles. A typical electric sedan uses 25 kWh to travel 100 miles, which is equal to $E_M = 8.53 \times 10^4$ Btu per 100 miles. Gasoline is known to have a fuel value of $E_G = 1.15 \times 10^5$ Btu/gallon; therefore the energy content of the electricity used to travel 100 miles is equivalent to 0.7417 gallons of gasoline. Thus, the electric car travels 100 miles on 0.7417 gallons of gasoline equivalent and achieves 135 mpg. In general, miles per gallon equivalent may be calculated as

$$MPGeq = \frac{((\text{total miles driven, d}) \times (\text{energy of one gallon of gasoline}) (E_G)}{(\text{total energy of all fuels consumed to cover distance } d)(E_M)} \quad (14.4)$$

Quality of energy used to make transportation fuels. In a debate about biofuels between Professors Bruce Dale and David Pimentel [5], in one argument given by Professor Dale he writes: "One megajoule of coal is simply not equivalent to 1 MJ of petroleum. If you doubt this, try grinding up some coal, put it in your gas tank, and see how far you can drive on coal's energy content. Pimentel's 'net energy' analysis is akin to adding 10 U.S. dollars, 10 Mexican pesos, and 10 Indian rupees to get $30 – an absurdity. Different energy carriers and different currencies have different qualities. They cannot simply be added together as Dr. Pimentel does."

Example 14.5 Calculate the miles per gallon equivalent (MPGeq) of a plug-in hybrid electric vehicle that travels 20 miles on the gasoline engine and 50 miles on the battery. The efficiency of each engine is 50 mpg and 30 kWh per 100 miles, respectively. A gallon of gasoline contains 115,000 Btu.

Solution

Consider the total distance traveled to be $d = 70$ miles. The total energy from gasoline and electricity to travel this distance is

$$E_M = \left(\frac{20}{50} \times 1.15 \times 10^5\right) + \left(\frac{30}{100} \times 50 \times 3412\right)$$

$$= 97,180 \text{ Btu}$$

Thus, using Equation 14.4, the miles per gallon equivalent is

$$MPGeq = \frac{70 \times 115,000}{97,180}$$

$$= 82.84 \text{ MPGeq}$$

Therefore, this vehicle achieves a fuel efficiency of 82.84 MPGeq.

Exergy and cumulative exergy are also appealing concepts for defining aggregate metrics, but before we do that we should understand the assumptions underlying aggregation. Such insight will help in defining meaningful and useful aggregate metrics, and in their proper interpretation and use.

The use of a common unit and aggregate metrics for comparing alternatives involves some implicit assumptions about substitutability between the flows being represented in a common unit. For example, adding the fuel value of coal, natural gas, and crude oil, as is done in the net energy analysis described in Chapter 12, implies that one joule of each resource is equivalent and substitutable. Similarly, if we determine the cumulative exergy of a process that uses coal, water, and iron ore, we implicitly assume that one joule of work available from each resource is equivalent. Unfortunately, this assumption of substitutability between resources in terms of their fuel value or exergy is often wrong, and the resulting aggregate metrics can be misleading, as argued by Professor Dale in Box 14.1.

A joule of coal fuel value is not equivalent to one joule of natural gas fuel value since producing 1 kWh of electricity requires 10,354 Btu (10.9242 MJ or 3.032 kWh) of natural gas or 10,089 Btu (10.6446 MJ or 2.955 kWh) of coal. In addition, natural gas is cleaner burning and more versatile in its applications. Similarly, one joule each of solar, wind, coal, and natural gas exergy are not comparable since, as shown in Table 14.3, one joule of each resource produces very different quantities of electrical energy [6].

Table 14.3 Efficiency and relative quality of selected resources based on exergy and emergy [6].

Energy source	Exergetic efficiency, η (%)	Relative quality based on η	Transformity, τ (sej/J)	Relative quality based on τ
Geothermal	13.8	5.0	20,300	8.0
Wind	2.8	1	2520	1
Hydropotential	51.5	18.5	27,764	11.0
Natural gas	21.0	7.6	171,000	67.6
Oil	36.1	13.0	148,000	58.5
Coal	22.2	8.0	87,100	38.4

The exergetic efficiency is that of converting the energy source to electricity. The transformity is that of the energy source. The unit sej refers to solar equivalent joules.

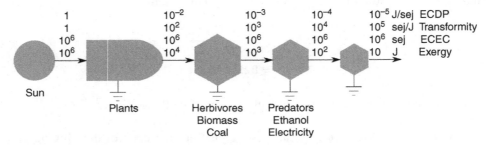

Figure 14.4 Exergy transformation in food or industrial chain.

Example 14.6 Represent 1 joule each of wind, coal, and natural gas in terms of electricity equivalent joules.

Solution

Since efficiency is the ratio of energy produced to energy used, data in the left three columns of Table 14.3 show that 1 J of wind results in 0.028 J of electricity, 1 J of coal results in 0.222 J of electricity, and 1 J of natural gas produces 0.21 J of electricity. Therefore, 1 J each of these three resources is equal to 0.028 + 0.222 + 0.21 = 0.46 of electricity equivalent joules.

An approach based on cumulative exergy consumption for quantifying resource quality may be devised by considering a simple food or industrial chain, shown in Figure 14.4. In this illustration, one million joules of sunlight is transformed into biomass by plants, which sustain herbivores, who in turn sustain predators. Other such chains may involve the conversion of plants to cellulosic biomass, which is converted to ethanol, or the conversion of plants to coal, which is converted to electricity. Clearly, one joule of ethanol or electrical exergy is not equivalent to one joule of solar exergy, since 10^5 joules of solar exergy are required per joule of ethanol exergy. Therefore, a joule of ethanol exergy is equivalent to 10^5 J of sunlight exergy. Thus, quality differences could be captured by considering

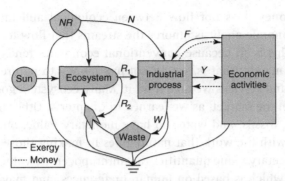

Figure 14.5 Exergy flow in industrial and ecological systems. NR indicates nonrenewable resources.

cumulative exergy consumption in terms of a common resource from which multiple products originate. The life cycle exergy input in this chain is one million joules of sunlight, as shown in Figure 14.4. The life cycle efficiency or cumulative degree of perfection is also shown in this figure. This quantity decreases with increasing quality. Thus, the CDP based on a common originating resource could provide a thermodynamic indication of resource quality. We will utilize this insight in Section 14.4 when we learn about emergy analysis.

14.3   Exergy Flow in Ecological and Economic Systems

The interaction of an industrial process or economic activity with other activities may be represented as an exergy flow, and is depicted in Figure 14.5. The continuous lines show the flow of exergy, while the dashed lines indicate the flow of money. The exergy of products manufactured by industry and sold to the economy is denoted by the continuous line Y. In return, there is a flow of money from the economy to industry denoted by the dashed line. The socioeconomic system provides exergy to enable industrial operation by means of goods and services such as labor and other materials. This is indicated by the continuous line F. In exchange, industrial processes pay money to society, indicated by the dashed line in the opposite direction. The physical and monetary interactions indicated by flows F and Y represent the worldview of conventional or neoclassical economics that we learned about in Figure 4.2. We also learned about a biophysical view of the economy, depicted in Figure 4.7, where in addition to the interaction between industry and markets, industrial processes also need inputs from nature. These include nonrenewable resources NR and renewable resources R_1. The waste from industry has exergy W. It returns to the ecosystems, and nature needs to do work R_2 to absorb it or bear its impact. Thus, Figure 14.5 represents the biophysical or scientific view depicted in Figure 4.7.

Notice that money does not flow between ecological and industrial systems. Conventional economic analysis ignores the streams that flow to and from nature in Figure 14.5. This is so because conventional economics tends to keep ecosystems outside the market; that is, we do not pay money to nature or to any other entity for goods and services we take from it. Many ecosystem goods and services are excluded from the market, as we learned in Chapter 4. Other natural resources such as minerals, fossils, and water do have monetary value, but this value may have little to do with the work that nature does in making those resources available. Instead, monetary value quantifies the anthropocentric value of a resource in the marketplace, which is based on human preferences, and may have little connection with physical aspects such as the work required to produce or maintain a resource.

Thus, all processes – ecological, industrial, and socioeconomic – involve the flow and transformation of exergy, making exergy a *common currency* that flows through all systems. If the flows in Figure 14.5 can be quantified, they can be used to define various kinds of metrics and support holistic decision-making while incorporating information about industrial, economic, and ecological flows in a scientifically sound manner.

Ecologists, geologists, and other natural scientists have relied on thermodynamics to understand ecological and planetary processes. Some of their insight is summarized in Figure 14.6 in the form of exergy flow and transformation in ecological systems [7]. This figure combines information from diverse disciplines collected over many decades. As we can see in the figure, planetary processes operate with three main independent exergy inputs: solar, crustal, and tidal. Solar exergy drives processes such as wind, water evaporation, photosynthesis, and surface heating and cooling. Tidal exergy drives tides, while crustal exergy drives geological cycles that cause movement of the Earth's crust and concentration of stored exergy in the form of minerals and ores. These processes in turn drive other ecological and economic activities. The availability of networks like those in Figures 14.5 and 14.6 implies that for a selected process, its exergy flows in industrial and ecological systems can be quantified, and may be used for defining holistic metrics. Next, we will learn about one such approach. Note that the field of environmental economics also addresses many of the issues that motivate this approach. We will learn more about this in Chapter 21.

14.4 Emergy Analysis

The approach of cumulative exergy consumption introduced in Section 14.1 considered exergy consumption in human activities. This is shown in Figure 14.7 by

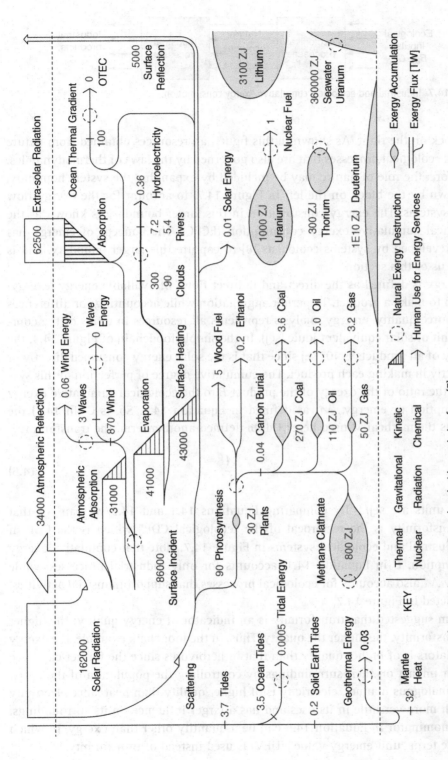

Figure 14.6 Exergy transformation in nature. Reproduced with permission from [7].

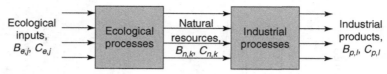

Figure 14.7 Industrial and ecological cumulative exergy consumption.

the block on the right. As shown in this figure, all resources obtained from nature rely on ecological processes that are also governed by the laws of thermodynamics. Therefore, the role of nature may be included by expanding the system boundary, as shown by the block on the left in Figure 14.7, to account for the exergy flow in ecosystems. The exergy consumption in this larger boundary is known as the ecological cumulative exergy consumption (ECEC). The concept of emergy has been developed by systems ecologists [8] to capture this larger boundary, and is the focus of this section.

Emergy is defined as the direct and indirect flows of available energy (exergy) needed to make a product. To permit aggregation while accounting for differences in resource quality, emergy analysis represents all resources in terms of the common unit of solar equivalent joule (sej). In the simple foodchain of Figure 14.4, the emergy of all products is 10^6 sej since that is the solar energy consumed directly or indirectly in making each product. The cumulative degree of perfection of this system is the ratio of the exergy of the product B to the ecological cumulative exergy consumption or emergy, $\mathcal{M}$, as defined in Equation 14.1. So, is C (in 14.1) the same as the mathcal M used here? Odum defined another term, the transformity:

$$\tau = \frac{\mathcal{M}}{B} \tag{14.5}$$

whose units are sej/J. By comparing Equations 14.1 and 14.5 we can see that the transformity is the reciprocal of the "ecological CDP," which is the CDP of the industrial and ecological systems in Figure 14.7. Note that cumulative exergy consumption, C in Equation 14.1, accounts for only industrial processes, while emergy $\mathcal{M}$ also accounts for ecological processes that support industrial activities, as depicted in Figure 14.7.

Odum suggested that transformity is an indicator of energy quality: the higher the transformity, the higher the quality. Thus, in the foodchain example, the exergy in predators is of higher quality than that in herbivores since the former can have a larger impact on their surroundings by controlling the population of the latter. In an analogous manner, electricity is of higher quality than heat since electricity is much more versatile in its uses and has a larger influence on its surroundings. The denominator in Equation 14.5 can be a quantity other than exergy, in which case the term "unit emergy value" (UEV) is used instead of transformity.

Table 14.4 Unit emergy values (UEV) of typical flows [9–11].

Item	UEV	Units
Primary planetary drivers		
Solar exergy	1.00E+00	seJ/J
Crustal heat	2.03E+04	seJ/J
Tidal energy	7.24E+04	seJ/J
Renewable resources		
Wind	2.52E+03	seJ/J
Rain water	3.06E+04	seJ/J
Ground water	3.98E+05	seJ/J
Wood	6.79E+08	seJ/g
Fossil resources		
Coal (anthracite and bituminous)	9.71E+04	seJ/J
Coal (sub-bituminous and lignite)	6.63E+04	seJ/J
Crude oil	1.48E+05	seJ/J
Natural gas	1.71E+05	seJ/J
Minerals		
Copper ($CuFeS_2$)	3.94E+06	seJ/g
Iron (Fe_2O_3)	2.81E+03	seJ/g
Aluminum (Al_2O_3)	1.23E+03	seJ/g
Gold (Au)	5.47E+10	seJ/g
Silicon (SiO_2)	3.19E+02	seJ/g
Fuels		
Gasoline	1.87E+05	seJ/J
Jet fuel	1.84E+05	seJ/J
Diesel	1.81E+05	seJ/J
LPG	1.70E+05	seJ/J

14.4.1 Emergy of Natural Resources

Representing all flows in solar equivalents requires aggregation of the three primary energy inputs to the planet: solar, crustal, and tidal. These three flows are of different quality; as shown in Table 14.3, converting crustal exergy into electricity is easier than converting solar exergy to electricity. The transformity of sunlight is taken as unity. Using knowledge about global energy flows and the relationship between them, the transformities (UEVs) of primary planetary flows have been calculated and are shown as the first three items of Table 14.4. The ecological work required to produce coal is depicted in Figure 14.8, and its transformity is shown in Table 14.4. This includes work done in the growing of biomass, its burial to produce peat, followed by coalification. Even though this work was done millennia ago, it is included in emergy accounting to give insight into the relative quality and role of ecosystems in producing this resource. Such methods have been used

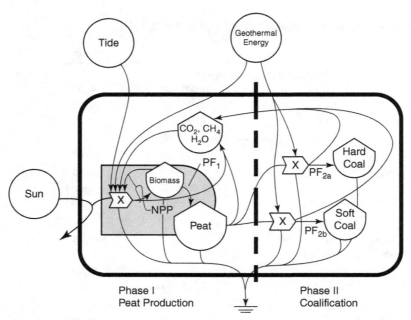

Figure 14.8 Exergy flow diagram representing work done in the generation of coal. Reproduced with permission from [9].

for calculating the unit emergy values (UEV) of natural resources, some of which are listed in Table 14.4.

Comparing the quality of the resources shown in Table 14.3 based on their transformities results in the numbers in the last three columns. Comparing these quality measures in solar equivalents with those in electrical equivalents shows the former quantify fossil resources to be of higher quality. This is so because fossil resources contain a large amount of embodied solar energy that is highly concentrated. In comparison, most renewable resources are relatively dilute, sunlight being the most dilute. Some implications of these differences in energy quality and concentration are discussed in Box 14.2.

BOX 14.2 Assessing Claims about Solar Energy

Today, many researchers are focusing on the development of artificial methods for photosynthesis. The following statement conveys the common motivation for this work [12]:

Among renewable energy resources, solar energy is by far the largest exploitable resource, providing more energy in 1 hour to the earth than all of the energy consumed by humans in an entire year.

This statement seems to imply that there is more than enough solar energy to satisfy human needs. Implicitly, it is assuming substitutability between solar energy and the types of energy (electricity, heat, fuel) that people actually use. As we have learned in this chapter, the quality of these energy sources is very different. Solar energy is of much lower quality than the forms of energy that satisfy human needs. In addition, as discussed in Box 12.1, sunlight also has a much lower power density than our density of use.

The transformity of solar electricity generated by a photovoltaic power plant is 89,200 seJ/J. This means that to produce 1 J of solar electricity, we need 89,200 J of solar energy input. Assuming this technology with all energy input to the process coming from the Sun, an hour of solar energy will produce $\frac{1}{89,200}$ th of the amount of electricity. This electricity will not power humans for an entire year, as stated in the quote, but for only $\frac{1}{89,200}$ th of a year, which is $\frac{1}{89,200} \times 365 \times 24 = 0.1$ hour or 6 minutes of global consumption. Even if the efficiency of solar energy tripled to 60 percent, an hour of sunlight would produce electricity for only about 20 minutes of human use. This calculation implicitly assumes that all the energy used to produce solar panels is from the Sun, which is not how solar panels are produced today. This calculation corresponds to a possible future based entirely on renewable energy, and conveys its challenges.

Another way of appreciating the misleading nature of the above statement is by considering the efficiency of solar panels, which, as mentioned, is about 20 percent today. This means that an hour of sunlight will generate enough electricity to satisfy only 20 percent of the annual global electricity consumption. Thus, a year of sunlight will be converted to electricity that is consumed in 2.4 months today. It will not satisfy a year's human requirement. This estimate does not account for the large amount of nonrenewable resources consumed in the production of solar panels, or any future increase in efficiency and electricity use.

Another argument can be made using the power density of sunlight, which is known to be less than that of most human uses by a factor of 1000, as can be seen in Box 12.1. So harnessing an hour of sunlight falling on a particular area of the Earth's surface will require 1000 times that area, which is not practical.

This analysis does not imply that research on solar energy conversion should not be carried out. However, ignoring differences in energy quality can result in misleading claims. Scientific breakthroughs could change this situation, but the dilute nature of solar energy as compared to fossil energy and wind will always and wind will always remain.

Example 14.7 Calculate the solar transformity for the processes in Figures 14.2a and 14.2b and compare these results with those in Examples 14.1 and 14.2. The transformity values for coal, fuel oil, and sunlight are 40,000 seJ/J, 54,000 seJ/J, and 1 seJ/J, respectively.

Solution

Using Equation 14.5, the solar transformity of coal-based electricity is

$$\tau = \frac{141.95 \times 40{,}000 + 7.05 \times 54{,}000}{34.51}$$
$$= 175{,}614 \text{ seJ/J}$$

For the solar plant, transformity of solar electricity is

$$\tau = \frac{270.82}{34.51}$$
$$= 7.85 \text{ seJ/J}$$

To compare these results with those based on the CEC, we can calculate the CDP as the reciprocal of these transformities. Then the ecological CDP values of the coal and solar systems are 0.006 percent and 12.7 percent, respectively. Note that the CDP of the coal process is much lower when ecosystems are included, while that of the solar process is unchanged. Thus, emergy analysis tells us that if we account for ecological work, the coal process is much less efficient than the solar process owing to the large amount of work done by nature in the generation of coal.

14.4.2   Emergy Algebra

Calculating the emergy of economic products is no different from determining the cumulative or life cycle flow of a resource. Thus, the methods for network analysis that we covered in Chapter 10 are also useful for emergy calculations. However, the allocation method used in emergy calculations is unique, and was devised to maintain the relevance of transformity to energy quality, and to avoid allocation in case of incomplete information. Emergy allocation treats "splits" and "coproducts" differently.

A *split* occurs when the products can be produced independently, such as in the distribution of water to multiple users or the separation of a desired mineral from the tailings. *Coproducts* can only be produced together, such as corn and stover or the yellow and white of an egg. In the case of splits, the allocation approach of partitioning is used as described in Section 10.1.2. Here, the emergy of products is proportional to a property of the products such as their exergy, as shown in Figure 14.9a. Notice that the transformity of both products is identical. For coproducts, the approach of no allocation is used, and the same emergy value is assigned to all coproducts, as shown in Figure 14.9b. Now, the transformity of the product that is made in smaller quantities is higher, implying that this product is likely to be of higher quality.

When coproducts are combined, their emergy values cannot be added, to avoid double counting. Their maximum value is assigned to the combined stream, as shown in Figure 14.9c. In general, nonrenewable resources are allocated as

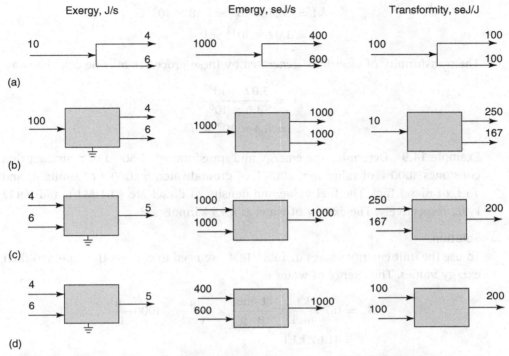

Figure 14.9 Allocation in emergy analysis. (a) Allocation in splits, (b) allocation for coproducts, (c) combining coproducts, (d) combining splits.

splits, while renewable resources are allocated as coproducts. If the inputs being used come from different time periods, then they may be added. For example, the emergy values of rainwater and groundwater may be added because stored groundwater is produced by emergy in an earlier time period as compared to rain-water, which is produced by current solar energy. Using this algebra, the emergy of a product may be calculated as follows:

$$\mathcal{M} = \max_i \left\{ \tau_i B_{R,i} \right\} + \sum_j \tau_j B_{R_t,j} + \sum_k \tau_k B_{N,k} \tag{14.6}$$

Here, subscript R is for renewables from the same time period, R_t is for renewables coming from different time periods, and N indicates nonrenewable resources.

Example 14.8 Continuing Example 10.4, determine the emergy and transformity of the electricity produced.

Solution
The emergy or ecological cumulative exergy consumption of 0.51 L of crude oil is the product of its exergy and transformity. We calcuated the exergy in Example 14.4, and the transformity of crude oil is obtained from Table 14.4:

$$M = 20.4 \times 10^6 \times 1.48 \times 10^5$$
$$= 3.02 \times 10^{12} \text{ seJ}$$

The transformity of electricity generated by these processes may be calculated as

$$\tau = \frac{3.02 \times 10^{12}}{3.6 \times 10^6}$$
$$= 8.4 \times 10^5 \text{ seJ/J}$$

Example 14.9 Determine the emergy and transformity of 500 J of a product that consumes 1000 L of rainwater, 2000 L of groundwater, 50,000 J of sunlight, and 15 L of diesel fuel. The fuel value and density of diesel are 43.1 MJ/L and 0.832 kg/L, respectively. The exergy of water is 0.75 kJ/mol.

Solution
To use the unit emergy values in Table 14.4, we need to convert the inputs to their exergy values. The exergy of water is

$$B_w = 0.75 \frac{\text{kJ}}{\text{mol}} \times \frac{1}{18} \frac{\text{mol}}{\text{g}} \times 1 \frac{\text{g}}{\text{cm}^3} \times 1000 \frac{\text{cm}^3}{\text{L}}$$
$$= 41.67 \text{ kJ/L}$$

To apply Equation 14.6, we need to classify the resources in terms of three categories: rainwater and sunlight are renewables that rely on solar exergy from the same time period (R); groundwater uses exergy mainly from previous years and may be added with other resources R_t; and diesel relies on ancient solar energy and is nonrenewable (N). Therefore, the emergy input to this process is

$$M = \max \{(50{,}000 \times 1), (1000 \times 41{,}670 \times 3.06 \times 10^4)\}$$
$$+ (2000 \times 41{,}670 \times 3.98 \times 10^5)$$
$$+ (15 \times 43.1 \times 10^6 \times 1.81 \times 10^5)$$
$$= (1.28 \times 10^{12}) + (3.32 \times 10^{13}) + (1.17 \times 10^{14})$$
$$= 1.51 \times 10^{14} \text{ seJ}$$

The transformity of the product is

$$\tau = \frac{1.51 \times 10^{14}}{500}$$
$$= 3 \times 10^{11} \text{ seJ/J}$$

Example 14.10 Data about inputs for constructing and operating a wind electricity plant are provided in the first three columns of Table 14.5. Calculate the emergy of each resource, and the transformity of wind electricity.

Table 14.5 Resources for wind electricity production in Italy. Reproduced with permission from [13].

Item	Unit	Amount	UEV (seJ/unit)	Emergy
Direct renewable inputs, $\mathcal{M}_R$				
Wind	J	4.85E+13	2.52E+03	1.22E+17
Nonrenewable and purchased inputs, $\mathcal{M}_N$ and $\mathcal{M}_F$				
Concrete	g	5.62E+06	2.59E+09	1.45E+16
Iron	g	2.65E+05	2.50E+09	6.63E+14
Steel	g	8.61E+05	5.31E+09	4.57E+15
Pig iron	g	1.56E+05	5.43E+09	8.47E+14
Copper	g	8.76E+04	3.36E+09	2.94E+14
Insulating and miscellaneous plastic material	g	1.01E+04	2.52E+09	2.55E+13
Lube oil	J	3.10E+09	1.11E+05	3.44E+14
Labor	Years	8.33E−02	6.32E+16	5.26E+15
Services	US$	2.94E+01	2.00E+12	5.88E+13
Electricity generated, $\mathcal{M}_Y$				
Electricity, with labor and services	J	1.35E+12	1.11E+05	1.49E+17
Electricity, without labor and services	J	1.35E+12	1.06E+05	1.44E+17

Solution

To calculate the emergy of each resource, we need their unit emergy values. These are listed in column 4 of Table 14.5. The resulting emergy for each resource is obtained by multiplying the numbers in columns 3 and 4, and is given in column 5. The total emergy for this wind electricity production system is obtained by using Equation 14.6 and adding the values in the last column. The emergy of wind electricity with and without accounting for labor and services is calculated to be 1.49×10^{17} sej and 1.44×10^{17} sej, respectively. The corresponding transformities of wind electricity may be calculated as the ratio of the corresponding emergy and exergy, giving $\frac{1.49E+17}{1.35E+12} = 1.1E + 05$ seJ/J and $\frac{1.44E+17}{1.35E+12} = 1.07E + 05$ seJ/J, respectively.

14.4.3 Aggregate Metrics

An important benefit of the thermodynamic methods we have covered so far is their ability to represent diverse types of flows in common units. This permits their aggregation to reduce dimensionality and define various metrics. For cumulative exergy and emergy analysis, metrics may be defined on the basis of the flows in Figure 14.5. A renewability index (RI) may be defined as the ratio of the emergy of the renewable resources to the total emergy of all resources:

$$RI = \frac{\mathcal{M}_R}{\mathcal{M}_Y} \qquad (14.7)$$

Sustainability requires processes to have a high renewability index.

The emergy yield ratio (EYR) is defined as

$$EYR = \frac{\mathcal{M}_Y}{\mathcal{M}_F} \qquad (14.8)$$

This metric captures the ratio of the emergy of products to the emergy required to convert raw materials into products. It may also be thought of as the ratio of the emergy provided to society to the emergy used from society, making it analogous to the concept of the energy return on investment (EROI) that we learned about in Chapter 12. However, unlike the EROI, the EYR can never be less than unity, since the numerator, $\mathcal{M}_Y = \mathcal{M}_F + \mathcal{M}_N + \mathcal{M}_R$ is always greater than or equal to the denominator. From a sustainability perspective, a higher EYR is preferred.

The environmental loading ratio (ELR) is defined in emergy analysis as the ratio of the nonrenewable emergy and inputs from the economy to the renewable emergy:

$$ELR = \frac{\mathcal{M}_N + \mathcal{M}_F}{\mathcal{M}_R} \qquad (14.9)$$

This metric represents the load on the environment due to the use of nonrenewable and non-local resources.

A sustainability index may be defined as the ratio of EYR and ELR:

$$SI = \frac{EYR}{ELR} \qquad (14.10)$$

A larger value of SI is preferred.

Example 14.11 Calculate the sustainability metrics for the wind electricity process using the data in Table 14.5. You may ignore the roles of labor and services.

Solution
Based on these data, the emergies of the various flows in Figure 14.5 are $\mathcal{M}_R = 1.22 \times 10^{17}$ seJ, $\mathcal{M}_F + \mathcal{M}_N = 2.2 \times 10^{16}$ seJ, and $\mathcal{M}_Y = 1.44 \times 10^{17}$ seJ. Using equations in this subsection, the emergy metrics may be calculated to obtain the results shown in Table 14.6.

As we can see from Table 14.6, the small values of EYR and RI demonstrate the large dependence on ecological work of oil-based power. Comparing the emergy metrics with metrics from energy analysis such as the EROI (Table 12.2), we can see that hydropower looks much better than wind energy in terms of their EROIs, but not in terms of emergy metrics. This difference is due to the fact that energy analysis ignores the role of materials and natural resources such as land and water, for which emergy does account.

Table 14.6 Comparing alternatives for generating electricity by emergy analysis [13].

Item	Oil plant (1280 MWe)	Wind (2.5 MWe)	Geothermal (20 MWe)	OTEC (4.57 MWe)	Hydro (85 MWe)
	Emergy indicators				
EYR	1.33	6.54	5.60	1.01	5.41
ELR	23.26	0.18	0.96	180	0.55
RI	0.26	0.85	0.78	0.13	0.70
SI	0.06	36.3	5.87	0.006	9.84

14.5 Summary

Consideration of the cumulative exergy consumption expands the benefits of exergy analysis to the life cycle. It is the direct and indirect exergy consumed in an activity or product. The industrial CEC developed in engineering ignores the work done in ecosystems, while the ecological CEC or emergy developed by systems ecologists includes the role of nature by considering exergy flow in natural systems. The cumulative degree of perfection is the second law efficiency of a life cycle, which is the reciprocal of the concept of transformity in emergy analysis. Methods based on CEC are able to account for the role of fuels, as in energy analysis, along with the role of materials. The resulting aggregate metrics can provide unique insight into factors such as renewability and environmental load in the life cycle of human activities.

Key Ideas and Concepts
- Cumulative exergy
- Industrial cumulative exergy consumption
- Ecological cumulative exergy consumption
- Renewability index
- Sustainability index
- Cumulative degree of perfection
- Emergy
- Transformity
- Emergy yield ratio
- Environmental loading ratio

14.6 Review Questions

1. Define cumulative exergy and cumulative degree of perfection.
2. Explain how electricity is a higher-quality resource than coal.
3. Explain why exergy is a common currency that flows through all systems.
4. What is a unit emergy value? What are its units?
5. Explain the emergy yield ratio. Why is its value always larger than or equal to 1?

Problems

14.1 For a light bulb, the exergy efficiency of fuel mining is 93 percent, of fuel conversion to electricity is 34 percent, and of the device is 12 percent. Calculate the cumulative degree of perfection for this bulb. Justify your approach.

14.2 Consider the coal to electricity process that we studied in the previous chapter. Calculate the industrial CDP and the ecological CDP for this process and compare the results with those from energy analysis. As shown in Figure 13.5, this process has outputs of 524 MW electricity, 62 MW flue gas, and 11 MW waste heat. Extracting 100 MJ of coal requires 6 MJ of fuel exergy, and 13 MW of make-up water. The exergy input required per megawatt of make-up water is negligible.

14.3 A gas–electric hybrid truck has an internal combustion engine that provides 24 mpg and a battery that uses 40 kWh per 100 miles. Calculate the miles per gallon equivalent for this vehicle.

14.4 Discuss whether there is a relationship between energy quality and the power or energy density of various renewable and non-renewable resources.

14.5 For the environmentally extended input–output data given in Table 14.7, answer the questions that follow given one million dollars of final demand from sector 2.
  1. Calculate the monetary throughput of each sector for the specified final demand.
  2. Determine the carbon footprint.
  3. If the energy return on investment of the product from sector 2 is 10, determine the fuel value of the product.

Table 14.7 Data for Problem 14.5.

	1	2	3	f	x	CO_2, t	SO_2, t	NO_x, t
1	200	300	200	300	1000	10	5	5
2	400	600	600	400	2000	10	20	10
3	300	900	200	100	1500	20	10	10
v	100	200	500					
x	1000	2000	1500					
NG, MJ	50	1000	200					
H_2O, L	0	100	0					

Table 14.8 Data for fossil route to produce 1,3-propanediol [14].

Input	NAICS sector code (2002)	Cost per mt PDO
Ethylene oxide	325190	$563.34
Syngas	325120	$933.58
Water	221300	$0.00
Natural gas (heating)	221200	$17.37
Electricity (cooling)	221100	$97.66
Cobalt catalyst	325188	$0.00
Nickel catalyst	331419	$0.00

4. Determine mid-point life cycle impact indicators for the given final demand. (Solve after covering Chapter 15.)
5. What is the transformity of the product from sector 2?

14.6 The supply chain of a plastic bottle involves transportation, bottle manufacturing, polymer manufacturing, and monomer manufacturing. The exergy consumed in each of the three manufacturing steps is 0.1 J of diesel in transportation, 0.5 J of coal, and 0.2 J of natural gas. Determine the ecological cumulative exergy consumed in this system per bottle. The transformities are 1.81E+05 seJ/J of diesel, 7E+04 seJ/J of coal, and 1.7E+05 seJ/J of natural gas. Calculate the renewability index of this bottle.

14.7 This question is based on the debate between Professors Dale and Pimentel about "The cost of biofuels" [5]. One statement made by Professor Dale is as follows.

> One megajoule of coal is simply not equivalent to 1 MJ of petroleum. If you doubt this, try grinding up some coal, put it in your gas tank, and see how far you can drive on coal's energy content. Pimentel's "net energy" analysis is akin to adding 10 U.S. dollars, 10 Mexican pesos, and 10 Indian rupees to get $30 – an absurdity. Different energy carriers and different currencies have different qualities. They cannot simply be added together as Dr. Pimentel does.

1. What is the common way of dealing with the aggregation of different currencies? Suggest a way to address the difference in resource quality.
2. Use emergy analysis to calculate metrics based on the data in Table 12.4. Which fuel looks better from this calculation?

14.8 1,3-propanediol (PDO) is an important intermediate in the production of various plastics and fibers. It may be derived from fossil or biomass resources. Tables 14.8 and 14.9 contain information about the cost of inputs to the fossil and biomass routes, respectively, to produce PDO. Using these data and the Eco-LCA tool, answer the following questions.

Table 14.9 Data for biomass route to produce 1,3-propanediol [14].

Input	NAICS sector code (2002)	Cost per mt PDO
Glucose	311221	$1086.61
Ammonia	315310	$2.25
Water	221300	$0.05
Natural gas (heating)	221200	$148.68
Electricity (cooling)	221100	$274.86

Table 14.10 Data about commercial cabbage cultivation in Kenya.

#	Item	Detail	Input	Unit	UEV	Unit	Type
Cabbage cultivation							
1	Solar radiation		62.80	TJ	1.00	seJ/J	R
2	Rain		9.75	Gg	84.82	ksej/g	R
3	Wind		0.14	TJ	1.43	ksej/J	R
4	Geothermal heat		26.78	GJ	6.00	ksej/J	R
5	Net topsoil loss		0.69	GJ	112.00	ksej/J	N
6	Tractor	Iron/steel	117.45	kg	3.38	Gsej/g	F
		Tire, rubber	52.20	kg	4.22	Gsej/g	F
7	Fuel	Diesel	144.86	kg	2.83	Gsej/g	F
8	Tools	Steel	22.05	kg	3.38	Gsej/g	F
		Wood	2.45	kg	1.72	Gsej/g	F
9	Fertilizer	N	120.00	kg	3.73	Gsej/g	F
		P	52.39	kg	3.83	Gsej/g	F
		K	99.57	kg	1.08	Gsej/g	F
10	Pesticide		240.00	kg	14.50	Gsej/g	F
11	Human work		1326	h	273.79	Gsej/h	
Output							
12	Cabbage		6.25	GJ	1.08	Msej/J	
			60.00	Mg	112.82	Msej/g	
Transport to market							
13	Transport		3.00	Ggkm	53.76	ksej g/km	F
Vegetable market							
14	Cabbage		6.25	GJ	1.11	Msej/J	
	(waste)		60.00	Mg	115.51	Msej/g	

R, renewable emergy; N, local non-renewable emergy; F, imported emergy.

1. Compare the life cycle CO_2 emissions, water use, and land use for the fossil and biomass routes to produce PDO.
2. Calculate the CDP and transformity of both processes. Is the insight from these metrics similar?

Table 14.11 Data related to a rain garden in Syracuse, New York.

Item	Unit	Raw data (units)	Emergy/unit (sej/unit)
Renewable			
Sun	J	7.55E+12	1
Rain (chemical)	J	8.53E+10	3.10E+04
Wind (kinetic)	J	1.45E+09	2.45E+03
Non-renewable local			
Net topsoil loss	J	2.51E+08	1.24E+05
Purchased			
Construction debris disposal	g	1.07E+09	1.94E+08
Stone	g	1.37E+08	1.64E+09
Topsoil	g	2.68E+08	1.68E+09
Fertilizers	g	2.34E+04	8.28E+09

3. Calculate the renewability index, environmental loading ratio, and sustainability index for both routes.
4. Discuss the sustainability of both routes.

14.9 Data about commercial cabbage cultivation in Kenya is provided in Table 14.10 [15]. Answer the following questions.

1. Calculate the emergy of each flow and the renewability index, environmental loading ratio, and sustainability index.
2. A parallel study was done for cabbage grown in Kenya by traditional methods. The renewability index was found to be 31 percent, the ELR was 2.18, and the EYR was 1.51. Compare these index values with those for commercial cabbage. Discuss the pros and cons of both farming methods in Kenya.

14.10 Data related to the installation and operation of a rain garden in Syracuse, New York are provided in Table 14.11. The total emergy of the purchased materials is 943,006 sej. Calculate the emergy metrics for this system. Identify ways of improving the sustainability characteristics of this rain garden.

References

[1] H. T. Odum. *Environment, Power, and Society for the Twenty-First Century.* Columbia University Press, 2007.

[2] J. M. Cullen and J. M. Allwood. Theoretical efficiency limits for energy conversion devices. *Energy*, 35(5):2059–2069, 2010.

[3] J. Szargut. *Exergy Method: Technical and Ecological Applications.* WIT Press, 2005.

[4] J. Szargut, D. R. Morris, and F. R. Steward. *Exergy Analysis of Thermal, Chemical and Metallurgical Processes*. Hemisphere Publishing, 1988.

[5] W. Schulz. The costs of biofuels. *Chemical and Engineering News*, 85(51):12–16, 2007.

[6] B. R. Bakshi, A. Baral, and J. L. Hau. Thermodynamic methods for resource accounting. In B. R. Bakshi, T. G. Gutowski, and D. P. Sekulic, editors, *Thermodynamics and the Destruction of Resources*. Cambridge University Press, 2011.

[7] W. A. Hermann. Quantifying global exergy resources. *Energy*, 31(12):1685–1702, 2006.

[8] H. T. Odum. *Environmental Accounting: EMERGY and Environmental Decision Making*. Wiley, 1996.

[9] M. T. Brown and S. Ulgiati. Updated evaluation of exergy and emergy driving the geobiosphere: a review and refinement of the emergy baseline. *Ecological Modelling*, 221(20):2501–2508, 2010.

[10] M. T. Brown, G. Protano, and S. Ulgiati. Assessing geobiosphere work of generating global reserves of coal, crude oil, and natural gas. *Ecological Modelling*, 222(3):879–887, 2011.

[11] N. Jamali-Zghal, O. Le Corre, and B. Lacarrière. Mineral resource assessment: compliance between emergy and exergy respecting Odum's hierarchy concept. *Ecological Modelling*, 272:208–219, 2014.

[12] N. S. Lewis and D. G. Nocera. Powering the planet: chemical challenges in solar energy utilization. *Proceedings of the National Academy of Sciences*, 103(43):15729–15735, 2006.

[13] M. T. Brown and S. Ulgiati. Emergy analysis and environmental accounting. In Cutler J. Cleveland, editor, *Encyclopedia of Energy*, volume 2. Knovel, 2004.

[14] R. A. Urban and B. R. Bakshi. 1,3-propanediol from fossils versus biomass: a life cycle evaluation of emissions and resource use. *Industrial and Engineering Chemistry Research*, 48(17):8068–8082, 2009.

[15] F. Saladini, S. A. Vuai, B. K. Langat, et al. Sustainability assessment of selected biowastes as feedstocks for biofuel and biomaterial production by emergy evaluation in five African countries. *Biomass and Bioenergy*, 85(Supplement C):100–108, 2016.

15 Life Cycle Impact Assessment

> Life cycle thinking is the prerequisite of any sound sustainability assessment. It does not make any sense at all to improve (environmentally, economically, or socially) one part of the system in one country or in one step of the life cycle if this "improvement" has negative consequences for other parts of the system which may outweigh the advantages achieved.
>
> Walter Klöpffer [1]

The sustainability assessment methods that we covered in previous chapters have included univariate footprint methods and thermodynamic aggregation methods. All these methods, except the carbon footprint method, focused on the use of resources such as water, fuel, and materials, while considering boundaries that range from including only human activities to including work done by nature as well. These methods do not consider emissions to the environment and their impact. The carbon footprint quantifies the potential impact of only greenhouse gases on climate change. In addition to resource consumption and greenhouse gas emissions, many other emissions such as volatile organic solvents, particulate matter, and ozone-depleting chemicals impact our well-being and that of the environment. The methods considered so far do not account for the impact of such emissions.

In this chapter we will learn about life cycle impact assessment (LCIA), which is the third step in the approach of life cycle assessment (LCA). We learned about the previous two steps of "goal definition and scope" and "inventory analysis" in Chapters 8 and 9. The overall approach of LCA is depicted in Figure 8.2. LCIA accounts for a large variety of emissions and resource use and combines many of the benefits of the sustainability assessment methods described in previous chapters.

15.1 Steps in Life Cycle Impact Assessment

As we learned in Chapter 9 and saw in a typical inventory in Table 9.1, life cycle inventories for a selected product or process may consist of the emissions of hundreds of chemicals and use of a large number of resources. Determining the impact on the basis of such data is challenging not just because of the high dimensionality of the inventory, but also because of the need to associate emissions and resource use with human and ecological impacts. The footprint methods in Chapter 11 dealt with this "curse of dimensionality" by focusing only on selected flows such as greenhouse gases or water, while ignoring the rest. Methods based on energy and exergy analysis address the dimensionality challenge by representing resources in terms of the fuel value of fossil resources or the exergy of a more diverse array of resources. Cumulative exergy and emergy analysis consider the work done in industrial and ecological systems for making various ecological goods and services available. However, none of these methods considers the large variety of molecules produced by human activities and their impact on human and ecological systems due to their release into society and the environment. Such impact assessment is an essential part of evaluating and developing sustainable solutions, but is a challenging task due to the thousands of chemicals in use today that may be emitted into the environment.

Life cycle impact assessment is a systematic approach for assessing the impact of emissions, and consists of four steps [2]:

1. classification
2. characterization
3. normalization
4. weighting

The first two steps are mandatory according to the ISO 14040 standards, while the last two are optional. In this chapter we will learn about all four steps and their application. An overview of these steps for a typical LCIA approach is shown in Figure 15.1.

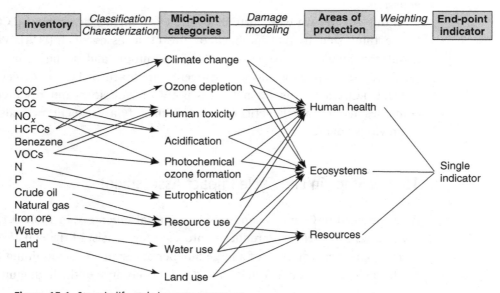

Figure 15.1 Steps in life cycle impact assessment.

Table 15.1 Typical classification categories in LCIA.

Category	Cause and contributors	Unit
Climate change	Release of greenhouse gases such as CO_2, CH_4, N_2O, CFCs	kg CO_2 equivalent
Ozone depletion potential	Emission of gases that cause loss of stratospheric ozone such as chlorofluorocarbons (CFCs) and hydrochlorofluorocarbons (HCFCs)	kg CFC-11 equivalent
Photochemical smog formation	Reaction between NO_x and volatile organic compounds in the presence of sunlight to form ground-level ozone	kg O_3 equivalent
Human health pollutants	Pollutants such as particulate matter and various toxics. These include hundreds of chemical compounds.	kg 1,4-dichlorobenzene equivalent
Ecotoxicity	Pollutants that are toxic to ecosystems including various chemicals.	kg 1,4-dichlorobenzene equivalent
Acidification	Addition of hydrogen ions (H^+) to the local environment. Caused by emissions such as NO_x, SO_x, NH_3	kg SO_2 equivalent
Eutrophication	Presence of excessive nutrients such as N and P compounds	kg N equivalent
Land use	Area of land used	m^2y
Resource depletion	Fossil fuel	MJ
	Mineral use	kg antimony equivalent

15.1.1 Classification into Impact Categories

The first step in reducing dimensionality in LCIA is to classify the emissions and resources into selected categories, such as those listed in Table 15.1. These categories belong to three broad "areas of protection":

- human impact
- ecological impact
- resource use

Classification of various chemicals relies on knowledge about their environmental impact. Such information has been developed through environmental and toxicological studies, and is included in LCIA databases such as those listed in Table 9.3.

15.1.2 Characterization into Common Units

Inventory data classified in the same category are multiplied by their corresponding characterization factor to represent all flows in the common unit

Table 15.2 Illustration of the characterization step in LCIA.

Emissions/resource	Quantity (kg)	GWP	ODP	HT	AP	RD
CO_2	1.792	×1	–	–	–	–
CO	6.7E–4	–	–	×0.012	–	–
NO_x	1.091E–3	–	–	×0.78	×0.7	–
SO_2	9.87E–4	–	–	×1.2	×1	–
Silver	0.6	–	–	–	–	×1.18
Total		1.792	0	2.04E–3	1.75E–3	7.1E–01

Note: GWP, global warming potential; ODP, ozone depletion potential; HT, human toxicity; AP, acidification potential; RD, resource depletion.

of the category. This step is represented by Equation 10.16, which is reproduced below:

$$h_i = \sum_j \phi_{ij} r_j$$

Here, h_i is the total impact for category i, r_j is the mass of inventory flow j, and ϕ_{ij} is the characterization factor for impact category i and flow j. This step represents the chemicals classified in the same group in terms of a common unit, to permit their addition. These common units are shown in Table 15.1, and typical characterization factors are given in Table 11.1 for greenhouse gases and in Table 15.2 for some selected chemicals. These factors have been determined from research about the environmental and human impact of each chemical that is emitted to the environment, and from estimated stocks for abiotic resources. The resulting impact categories are often referred to as "mid-point" indicators, as shown in Figure 15.1.

Example 15.1
Classify and characterize the following pollutants and resources from a selected life cycle inventory: 1.792 kg of CO_2, 6.7×10^{-4} kg of CO, 1.091×10^{-3} kg of NO_x, 9.87×10^{-4} kg of SO_2, and 0.6 kg of silver. The relevant characterization factors and mid-point indicators for these compounds are shown in Table 15.2.

Solution
The mid-point impact indicator for each flow may be calculated using Equation 10.16. For the human toxicity impact category,

$$h_i = (0 \times 1.792) + (0.012 \times 6.7 \times 10^{-4}) + (0.78 \times 1.091 \times 10^{-3})$$
$$+ (1.2 \times 9.87 \times 10^{-4}) + (0 \times 0.6)$$
$$= 2.04 \times 10^{-3} \text{ kg 1,4-dichlorobenzene equivalent}$$

Using matrix notation and Equation 10.17,

$$h = \begin{bmatrix} 1 & 0 & 0 & 0 & 0 \\ 0 & 0 & 0 & 0 & 0 \\ 0 & 0.012 & 0.78 & 1.2 & 0 \\ 0 & 0 & 0.7 & 1 & 0 \\ 0 & 0 & 0 & 0 & 1.18 \end{bmatrix} \begin{bmatrix} 1.792 \\ 6.7 \times 10^{-4} \\ 1.091 \times 10^{-3} \\ 9.87 \times 10^{-4} \\ 0.6 \end{bmatrix}$$

$$= \begin{bmatrix} 1.792 \\ 0 \\ 2.04 \times 10^{-3} \\ 1.75 \times 10^{-3} \\ 7.1 \times 10^{-1} \end{bmatrix} \begin{matrix} \text{kg CO}_2 \text{ eq} \\ \text{kg CFC-11 eq} \\ \text{kg 1,4 DB eq} \\ \text{kg SO}_2 \text{ eq} \\ \text{kg Sb eq} \end{matrix}$$

The solution is also shown in Table 15.2.

Example 15.2 Continuing Example 10.4, determine the mid-point indicators for producing 1 kWh electricity.

Solution
Only two mid-point impact indicators are relevant to the emissions from these processes. The global warming potential of the life cycle is identical to the carbon footprint, and is 0.7 kg CO_2 eq/kWh. Acidification potential is $(0.28 \times 1) + (0.13 \times 0.7) = 0.37$ kg SO_2 eq.

Choosing between alternatives based on mid-point impacts requires their comparison. Even with the smaller number of dimensions of mid-point indicators as compared to the original emissions data, this can still be a daunting task since the impacts are in diverse units. Furthermore, when comparing multiple products, only in rare cases do all the indicators for one product outperform the indicators for another product.

15.1.3 Normalization and Weighting
Various approaches have been suggested for analyzing the mid-point impact assessment results. A common approach is to normalize each impact h_i by a reference value $h_{i,r}$. Thus, the normalized value is

$$\hat{h}_i = \frac{h_i}{h_{i,r}} \tag{15.1}$$

Common reference values include the total impact in a selected region and the largest value among alternatives. Thus, the global warming potential from a product being analyzed would be divided by the global warming potential of emissions in the entire region where the product is produced. Typical normalization factors

Table 15.3 Normalization in LCIA.

Impact category	Quantity, h_i	Normalization factor, $h_{i,r}$	Normalized value, $\hat{h}_i$
GWP	1.792	37.7E+12	47.5E–15
ODP	0	1E+09	0
HT	2.04E–03	576E+09	3.54E–15
AP	1.75E–03	286E+09	6.12E–15

Note: The units of each category and its normalization factors are given in Table 15.1.

for this approach are shown in Table 15.3. Another approach for normalization is to divide all numbers in an impact category by the largest quantity so that the largest value is equal to unity for all categories. The resulting dimensionless numbers allow convenient comparison.

Example 15.3

Some mid-point indicators for various types of grocery bags are listed in Table 15.4. Choose the best option based on these results.

Solution

In each column of Table 15.4, the smallest value in each category is indicated with italics. A close look shows that no bag is better in all categories. For example, on the basis of the global warming potential, the best bag is the reusable polypropylene bag, while on the basis of photochemical oxidant formation, the best bag is the biodegradable bag. From the numbers in the table, it is very difficult to identify the bag with the lowest impact. For easier visualization of the trade-offs, the indicators are plotted in Figure 15.2 after scaling the largest value to be 100. Here, the reference value for normalization is the largest value in each impact category.

A glance at this plot shows that a paper bag has many more tall bars. Closer inspection reveals that it has the highest impact in all categories except eutrophication and mineral depletion. At the other extreme is the reusable PET bag owing to its lowest values in many categories. For categories in which the values for the reusable PET bags are not the smallest, the values are quite close to the lowest values. Thus, among all the bags considered, the reusable PET bag seems to be best. However, no conclusion can be reached unambiguously.

As this example illustrates, after applying various mid-point assessment methods, it is still common to face the challenge of having to choose between a higher impact in one category versus another. To address this challenge, end-point assessment methods have been developed for further aggregation to a smaller number of indicators, and even to a single indicator.

Table 15.4 Mid-point indicators for different types of grocery bags. Reproduced with permission from [3].

Impact category	Unit	HDPE plastic bag 100 percent virgin	HDPE plastic bag with recycled content	Biodegradable bag	Oxo-degradable bag	Paper bag	Reusable PET bag	Reusable PP bag
GWP	kg CO$_2$eq	7.52	7.35	9.19	6.69	44.74	6.47	5.43
POCP	kg C$_2$H$_4$eq	0.045	0.038	−0.001	0.036	0.072	0.005	0.004
EP	kg PO$_4$eq	0.005	0.005	0.278	0.004	0.033	0.003	0.003
LU	Ha/yr	6.6E-6	5.9E-6	3.5E-4	9.2E-7	2.0E-3	2.0E-6	1.3E-6
WU	kL H$_2$O	0.013	0.053	0.050	0.012	0.423	0.038	0.016
SW	kg	2.74	3.31	0.83	2.56	3.24	0.81	3.27
Fossils	MJ s	19.93	19.07	9.96	17.94	44.77	6.46	6.65
Minerals	MJ s	8.44E-5	7.79E-4	1.02E-2	1.73E-3	7.42E-3	3.39E-5	2.50E-5

The smallest values in each impact category are in italics.

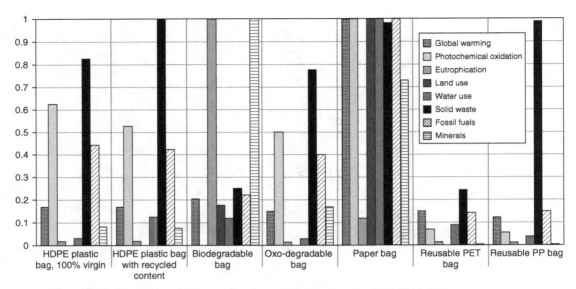

Figure 15.2 Comparison of indicators for grocery bags based on the data in Table 15.4.

15.1.4   End-Point Assessment

The framework and steps in end-point assessment are depicted in the right half of Figure 15.1. As shown, the mid-point impact categories are further aggregated into damage categories such as human health, ecosystem damage, and resource depletion by the following formula:

$$E_k = \sum_j \theta_{j,k} h_j \tag{15.2}$$

Here, $\theta_{j,k}$ is the characterization factor that converts the jth mid-point indicator h_j to an end-point indicator E_k.

Several methods have been developed for arriving at end-point indicators. Some result in a single indicator by using weighting factors like those in Table 15.5 to convert midpoint indicators into a common unit that can be added. Here, the units of each factor are the reciprocal of the corresponding impact category.

Example 15.4

Using the weighting factors in Table 15.5, represent the mid-point indicator values in Table 15.2 by an end-point indicator.

Solution

Using Equation 15.2,

$$\begin{aligned}
E_k &= (1.792 \times 0.240) + (2.04 \times 10^{-3} \times 0.039) + (1.75 \times 10^{-3} \times 0.039) \\
&\quad + (7.1 \times 10^{-1} \times 0.361) \\
&= 0.69
\end{aligned}$$

With such a single score, it becomes possible to easily compare alternatives.

Table 15.5 Typical weighting factors for converting mid-point into end-point indicators in LCIA.

Impact category	Weight
Resource depletion potential	0.361
Global warming potential	0.240
Photochemical oxidant chemical potential	0.113
Acidification potential	0.039
Human toxicity potential	0.039
Ecotoxicity, aquatic	0.106
Eutrophication potential	0.102

Table 15.6 Mid-point and end-point contributions of Products A and B.

Category	Mid-point, A	Mid-point, B	Weight	End-point, A	End-point, B
GWP	0.14	0.25	0.240	3.4E–02	6.0E–02
POCP	0.55	0.72	0.113	6.2E–02	8.1E–02
AP	0.43	0.15	0.039	1.7E–02	5.9E–03
ET	0.4	0.3	0.106	4.2E–02	3.2E–02
Total				1.55E–01	1.79E–01

Example 15.5 Choose between products A and B on the basis of the data in Table 15.6.

Solution

The mid-point indicators for products A and B are plotted in Figure 15.3a. Choosing between the two products is difficult on the basis of these results since no product dominates in all categories. End-point indicators are shown in Figure 15.3b obtained using the weights in Table 15.5. Owing to the weighting scheme, the impact of photochemical oxidation dominates, followed by that of global warming. In terms of the end-point indicator, product A has a smaller impact and should be preferred.

Other approaches use damage models as summarized in Figure 15.1, and combine mid-point indicators into categories such as human impact, ecological impact, and resource depletion, which are area-of-protection (AoP) indicators. These may be further combined into a single indicator.

The human health conversion factor commonly relies on the approach of disability adjusted life years (DALY), developed by the World Health Organization to quantify the human impact of emissions. It is quantified from public health studies as the years lived disabled (YLD) or years of life lost (YLL) due to an impact. For

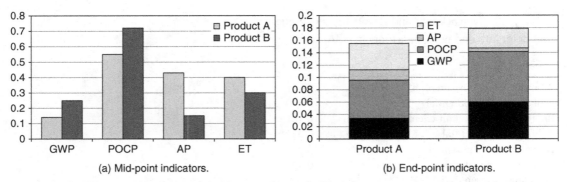

(a) Mid-point indicators.   (b) End-point indicators.

Figure 15.3 Mid-point and end-point indicators for life cycle impact assessment in Example 15.5.

the jth mid-point indicator, DALY is calculated as

$$\theta_{j,H} = \text{YLL}_j + \text{YLD}_j \tag{15.3}$$

Ecological impact may be quantified in terms of the potentially disappeared fraction (PDF) of an ecosystem due to the impact of an emission. This characterization factor for the jth mid-point indicator, $\theta_{j,E}$, may be calculated as

$$\theta_{j,E} = \text{PDF}_{j,terr}\text{SD}_{j,terr} + \text{PDF}_{j,fw}\text{SD}_{j,fw} + \text{PDF}_{j,mw}\text{SD}_{j,mw} \tag{15.4}$$

where SD stands for the species density in the affected region. The three terms add the impact on terrestrial, fresh water, and marine water ecosystems.

Resource depletion is quantified in various ways. One approach is based on considering the increase in resource cost due to depletion. This marginal cost increase (MCI) is the third characterization factor, and is calculated as the ratio of the change in cost to the change in consumption:

$$\theta_{j,R} = \frac{\Delta Cost_j}{\Delta Yield_j} \tag{15.5}$$

Other approaches for quantifying resource depletion rely on measures such as cumulative exergy, which we learned about in Chapter 14. Alternatively, the additional energy required to extract the next kilogram of resource may be used.

Typical end-point characterization factors based on the ReCiPe2016 method [4] are listed in Table 15.7. As we can see, societal preferences are already included in these factors.

Example 15.6

Represent the results from Example 15.1 in terms of end-point indicators using the ReCiPe method with hierarchist values (see the discussion after Table 15.8). For silver, the surplus ore potential is 2000 kg ore/kg Ag, which may be converted into Cu equivalents by the hierarchist value, 153 kg Cu eq/kg ore.

Table 15.7 Mid-point to end-point conversion factors from ReCiPe2016 [4].

Impact category	Unit	End-point CF, θ		
		Individualistic	Hierarchist	Egalitarian
Human health				
Global warming	DALY/kg CO_2eq	8.12E–08	9.28E–07	1.25E–05
Stratospheric ozone depletion	DALY/kg CFC-11eq	2.37E–04	5.31E–04	1.34E–03
Ionizing radiation	DALY/kBq Co-60 emitted to air eq	6.80E–09	8.50E–09	1.40E–08
Fine particulate matter formation	DALY/kg PM2.5eq	6.29E–04	6.29E–04	6.29E–04
Photochemical ozone formation	DALY/kg NO_xeq	9.10E–07	9.10E–07	9.10E–07
Toxicity (cancer)	DALY/kg 1,4-DCB emitted to urban air eq	3.32E–06	3.32E–06	3.32E–06
Toxicity (non-cancer)	DALY/kg 1,4-DCB emitted to urban air eq	6.65E–09	6.65E–09	6.65E–09
Water consumption	DALY/m^3 consumed	3.10E–06	2.22E–06	2.22E–06
Terrestrial ecosystems				
Global warming	Species yr/kg CO_2eq	5.32E–10	2.80E–09	2.50E–08
Photochemical ozone formation	Species yr/kg NO_xeq	1.29E–07	1.29E–07	1.29E–07
Acidification	Species yr/kg SO_2eq	2.12E–07	2.12E–07	2.12E–07
Toxicity	species yr/kg 1,4-DBC emitted to industrial soil eq	5.39E–08	5.39E–08	5.39E–08
Water consumption	species yr/m^3 consumed	0.00E+00	1.35E–08	1.35E–08
Land use or occupation	Species yr/annual crop eq	8.88E–09	8.88E–09	8.88E–09
Freshwater ecosystems				
Global warming	Species yr/kg CO_2 eq	1.45E–14	7.65E–14	6.82E–13
Eutrophication	Species yr/kg P to fresh water eq	6.10E–07	6.10E–07	6.10E–07
Toxicity	species yr/kg 1,4-DBC emitted to fresh water eq	6.95E–10	6.95E–10	6.95E–10
Water consumption	species yr/m^3 consumed	6.04E–13	6.04E–13	6.04E–13

Table 15.7 (cont.)

Marine ecosystems				
Toxicity	species yr/kg 1,4-DBC emitted to seawater eq	1.05E–10	1.05E–10	1.05E–10
Resources				
Mineral resource scarcity	USD2013/kg Cu	1.59E–01	2.31E–01	2.31E–01
Fossil resource scarcity				
Crude oil	USD2013/kg	0.46	0.46	0.46
Hard coal	USD2013/kg	0.03	0.03	0.03
Natural gas	USD2013/N m^3	0.30	0.30	0.30
Brown coal	USD2013/kg	–	–	0.03
Peat	USD2013/kg	–	–	0.03

Solution

Use the mid-point to end-point conversion factors in Table 15.7 to get the following results. By comparing the mid-point indicators in the last row of Table 15.2 with the impact categories in Table 15.7 we can see that the impacts relevant to human health are global warming and toxicity (non-cancer):

$$E_H = (1.792 \times 9.28 \times 10^{-7}) + (2.04 \times 10^{-3} \times 6.65 \times 10^{-9})$$
$$= 1.66 \times 10^{-6} \text{ DALY}$$

The impact on terrestrial and fresh water ecosystems is due to global warming and acidification and may be calculated as follows:

$$E_E = ((2.80 \times 10^{-9} + 7.65 \times 10^{-14}) \times 1.792) + (2.12 \times 10^{-7} \times 1.75 \times 10^{-3})$$
$$= 5 \times 10^{-9} \text{ species} \cdot \text{years}$$

The impact on resources is calculated separately for the use of mineral and fossil resources:

$$E_{R,m} = 0.6 \text{ kg Ag} \times 2000 \frac{\text{kg ore}}{\text{kg Ag}} \times 153 \frac{\text{kg Cu eq}}{\text{kg ore}} \times (2.31 \times 10^{-1})$$
$$= 4.24 \times 10^4 \text{ USD2013}$$

Further aggregation to a single indicator is subjective and relies on information about human preferences. Weights for combining indicators may be determined by converting each impact into its monetary value. Various methods are available in environmental economics for such monetization. Alternatively, surveys of a

Table 15.8 Weights for combining impact categories by cultural theory.

	Individualist	Egalitarian	Hierarchist	Average
Ecosystem quality	25	50	40	40
Human health	55	30	30	40
Resources	20	20	30	20

representative population may be conducted to determine the relative importance the members of the population assign to various impact categories.

In one such approach for quantifying social preferences, people are categorized in four broad categories:

- Individualists, who consider nature to be capable of recovering from any disruption, and are driven by short-term self-interest with high technological optimism.
- Egalitarians, who are almost the exact opposite of individualists and adopt the most precautionary approach with the longest time frame. They consider nature to be fragile and require us to use it carefully.
- Hierarchists, who are in the middle. They consider nature to be capable of dealing with some disruption, and prefer that economic activity be kept within limits.
- Fatalists, who consider nature and people to be fickle and untrustworthy.

The development of a single indicator is made possible by choosing a category and weighting the three end-point indicators, as shown in Table 15.8.

We have learned about methods for reducing the dimensionality from hundreds of flows in a typical life cycle inventory to about 20 mid-point indicators, to three AoP indicators, to a single indicator. The attraction of dimensionality reduction is quite obvious since it simplifies the task of choosing between alternatives and of making the results easier to communicate. However, there are some significant disadvantages of aggregation, which we learned about in Section 14.2. These are the implicit assumption of substitutability and the loss of information. When we convert multiple flows into a common unit, it is essential to use appropriate conversion factors for a fair comparison. The mid-point characterization factors represent such conversion factors. Even after proper conversion, aggregation implicitly assumes the flows to be substitutable. Also, by aggregation we lose details about individual flows, which can make it difficult to identify opportunities for improvement by changing these flows. There is also a strong trade-off between reducing dimensionality and increasing uncertainty in the results. As the dimensionality is reduced, the results become less certain. In particular, end-point assessment methods can be highly subjective, owing to which many life cycle

Table 15.9 Software for life cycle assessment.

Name	Type
SimaPro	Process LCA, commercial
GaBi	Process LCA, commercial
OpenLCA	Process LCA, free
EIOLCA	Input–output LCA, free
Eco-LCA	Input–output LCA, free
USEEIO	Input–output LCA, free

studies prefer to stop at mid-point indicators or develop a hierarchy of indicators where the end-point and single indicators are determined only if necessary.

15.2   Software for Sustainability Assessment

Applying the various sustainability assessment methods to practical problems is difficult and tedious owing to the need for obtaining and manipulating a large amount of data. Therefore, several software packages are available for performing LCA, some of which are listed in Table 15.9. Most of the packages focus on process LCA with the help of commercial and free inventory databases like those listed in Table 9.3. Other packages are useful for input–output LCA at the economy scale. The companion website to this book contains information, tutorials, links, and case studies based on some of these tools.

15.3   Summary

Life cycle assessment is the most comprehensive approach for determining the environmental impact of different types of activities. It considers a large system boundary, many types of resources and emissions, and their environmental impacts. This approach is popular and many databases and software packages are available. It is also an active area of research since many requirements of sustainability are not yet satisfied.

Current research is developing methods for life cycle sustainability analysis to consider economic, environmental, and social aspects over the life cycle. A brief overview of this approach is provided in Box 15.1. Methods are also being developed for considering the dynamics of decisions based on LCA. This aims to account for possible economic feedback and rebound effects, and is called consequential LCA. Other efforts are accounting for the impact of human activities on ecosystem services and biodiversity. The inventory for LCA is also being used to understand the nexus between multiple flows such as food, energy, and water.

BOX 15.1 Life Cycle Sustainability Analysis

Life cycle sustainability analysis (LCSA) is an approach to account for the direct and indirect impacts of an activity on environmental, economic, and societal aspects [1]. As described in this chapter, LCA focuses on environmental aspects, and can reduce the chance of shifting environmental impacts along the life cycle. It aims to reduce the chance of burden-shifting for economic and societal impacts as well. Economic impacts are determined by the approach of life cycle costing (LCC), while societal impacts are determined by societal LCA (SLCA).

- LCC is an approach that predates LCA. It involves determining the monetary costs in each step of a life cycle. Such data are usually quite readily available. Comparing the life cycle costs of alternatives can help choose the one that is cheaper and more economically sustainable. For decisions about specific products or processes, engineering cost analysis may be used instead of life cycle costs. We will learn more about this in Chapter 17.
- SLCA is an emerging approach that considers the impacts of an activity on society. Categories of stakeholders include workers, local community, society, consumers, and value chain actors. Impact categories include human rights, working conditions, health and safety, cultural heritage, governance, and socioeconomic repercussions. Obtaining such data is a challenge for the large diversity of products across the world. In many engineering studies, it is common to consider also the societal impacts such as jobs created and safety of industrial processes.

LCSA is often represented as

$$LCSA = LCA + LCC + LCSA$$

It is an active area of research.

Key Ideas and Concepts

- Mid-point indicator
- End-point indicator
- Characterization
- Normalization
- Disability adjusted life year
- Potentially destroyed fraction
- Hierarchist approach
- Egalitarian approach

Table 15.10 Characterization factors for Problem 15.4.

Chemical	GWP (kg CO$_2$eq /kg)	Particulate air (PM2.5 eq/kg)	Eutrophic. air (kg N eq/kg)	Eutrophic. water (kg N eq/kg)	Ecotoxicity water (CTU eco/kg)	Human health (CTU noncancer/kg)
Ammonia	0.00E+00	6.67E–02	1.19E–01	7.79E–01	na	na
Carbon dioxide	1.00E+00	na	na	na	na	na
Carbon monoxide	0.00E+00	3.56E–04	na	na	na	na
Methanol	0.00E+00	0.00E+00	0.00E+00	0.00E+00	2.26E–01	1.63E–08

15.4 Review Questions

1. What are the four steps in life cycle assessment?
2. What are the steps in life cycle impact assessment?
3. What are the units of eutrophication and ecotoxicity?
4. Why does end-point impact assessment rely on subjective weighting factors?
5. What are the advantages and disadvantages of representing life cycle impact in terms of a single indicator?

Problems

15.1 Emissions from a life cycle include 1000 kg of CO$_2$, 200 kg of SO$_2$, 100 kg of NO$_2$, and 10 kg of CH$_4$. Convert this information to mid-point indicators.

15.2 Using the life cycle assessment results for grocery bags in Table 15.4, normalize and create the bar plot shown in Figure 15.2.

15.3 For the partial life cycle inventory of electricity production from natural gas given in Table 9.1, apply life cycle impact assessment to determine the mid-point and end-point indicators.

15.4 The cradle-to-gate inventory of acetic acid includes the following emissions to air: 5.70E-04 kg of ammonia, 1.76E-03 kg of carbon dioxide, 3.97E-03 kg of carbon monoxide, and 4.00E-05 kg of methanol. Emissions to water include 9.60E-04 kg of acetic acid and 5.20E-05 kg of ammonia. Using the information in Table 15.10, determine the mid-point environmental impact of this life cycle inventory.

15.5 The life cycle inventory data for soybean oil after allocation are given in Table 15.11. Answer the following questions using the data available here. Discuss the shortcomings of each step in brief.

Table 15.11 Life cycle inventory of soybean oil. Adapted from [5].

	Soy bean production	Steam production	Electricity production	Transport by road	Bean oil production	TOTAL
Contribution factors						
	2.14	0.51	75	28.6	0.43	
Energy resources (GJ)						
	1.24	1.77	0.68	0.25	–	3.95
Emissions to air (kg)						
CO_2	97.2	83.3	46.9	23.1	–	250.5
CO	0.021	0.021	0.00229	0.057	–	0.101
Hydrocarbons	0.01	1.03	0.004	0.04	–	1.08
NO_x	0.15	0.26	0.0009	0.34	–	0.75
SO_2	0.71	0.07	0.0058	0.03	–	0.81
Particles	0.032	0.0006	–	–	–	0.033
Liquid particles in air	–	–	0.001	0.023	–	0.025
Emissions to water (kg)						
Nitrogen	212	–	–	–	–	212
Phosphates	60	–	–	–	–	60
Potassium oxide	120	–	–	–	–	120
Calcium oxide	45	–	–	–	–	45
Magnesium oxide	36.4	–	–	–	–	36.4
Insecticides	25.7	–	–	–	–	25.7
Herbicides	3.54	–	–	–	–	3.54
Oil	–	0.0013	–	–	–	0.0013
Hexane	–	–	–	–	1.29	1.29
Solid waste (kg)						
High risk	–	–	0.00013	–	–	0.00013
Industrial	–	0.12	0.000052	–	–	0.12

1. Perform the classification and characterization of these emissions and resource use into various life cycle impact categories. Can you combine or compare the results of different impact categories? Why or why not?
2. Normalize the results of characterization using the figures for global emissions given in the table. What is the benefit of this normalization step? Can the impact categories be combined now? Justify your answer.

15.6 As we learned in Chapter 3, the availability of ecosystem services is essential for sustaining all activities on Earth. Compare the methods of emergy (ECEC) analysis, life cycle assessment, and water footprint analysis by answering the following questions. You may benefit from referring to earlier chapters on each of these methods.

Table 15.12 Inventory for polyethylene terephthalate life cycle.

Parameter	Unit	Quantity
Functional unit	volume (oz.)	12
Monomer (PET resin prod. – RNA)[a]	grams	18
Electricity consumption (US grid, 2010)	mj	0.1
Truck transportation (single unit truck, diesel) – RNA[a]	ton km	3.5
Train transportation (Diesel) – RNA[a]	ton miles	1.2
Cost	$	1.50

Note: [a] Rest of North America.

1. Consider the ability to account for the role of ecosystem services in the life cycle. Which ecosystem service is each method able to account for?
2. Summarize the approach each method uses to account for ecosystem services, and discuss its pros and cons.
3. Does each method account for the carrying capacity of ecosystems? What are the pros and cons of each approach?

15.7 Calculate the carbon footprint for manufacturing a cup made of polyethylene terephthalate using the data in Table 15.12.

1. Use the openLCA tool to determine the carbon footprint from the information provided in this table. You do not have to perform a complete life cycle analysis for this. You can use the "Quick results" option to calculate the carbon dioxide emissions for the production process. Identify the sectors that contribute the most to carbon dioxide emissions. (Hint: Create a new process called "Cups manufacturing, plant, US" and use this as a starting point to build your model.)
2. Now use an environmentally extended input–output model to determine the carbon footprint of this cup. Polyethylene terephthalate belongs to sector 325211. Comment on the results.
3. Explain the differences between a process-based analysis of calculating the carbon footprint and an input–output-based approach to calculating the carbon footprint.

15.8 Ernie Engineer has developed a new plug-in electric vehicle with a 24 kWh battery pack. Charging the pack to its full capacity requires a total of 30 kWh of electricity from the charging station. Two-thirds of this electricity comes from fossil fuels (assume bituminous coal) and the remaining from biomass. Only 90 percent of the electricity from the grid reaches the charging station owing to inefficiencies in the transmission system, thus requiring more primary fuel to generate this electricity. The car has a fuel efficiency of 130 kWh/100 miles. Determine the well-to-wheel emissions and their life cycle impact for traveling a

Table 15.13 Life cycle impact assessment of cheesecake packaging systems.

Impact category	Unit	XrPet	AerPack
Climate change human health	DALY	1.22E–05	9.48E–06
Ozone depletion	DALY	1.16E–09	8.61E–10
Human toxicity	DALY	1.91E–05	1.36E–05
Photochemical oxidant formation	DALY	4.35E–10	3.04E–10
Particulate matter formation	DALY	1.98E–06	1.53E–06
Ionizing radiation	DALY	7.49E–09	5.87E–09
Climate change ecosystems	species yr	6.53E–08	5.05E–08
Terrestrial acidification	species yr	5.32E–10	4.33E–10
Fresh-water eutrophication	species yr	3.07E–11	2.36E–11
Terrestrial ecotoxicity	species yr	1.92E–09	1.58E–09
Fresh-water ecotoxicity	species yr	3.04E–11	2.27E–11
Marine ecotoxicity	species yr	4.22E–09	3.06E–09
Agricultural land occupation	species yr	3.14E–08	2.68E–08
Urban land occupation	species yr	8.15E–10	5.04E–10
Natural land transformation	species yr	2.91E–08	2.23E–08
Metal depletion	$	1.09E–02	6.93E–03
Fossil depletion	$	1.63E–01	1.23E–01

distance of 100 miles. You do not have to account for the manufacturing of the car here. You may use OpenLCA with the TRACI 2.1 impact assessment method.

15.9 Food packaging can reduce food spoilage but also has a large environmental impact owing to the nature of the packaging materials. In an effort to reduce the environmental impact of food packaging, two types of packaging are being compared with regards to their ability to reduce spoilage and the impact of the materials used. Usually, cheesecake is packed in low-density polyethylene film, and has a short shelf life. This study considers two packaging materials that can extend shelf life: recycled polyethylene terephthalate (XrPet) and a multilayer gas and water barrier film (AerPack). The results of comparing these two life cycles are summarized in Table 15.13. Note that DALY is a measure of the overall severity of a disease, expressed as the number of years lost due to illness, disability or premature death; species yr is a measure of the extinction rate, which is the loss of species during a year; $ is a measure of the surplus costs of future resource production over an infinite time frame, considering a 3 percent discount rate. The unit is $2000.

1. Compare the two packaging systems using the results in Table 15.13 without further aggregation.
2. Represent the impact assessment results in terms of a single indicator and compare the two packaging systems.

15.10 Ionic liquids form a new class of solvents that are expected to be more environmentally friendly than conventional solvents owing to their low volatility and high stability. Since these solvents are still being studied in research laboratories, their performance in an industrial setting is not known. However, assessing the sustainability of such emerging technologies at early stages is important to guide research and development. The photochemical ozone creation potential (POCP) in a life cycle is a function of the quantity of volatile organic compounds used in each stage. For an ionic liquid, $POCP = 0.1x_1 + 0.5x_2 + 0.3x_3$ in units of kg C_2H_4eq. The variable x_1 is the quantity of ionic liquid used. For the base case, the quantities of solvents used are $x_1 = 105$ kg, $x_2 = 20$ kg, and $x_3 = 1000$ kg. For this case, the ionic liquid is used only once and then discarded, as is done in the laboratory. However, in an industrial setting, the ionic liquid is expected to be used up to 100 times. Calculate the effect of this uncertainty on the POCP value for the ionic liquid process. You may assume that the POCP due to additional processing in the reuse is negligible. A conventional process has a POCP of 200 kg C_2H_4eq. Is the ionic liquid-based process better than the conventional process in terms of life cycle POCP values?

References

[1] W. Klöpffer. Life-cycle based methods for sustainable product development. *The International Journal of Life Cycle Assessment*, 8(3):157, 2003.

[2] D. W. Pennington, J. Potting, G. Finnveden, et al. Life cycle assessment part 2: current impact assessment practice. *Environment International*, 30(5):721–739, 2004.

[3] H. Lewis, K. Verghese, and L. Fitzpatrick. Evaluating the sustainability impacts of packaging: the plastic carry bag dilemma. *Packaging Technology and Science*, 23(3):145–160, 2010.

[4] M. A. J. Huijbregts, Z. J. N. Steinmann, P. M. F. Elshout, et al. ReCiPe2016: a harmonised life cycle impact assessment method at midpoint and endpoint level. *The International Journal of Life Cycle Assessment*, 22(2):138–147, 2017.

[5] R. Heijungs. *Life Cycle Assessment: What It Is and How to Do It.* United Nations Environment Programme, 1996.

16 Ecosystem Services in Sustainability Assessment

> Does the educated citizen know he is only a cog in an ecological mechanism? That if he will work with that mechanism, his mental wealth and his material wealth can expand indefinitely? But if he refuses to work with it, it will ultimately grind them to dust. If education does not teach us these things, then what is education for?
>
> Aldo Leopold [1]

We learned in Chapter 3 that to sustain all human activities, the availability of ecosystem goods and services is essential. For their continued availability, human activities need to be within the carrying capacity of ecosystems that provide the necessary goods and services. All the methods we have learned about so far in Part III account for the impact of human activities on ecosystems. For example, the carbon footprint and life cycle analysis (LCA) account for emissions and their impact, while thermodynamic methods (energy, exergy, and emergy), the water footprint, and LCA account for the use of resources. Emergy also accounts for the work done by ecosystems to provide resources. However, none of these methods considers the capacity of ecosystems to absorb the emissions or provide the resources. In addition, these methods are best for comparing alternative products and processes, resulting in an assessment of "relative sustainability." They encourage the reduction of life cycle environmental impact, or doing "less bad," resulting in efforts toward continuous improvement. While this is certainly beneficial and should be done, what is also needed is to address the underlying reasons for unsustainability, such as exceeding nature's capacity to provide resources and mitigate emissions.

In this chapter we will learn about approaches that account for the demand and supply of ecosystem services, and see how the resulting methods can encourage environmental sustainability not just by comparing and choosing between alternatives, but also by comparing our demand for ecosystem services with the capacity of nature to supply these services. The resulting metrics provide information about "absolute sustainability" by quantifying the extent to which human activities exceed nature's capacity. Such approaches encourage technological improvements, as is done by conventional methods, but also encourage protection and restoration of ecosystems. Thus, these methods can encourage doing both "less bad" and "more good."

Table 16.1 Demand, supply, and serviceshed for ecosystem services [2].

Ecosystem service, k	Demand, $D_{i,j,k}$	Supply, $S_{i,j,k}$	Serviceshed scale, J
Climate regulation	CO_2 emission	Ecological sequestration capacity	Global
Water provisioning	Water withdrawal	Renewable water from rivers, aquifers, etc.	Watershed
Air quality regulation	Air pollutants	Cleaning capacity of trees, wind, etc.	Regional
Water quality regulation	Water pollutants	Cleaning capacity of aquatic bodies (wetland, river, lake)	Watershed
Abiotic materials provisioning	Abiotic materials use	Material production in nature	Mine

16.1   Synergies Between Human and Natural Systems

The approach for including ecosystems in sustainability assessment is motivated by the need for greater harmony between human and natural systems. As illustrated in Figure 1.10, human activities benefit from ecosystems since they absorb or mitigate emissions from human activities. Ecosystems also provide resources that support these activities. Ecosystems can benefit from human activities by obtaining protection and resources for their sustenance.

Emissions and resources use by human activities may be interpreted as a demand for ecosystem goods and services. For example, the emission of CO_2 indicates the quantity of climate regulation ecosystem service demanded by this activity, while the quantity of water withdrawn indicates the demand for the water provisioning service. Other examples are in Table 16.1. Also shown in this table is the supply of ecosystem services, or its capacity to provide goods and services. Sustainability assessment methods that we have learned about in previous chapters focus on reducing the demand for ecosystem services imposed by human activities. They ignore the supply of ecosystem services. Therefore, such methods may reduce the demand, while the supply may remain unchanged or even decrease. This is illustrated in Figure 16.1a. By explicitly including the role of ecosystems, as we will learn in this chapter, this approach can also encourage efforts toward increasing the availability of ecosystem services, as illustrated in Figure 16.1b.

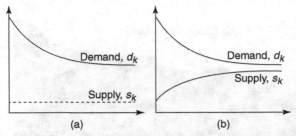

Figure 16.1 (a) Most methods focus on reducing the demand for ecosystem services, while ignoring the supply; (b) explicitly accounting for the role of ecosystems encourages a reduction in demand and an increase in the supply of ecosystem services.

This bears similarities to the approach of ecological footprints, which considers the demand imposed by human activities on planetary resources, and the availability of biological capacity to meet this need. It is summarized in Box 16.1.

BOX 16.1 Ecological footprint

The ecological footprint (EF) is a unique approach that quantifies the impact of human activities in terms of land area and compares it with the available land area, resulting in an indication of the degree to which human activities may be overshooting Earth's biocapacity [3].

Quantifying human impact as land area. The land use types that this approach considers are: (1) land for the provisioning of agricultural products of food and fiber; (2) grazing land and cropland for livestock products; (3) grounds for marine and inland fishery and aquaculture; (4) forestland for timber products; (5) land to neutralize emissions; and (6) land that is built up for shelter and other infrastructure. The only emissions considered in this approach currently are anthropogenic carbon dioxide emissions. These emissions are considered by calculating the land required to neutralize them.

The quantity of goods produced by these categories is converted into land area by the following approach. Given the production from the first four categories in physical units, it is converted to land units by dividing by the global yield of that product. For example, one ton of rice is converted to land area by dividing it by the world rice yield in tons/acre. Since the productivity of land all over the world for rice varies, this difference in land quality is accounted for by an equivalency factor. This factor keeps the footprint unchanged for average land quality, and magnifies it for land that has a higher yield than the average.

Quantifying biocapacity. In terms of land area, the biocapacity of our planet consists of only 22 percent of the total surface area. The reason is that only 18 percent of the land and 4 percent of the ocean are biologically productive. The rest of the surface is composed of low-productivity ocean (67 percent) and deserts, ice caps, and barren land (11 percent).

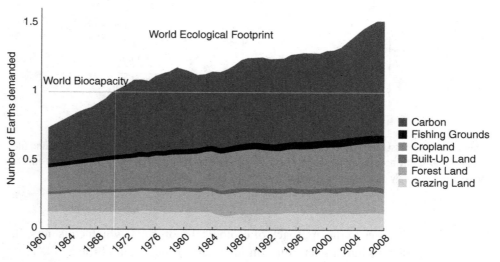

Figure 16.2 Change in global ecological footprint. Reproduced with permission from [4].

Insight from ecological footprint calculations. The EF method has been most popular for regional analysis, particularly of individual countries and the whole planet. At the planetary scale, the change in this footprint over the last half-century is shown in Figure 16.2. This figure shows that at the global scale, the EF exceeded the global biocapacity around 1970 and has continued to overshoot, mainly due to the increase in forest area needed to sequester carbon dioxide emissions.

Ecological footprint calculations for each country have been documented as well, and typical results based on these are presented in Figure 16.3. This figure plots the EF in global hectares per person versus the Human Development Index (HDI), and shows that a high HDI (>0.8) is possible with footprints from less than 2 global hectares per person to nearly 10 global hectares per person. In order to move up from a low HDI, increasing the EF seems essential, but once a high enough HDI is reached, the marginal increase in HDI for a marginal increase in EF need not be much. This observation is similar to the relationship between per capita energy use and HDI that is shown in Figure 12.1.

16.2   Ecosystem Services in Life Cycle Assessment

The approach for including ecosystem services in LCA is called techno-ecological synergy in LCA (TES-LCA) because it assesses the extent to which synergies exist between technological and ecological systems. The goal of TES-LCA is not just to account for the role of ecosystems, but also to evaluate and encourage harmony

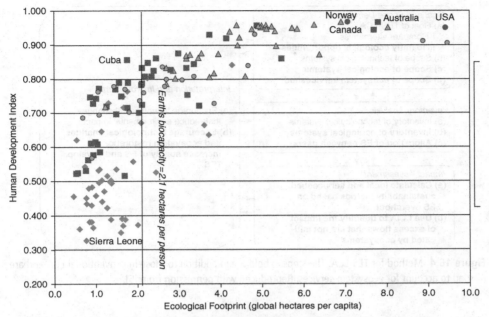

Figure 16.3 Ecological footprint versus Human Development Index for various countries.

between human and natural systems. In this chapter, we will focus on the use of TES for LCA, while in Chapter 20 we will learn about using the ideas of TES to design sustainable systems. The overall approach of TES-LCA is similar to the approach depicted in Figure 8.2 for other sustainability assessment methods, but with some additional steps that are shown in bold in Figure 16.4 and described in this section.

16.2.1   Goal and Scope Definition

As we learned in Chapter 8, the goal of sustainability assessment methods is to choose between alternatives to reduce their life cycle environmental impacts. TES-LCA has the additional goal of identifying ecological opportunities for enhancing the synergy between technology and ecosystems.

Determining the scope of conventional sustainability assessment involves defining the system boundary. We learned about the process, input–output, and hybrid models as common approaches for defining the boundary. In TES-LCA, in addition to defining this technological boundary, we also need to define a boundary of ecosystems at two scales: local and serviceshed.

- The local scale includes ecological systems in the immediate vicinity of the human activity in the life cycle. The boundary of the local ecosystem is chosen such that only the selected human activity benefits from the services provided

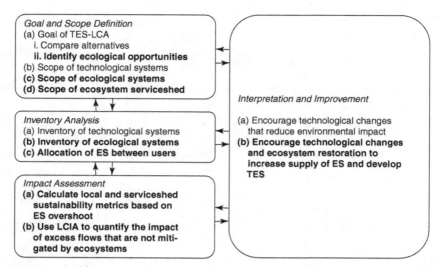

Figure 16.4 Method for TES-LCA. The steps in bold are in addition to those in conventional LCA and are needed to account for ecosystem services. Reproduced with permission from [5].

by the local ecosystem. Thus, the local ecosystem for a house can be its yard, for a manufacturing site it can be the ecosystems on this site, and for a city the ecosystems within the city limits.

• The serviceshed is the region from which ecosystem services are available to an activity. For example, the serviceshed for the carbon sequestration ecosystem service is the entire planet, since emitted greenhouse gases from an activity can travel and be sequestered anywhere in the world. For the water provisioning service, the serviceshed is the watershed since it is the region from which water is available to beneficiaries in the region.

Typical local- and serviceshed-scale ecosystems for manufacturing and farming processes are listed in Table 16.1 and depicted in Figure 16.5.

We learned in Chapter 9 that life cycle inventory data are usually averaged over a selected region. To use such LCI data in TES-LCA, if the LCI data are from multiple servicesheds, data about the supply of ecosystem services may be obtained as the average of servicesheds that correspond to the processes that constitute the LCI.

Example 16.1   Consider a cradle-to-gate life cycle of a biofuel that consists of two steps: farming and manufacturing, as shown in Figure 16.5. Define the goal and scope for this system to perform a TES-LCA with emphasis on the ecosystem services of water provisioning and climate regulation.

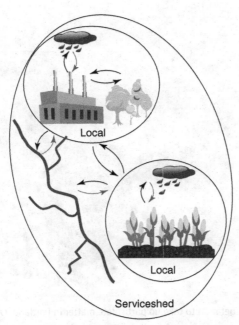

Figure 16.5 Illustration of ecosystems at local and serviceshed scales for the water provisioning ecosystem service.

Solution

Goals. The goals are to evaluate the extent to which the biofuel life cycle operates within nature's capacity, and to identify improvement opportunities by selecting technological and ecological options.

Scope. The scope of the technological system consists of the farm and manufacturing systems. The scope of the ecological system consists of the local ecosystems surrounding the two technological systems, as depicted in the figure, and the serviceshed for the selected ecosystem services. For water provisioning, the serviceshed is the watershed within which both processes are shown to lie, and for climate regulation the serviceshed is global.

16.2.2 Inventory Analysis

As we learned in previous chapters, sustainability assessment requires data about inputs and outputs of all human activities in the life cycle. As described in Section 16.1, this information can be interpreted as the demand for ecosystem services. In addition to this information, TES-LCA also requires data about the supply of ecosystem services at selected scales such as the locality and for the serviceshed. This information may be obtained from ecological models and from measured data from local or satellite measurements. For example, Figure 16.6 shows the supply of the air quality regulation ecosystem service in the city of Portland, Oregon

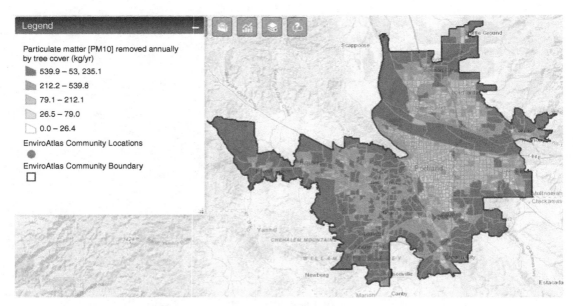

Figure 16.6 Capacity of vegetation to take up particulate matter in Portland, Oregon.

that is provided by vegetation. This service is the capacity of vegetation to uptake particulate matter of 10 µm size (PM10) and smaller.

Allocation of ecosystem services. Information about the supply of ecosystem services is needed at local and serviceshed scales. Data at the local scale may be considered to belong to the activity that is demanding it, particularly if the service is provided from land owned by the activity. However, the supply of ecosystem services from publicly owned land in the serviceshed belongs to all users in the serviceshed and needs to be allocated between them. As we learned in Chapter 9 about allocation between multiple products, the allocation of ecosystem services between multiple users is also a subjective exercise. The supply of ecosystem services may be partitioned on the basis of criteria such as quantity of the ecosystem service used by each activity, the monetary value of economic activity, the number of people employed or affected by the activity, etc. The selected allocation criterion is used to calculate the allocation weights using equations of the same form as Equations 9.1 and 9.2. As depicted in Figure 16.7, ecosystem service S_3 available from public land in the relevant serviceshed is partitioned into $w_{1,3}S_3$ and $w_{2,3}S_3$. If the partitioning is done in proportion to the demand of the ecosystem service, then for the ith activity and kth ecosystem service, the weighting factor may be determined as

$$w_{i,k} = \frac{d_{i,k}}{\sum_i d_{i,k}} \tag{16.1}$$

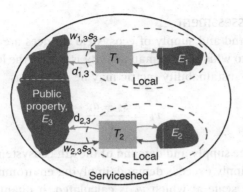

Figure 16.7 Allocation of ecosystem services in public property to activities in serviceshed that demand them.

If markets exist for the ecosystem service, then the quantity of ecosystem service purchased by each user is the allocated quantity. Thus, markets can play the role of optimally allocating ecosystem services between suppliers and users. We will learn more about markets for ecosystem services in Chapter 21.

Example 16.2 An urban area consists of three localities and a public biodiversity conservation area. The emission of NO_x in each locality is 5000 t/yr, 1000 t/yr, and 3000 t/yr. The areas of each locality are 2 km^2, 3 km^2, and 4 km^2, and their capacities to sequester NO_x are 2000 t/yr, 1200 t/yr, and 1500 t/yr. The NO_x capture capacity of the public area is 20,000 t/yr and its area is 2.5 km^2. Emissions from this area are negligible. Allocate the capacity of the public area to the three localities in proportion to their demand and land area.

Solution
The sequestration capacity of 20,000 t/yr may be allocated in proportion to the demand for the NO_x sequestration ecosystem service as follows:

$$s_{L1,a} = \frac{5000}{5000 + 1000 + 3000} \times 20,000$$
$$= 11111.1 \text{ t/yr}$$

$$s_{L2,a} = \frac{1000}{5000 + 1000 + 3000} \times 20,000$$
$$= 2222.2 \text{ t/yr}$$

$$s_{L3,a} = \frac{3000}{5000 + 1000 + 3000} \times 20,000$$
$$= 6666.7 \text{ t/yr}$$

Similarly, allocation based on land area results in the allocated supply for the three localities to be 4444.4 t/yr, 6666.7 t/yr, and 8888.9 t/yr.

16.2.3   Impact Assessment

Data about the demand and supply of ecosystem services are used in TES-LCA to calculate the extent to which the human activity is within the local and serviceshed scale capacities. The sustainability metric may be defined as

$$v_k = \frac{s_k - d_k}{d_k} \tag{16.2}$$

Here, s_k and d_k are the supply and demand of the kth ecosystem service. A value of $v_k \geq 0$ means that supply exceeds demand, implying environmental sustainability.

Depending on the scale at which v_k is calculated, it quantifies different types of environmental sustainability: local, serviceshed, and average. The local metric is for the process and its immediate surroundings; the serviceshed metric is defined as

$$v_k^* = \frac{s_k^* - d_k^*}{d_k^*} \tag{16.3}$$

where the * superscript denotes a serviceshed value. If

$$v_k^* \geq 0 \tag{16.4}$$

then we can claim "absolute environmental sustainability" since the activities in the serviceshed are within the carrying capacity.

Depending on the signs of the local and serviceshed metrics, the system may lie in one of the four quadrants shown in Figure 16.8. A system with both metrics in Quadrant I is sustainable at the local and serviceshed scales. This is the best-case scenario since the system is within ecological limits at both scales. Quadrant II implies sustainability in the serviceshed but not at the local scale. Such a situation could arise if a manufacturing process withdraws more water than is available within its location, but, in the overall watershed, the water withdrawal is within the capacity of natural processes. Quadrant III is the worst-case scenario since the system exceeds ecosystem capacity at both local and serviceshed scales. Quadrant IV represents an island of sustainability since the system is locally sustainable, but unsustainable in the serviceshed. Systems in Quadrants I and II satisfy the requirement of absolute sustainability given by Equation 16.4.

Depending on the demand and supply of ecosystem services, there may be emissions that cannot be taken up by ecosystems, or resources that exceed nature's ability to supply them in a renewable manner. These flows are likely to have a negative impact on the environment. In TES-LCA, the impact of these flows is quantified by the conventional life cycle impact assessment methods that we learned about in Chapter 15.

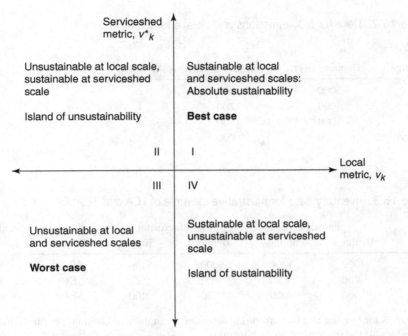

Figure 16.8 Local and serviceshed sustainability.

Example 16.3 Calculate the local and absolute sustainability metrics for NO_x sequestration in Example 16.2. You may assume the city to be the serviceshed for NO_x regulation.

Solution

The demand and supply of the air quality regulation ecosystem service (the part that takes up NO_x) are listed in Table 16.2. The local sustainability metrics may be calculated by Equation 16.2 as

$$v_{L1} = \frac{2000 - 5000}{5000}$$

$$= -0.6$$

Similarly, $V_{L2} = 0.2$ and $V_{L3} = -0.5$.

At the scale of the serviceshed, the total supply is $s_k^* = 24,700$ t/yr, and the total demand is $d_k^* = 9000$ t/yr:

$$v_k^* = \frac{24,700 - 9000}{9000}$$

$$= 1.74$$

Thus, this system is locally sustainable for locality 2 and is absolutely sustainable for the serviceshed. All metrics lie in Quadrant I or II. Localities 1 and 3 are "islands of unsustainability."

Table 16.2 Data for NO_x emissions and uptake for Example 16.3.

Locality	Demand (t/yr)	Supply (t/yr)	Area (km^2)
1	5000	2000	2
2	1000	1200	3
3	3000	1500	4
Reserve	0	20,000	2.5

Table 16.3 Inventory data for illustrative example of LCA and TES-LCA.

Flow	Farming		Manufacturing		Serviceshed	
	Demand	Supply	Demand	Supply	Demand	Supply
CO_2	1000	1300	5000	100	8E+12	4E+12
SO_2	200	20	700	50	3E+3	3.5E+3
H_2O	5000	7000	200	100	5E+6	8E+6

Note: Values for farming and manufacturing activities are in units of kilogram per functional unit, while values for the serviceshed are in kilograms.

16.2.4 Interpretation and Improvement

In this last step, TES-LCA focuses on technological and ecological changes that can be made to reduce ecological overshoot. Like conventional LCA, modifications to technological systems are considered that reduce emissions and resource use and their impact. However, as shown in Figure 16.4, TES-LCA also considers ways of continuing and enhancing goods and services from nature by restoring and protecting ecological systems.

The interpretation and improvement step considers alternatives for increasing v_k, which is possible by reducing demand (d_k) and increasing supply (s_k). Thus, in addition to technological modifications, TES-LCA also considers alternatives such as ecological protection and restoration. Insight about the sustainability of the serviceshed also directs attention toward improvements in this larger region by interaction with other human and ecological activities in the serviceshed.

Example 16.4 The boundary of a conventional LCA consists of a farming step and a manufacturing process. Data for the relevant technological and ecological systems are provided in Table 16.3. Perform a conventional LCA and a TES-LCA for this system.

Solution

Conventional LCA quantifies the environmental impact due to emissions and resource use, which are given in Table 16.3 under the "Demand" columns. The mid-point indicators are 6000 kg CO_2eq of global warming potential, 900 kg SO_2eq of acidification potential, and 5200 kg of water use. The last step, of interpretation and improvement, will consider ways of reducing the environmental impact, usually by considering modifications to technologies or human activities.

For *TES-LCA*, the LCA boundary needs to be extended by including local and serviceshed scale ecosystems that provide the selected ecosystem services. This boundary for a water provisioning ecosystem service is depicted in Figure 16.5. The local scale for the manufacturing can be its site or campus, and for farming could be the farm itself. The serviceshed for water provisioning is the watershed, as shown in the figure, while for carbon sequestration the serviceshed is global.

TES-LCA calculates the sustainability metric v_k by using Equation 16.2:

$$v_1 = \frac{(1300 + 100) - (1000 + 5000)}{1000 + 5000}$$

$$= -0.77$$

Similarly, $v_2 = -0.92$, $v_3 = 0.36$. Thus, in terms of local environmental sustainability, this life cycle is locally sustainable only for water and not for carbon sequestration and air quality regulation.

Applying *impact assessment* to emissions that exceed nature's capacity results in the following mid-point indicators: 4600 kg CO_2eq, 830 kg SO_2eq, and -1900 kg H_2O use. A negative impact means that the system provides resources or sequestration capacity. Thus, this system makes 1900 kg H_2O available to the serviceshed.

The last two columns represent the supply and demand for the servicesheds of each ecosystem service. Using these data, the absolute sustainability metrics may be calculated to be $v_1^* = -0.5$, $v_2^* = 0.14$, $v_3^* = 0.6$. A plot of the local and absolute sustainability metrics for this example is shown in Figure 16.9. Comparing this with Figure 16.8 conveys that the water provisioning ecosystem service shows absolute sustainability, air quality regulation is an island of unsustainability, and carbon sequestration is unsustainable at local and serviceshed scales.

16.3 Computational Structure

We learned about the computational structure of sustainability assessment methods in Chapter 10. In this section, we will extend this framework to include ecosystems and enable TES-LCA [6]. Let us consider the general technological and ecological systems depicted in Figure 16.10. Here, a_{ij} represents the flow of the jth economic good or service to or from process i. Similarly, d_{ij} is the interaction between the technological and ecological system and s_{ij} is the capacity of the

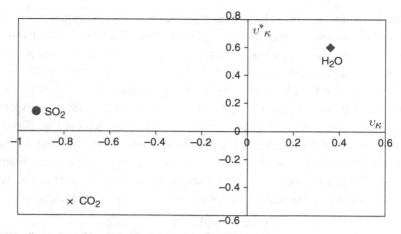

Figure 16.9 Illustration of local versus absolute sustainability.

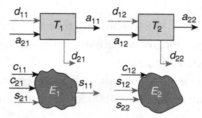

Figure 16.10 Technological and ecological systems.

jth ecosystem to provide the ith flow. Note the change in notation as compared to Chapter 10. Balance equations analogous to Equation 10.4 for economic flows are

$$a_{11}m_1 - a_{12}m_2 = f_1 \tag{16.5}$$

$$-a_{21}m_1 + a_{22}m_2 = f_2 \tag{16.6}$$

Here, m_j are the scaling factors or multipliers, and f_i are the final demands. In matrix notation, we can write

$$Am = f \tag{16.7}$$

where $A = \begin{bmatrix} a_{11} & -a_{12} \\ -a_{21} & a_{22} \end{bmatrix}$ is the technology matrix, $m = \begin{bmatrix} m_1 \\ m_2 \end{bmatrix}$ is the scaling factor, and $f = \begin{bmatrix} f_1 \\ f_2 \end{bmatrix}$ is the final demand. The environmental flows are represented as

$$Dm = r \tag{16.8}$$

where $D = \begin{bmatrix} -d_{11} & -d_{12} \\ d_{21} & d_{22} \end{bmatrix}$ is the environmental intervention matrix. Note that this is also the demand matrix of the ecosystem services used by the technological systems.

To expand this framework for including ecosystems, we will consider ecosystems in a manner analogous to technological systems. For the technological and ecological systems in Figure 16.10, we can write

$$a_{11}m_1 - a_{12}m_2 - c_{11}m_{1e} - c_{12}m_{2e} = f_1 \tag{16.9}$$

$$-a_{21}m_1 + a_{22}m_2 - c_{21}m_{1e} = f_2 \tag{16.10}$$

Here, c_{ij} represents the flow of an economic quantity from the ith technological system to the jth ecological system. These coefficients represent the economic flows needed to manage the ecosystem. For example, a lawn will require inputs such as fertilizers.

Similar balance equations may also be written for ecological flows, as shown below for the systems in Figure 16.10:

$$-d_{11}m_1 - d_{12}m_2 + s_{11}m_{1e} - s_{12}m_{2e} = f_{1e} \tag{16.11}$$

$$d_{21}m_1 + d_{22}m_2 - s_{21}m_{1e} - s_{22}m_{2e} = f_{2e} \tag{16.12}$$

Here, m_{ie} is the scaling factor for the ith ecosystem and f_{ie} is the final demand from ecosystems or the net flow to society of the ith flow. The previous four equations may be written in matrix form as

$$\left[\begin{array}{c|c} A & C \\ \hline D & S \end{array} \right] \left[\begin{array}{c} m \\ m_e \end{array} \right] = \left[\begin{array}{c} f \\ f_e \end{array} \right] \tag{16.13}$$

with $D = \begin{bmatrix} -d_{11} & -d_{12} \\ d_{21} & d_{22} \end{bmatrix}$ and $S = \begin{bmatrix} s_{11} & -s_{12} \\ -s_{21} & -s_{22} \end{bmatrix}$. Here, A is the technology matrix, C is the environmental management matrix, D is the environmental intervention matrix, and S is the ecosystem matrix. Matrices A and C represent economic flows, while D and S are ecosystem flows. Equation 16.13 may be written as

$$A_{te}m_{te} = f_{te} \tag{16.14}$$

where A_{te} is the techno-ecology matrix. The matrix f_e may be multiplied by the characterization matrix Φ for assessing the impact of the emissions:

$$h_e = \Phi f_e \tag{16.15}$$

Thus, TES-LCA performs conventional life cycle impact assessment on the net emissions or resource use that exceeds the capacity of the ecosystems under consideration.

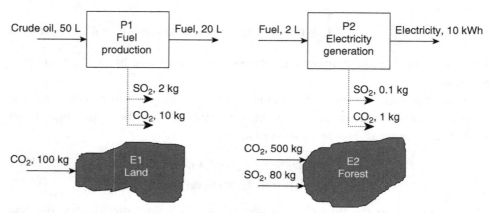

Figure 16.11 Technological modules and surrounding ecosystems for Example 16.5.

Example 16.5   Figure 16.11 shows the flows in the life cycles of fuel production and power generation, and the surrounding ecosystems. The fuel production facility owns land whose soil sequesters 100 kg of CO_2. The power plant for generating electricity is on a forested property that can sequester 500 kg of CO_2 and 80 kg of SO_2. Write the balance equations to determine the technology and techno-ecological matrices for this system.

Solution

The balance equations for the technologies are

$$20m_1 - 2m_2 = f_1$$
$$10m_2 = f_2$$

So the technology matrix is $A = \begin{bmatrix} 20 & -2 \\ 0 & 10 \end{bmatrix}$.

The balance equations for the ecosystem flows are

$$10m_1 + m_2 - 100m_{1e} - 500m_{2e} = f_{1e}$$
$$2m_1 + 0.1m_2 - 80m_{2e} = f_{2e}$$

The demand and supply matrices can be seen from these equations to be $D = \begin{bmatrix} 10 & 1 \\ 2 & 0.1 \end{bmatrix}$ and $S = \begin{bmatrix} -100 & -500 \\ 0 & -80 \end{bmatrix}$. Since there is no economic input to the ecosystems, the management matrix C is zero. Thus, the techno-ecological matrix is

$$A_{te} = \left[\begin{array}{cc|cc} 20 & -2 & 0 & 0 \\ 0 & 10 & 0 & 0 \\ \hline 10 & 1 & -100 & -500 \\ 2 & 0 & 0 & -80 \end{array} \right]$$

Example 16.6 If the total production of electricity is 1000 kWh and the total ecosystem is as specified in Example 16.5, determine the TES sustainability metrics for both sites.

Solution

The final demand for electricity is $f_2 = 1000$, and that for fuel is $f_1 = 0$. Since the total ecosystem capacity is provided, it means that the ecosystem scaling factors are equal to unity. Thus, $m_{1e} = m_{2e} = 1$. Substituting these values in the equations developed in Example 16.5, we get

$$20m_1 - 2m_2 = 0$$
$$10m_2 = 1000$$
$$10m_1 + m_2 - 100 - 500 = f_{1e}$$
$$2m_1 + 0.1m_2 - 80 = f_{2e}$$

Solving these four equations for the four unknowns results in $m_1 = 10$, $m_2 = 100$, $f_{1e} = -400$, $f_{2e} = -50$. Thus, the net emissions of CO_2 and SO_2 are -400 kg and -50 kg, respectively. The negative signs indicate that these are being sequestered. Knowing that the demand $d_1 = 10(10) + 100 = 200$ kg for CO_2 and the demand $d_2 = 2(10) + 0.1(100) = 30$ kg for SO_2, the sustainability metrics are

$$v_1 = \frac{600 - 200}{200}$$
$$= 2$$
$$v_2 = \frac{80 - 30}{30}$$
$$= 1.67$$

Thus, this system is environmentally sustainable for the provided flows.

16.4 Satisfying the Requirements for Sustainability

In this section we will consider the various sustainability assessment methods that we have learned about in terms of their ability to satisfy the requirements for sustainability described in Section 3.4 and Box 3.3. The methods we have covered include carbon and water footprinting, energy analysis, cumulative exergy and emergy analysis, LCA, and TES-LCA. All of these methods are similar in their consideration of spatial scales, temporal interactions, and cross-disciplinary effects. Methods for defining the network and the network analysis that we covered in

Chapters 8 and 10 may be used for each method, resulting in their ability to account for spatial scales from individual processes to their life cycles and national or global economies. None of the methods considers temporal interactions, owing to their static nature, and they are limited in their consideration of effects across disciplines such as economics. Only TES-LCA explicitly accounts for the role of ecosystems. In terms of multiple stocks and flows, LCA is the most advanced since life cycle impact assessment methods account for resource flows and emissions. TES-LCA inherits these benefits of LCA since it uses the same impact assessment approach for flows that exceed nature's capacity. Footprint methods only allow one type of flow: greenhouse gases for the carbon footprint and water for the water footprint. Exergy-based methods take into consideration diverse material and resource flows, but not the flow of emissions and their impact. In terms of ecosystem goods and services, all methods except TES-LCA account only for the demand of ecosystem goods such as water, minerals, and fossils. Emergy analysis also accounts for many ecosystem services by including the work done in nature for making resources available. Accounting for the supply of ecosystem services is not a strength in any of these methods except TES-LCA. Applying TES-LCA is more challenging than other methods since obtaining data about ecosystems involves additional effort.

16.5   Summary

One of the requirements for sustainability assessment methods is to take into account the demand and supply of ecosystem goods and services. In this chapter, we learned about methods that satisfy this requirement while considering the life cycle. The approach of TES-LCA modifies all the steps of conventional LCA to include ecosystems in the analysis boundary and to calculate local and absolute sustainability metrics based on the gap between the demand and supply of selected ecosystem services. This provides information about the extent to which the selected activity exceeds nature's capacity and helps identify opportunities for improving technological and ecological systems.

Key Ideas and Concepts

- Techno-ecological synergy (TES)
- Supply of ecosystem services
- Absolute sustainability
- Demand of ecosystem services
- Serviceshed
- Allocation of ecosystem services

16.6 Review Questions

1. How is the goal and scope step in TES-LCA different from conventional LCA?
2. Define the TES-LCA absolute sustainability metric.
3. What is an island of sustainability?
4. Give an example of how TES-LCA can encourage ecosystem restoration.
5. What is a serviceshed?

Problems

16.1 Describe how you may determine the serviceshed of the following ecosystem services: insect pollination, flood regulation, crude oil provisioning.

16.2 A watershed has a land area of 10,000 m² and receives annual precipitation of 100 m³. Two towns occupy areas of 1000 m² and 800 m². Farmland occupies 6000 m², and the remaining area is publicly owned. Assuming the precipitation to be uniformly distributed over the watershed, calculate the direct supply of water in each of these regions. If 50 percent of the precipitation on the publicly owned land is available for human use, allocate this water in proportion to land area between the towns and farmers.

16.3 The information in Table 16.4 is for three companies that share the same serviceshed for air quality regulation and water provisioning ecosystem services. Within this serviceshed also lies 1500 acres of public land. Assume that all the land in the serviceshed has a carrying capacity of 1.5 L of H_2O/acre for water provisioning and 0.15 kg of SO_2/acre and 0.1 kg of NO_x/acre for air quality regulation services.

Table 16.4 Data for Problem 16.3.

Company	A	B	C
Water use (L)	2000	3250	1800
SO_2 emissions (kg)	200	300	120
NO_x emissions (kg)	250	225	100
Land owned that can provide ecosystem services (acres)	800	2000	1000
Monetary value of products produced ($)	1,000,000	2,000,000	780,000

Table 16.5 Inventory for partial life cycle of corn ethanol with till and no-till farming.

	Farming		Biofuel		N fertilizer		P fertilizer		Diesel	
	d_k	s_k	d_k	s_k	d_k	s_k	d_k	s_k	d_k	s_k
					With tillage					
CO_2	37.4	93.8	11.8	0.003	2.5	0.19	0.27	5.2E–05	0.08	3.6E–05
H_2O	69.4	70.9	0.07	0.05	0.02	0.03	0.005	0.002	0.001	0.001
					Without tillage					
CO_2	41.8	111.7	11.8	0.003	2.9	0.22	0.24	4.55E–05	0.04	1.99E–05
H_2O	70.5	72.2	0.07	0.05	0.02	0.03	0.004	0.002	7.7E–04	7.2E–04

1. Calculate the sustainability metric of all activities in the serviceshed without allocation. This approach is equivalent to calculating metrics at the scale of the serviceshed.
2. Now calculate sustainability metrics for each company in the serviceshed using monetary allocation.
3. Repeat the previous part with allocation in proportion to the demand for ecosystem services.
4. Discuss the effect of these allocation approaches on the sustainability of each company.

16.4 A partial life cycle of corn ethanol accounts for the farming, ethanol production, nitrogen and phosphorus fertilizer manufacturing, and diesel production steps. The local demand and supply of CO_2 and water for each step are given in Table 16.5. Two types of farming are considered: with tillage of the soil and without tillage. Calculate the local sustainability metrics for each activity, and the global warming potential of the net CO_2 emissions. Compare the improvement opportunities that would be identified by conventional LCA versus TES-LCA.

16.5 Consider the systems shown in Figure 16.12. Answer the following questions. (Developed by Henry Cooper.)
1. What are the intervention, technology, ecology, and management matrices for this system?
2. The total final demand for this system is 100 kWh of electricity. Develop the $A_{te}m_{te} = f_{te}$ matrix for this situation.
3. Knowing that the ecosystems represent all the land available for each process, calculate the sustainability metrics for H_2O and CO_2. Are the facilities locally sustainable?
4. Suggest possible interventions that could be made to improve the sustainability of these facilities.

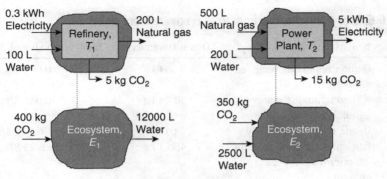

Figure 16.12 Technological and ecological systems for Problem 16.5.

16.6 The NREL life cycle inventory database contains information about the manufacture of phosphate fertilizer, which quantifies the demand for the air quality regulation ecosystem service in the form of SO_2 emissions. To produce 1 kg of phosphate fertilizer, 5.53×10^{-3} kg SO_2 is emitted. However, information about the supply of this ecosystem service is not available in this inventory, which may be specific to the phosphate fertilizer manufacturing site. From the National Emissions Inventory database, a phosphate manufacturing facility is found to be located in Illinois. Using the current forest cover recorded in the National Land Cover Database, the capacity of the site for taking up emissions is 2.51×10^{-3} kg SO_2/kg fertilizer. Calculate the TES sustainability metric v_k for the ecosystem service of mitigating SO_2 emissions at this site. If land is restored, the capacity of the site for taking up emissions is increased to 5.44×10^{-3} kg SO_2/kg fertilizer; what will be the value of v_k?

16.7 A small water catchment has three activities nested within it, each requiring water for their operations. The water demands from the three activities are 100 m³/year, 150 m³/year, and 250 m³/year. The areas of their sites are 30 hectares, 20 hectares, and 50 hectares, respectively. The water supply available in the small catchment is 1000 m³/year. Allocate the capacity of the catchment to the three activities in proportion to their demand and land area.

16.8 A residential system uses 61.64 GJ of electricity per year for cooling during summer, in the absence of any trees around the house. If shading trees were present, they would reduce electricity consumption by decreasing the cooling load to 60.10 GJ. The price of electricity is 8 cents/kWh. The initial cost of planting the trees is $200. How many years will it take for the cost and benefits of trees to break even? You may use discount factors of 0 percent and 3 percent.

Table 16.6 Ecosystem services from trees on the refinery site.

Abbr.	Benefit description	Quantity removed	Monetary value of removal
CO	Carbon monoxide removed annually	21.04 kg	0.084 $/kg
NO_2	Nitrogen dioxide removed annually	98.24 kg	0.033 $/kg
O_3	Ozone removed annually	1.18 t	0.033 $/t
PM2.5	Particulate matter less than 2.5 microns removed annually	45.13 kg	6.25 $/kg
SO_2	Sulfur dioxide removed annually	64.69 kg	0.0066 $/kg
PM10	Particulate matter less than 10 microns removed annually	334.25 kg	0.375 $/kg
CO_2seq	Carbon dioxide sequestered annually in trees	273.33 t	39 $/t
CO_2stor	Carbon dioxide stored in trees (note that this is not an annual rate)	5.71 kt	39 $/t

16.9 Consider rice production activities in three farms located in China, India, and the USA. The runoff of nitrate fertilizer requires the ecosystem service of water quality regulation. The nitrate concentrations in the runoff from the three farms are 14.35, 1.13, and 4.42 mg/L, respectively, and the hydraulic loading rates are 11.41, 8.68, and 7.61 m^3/day, respectively. Consider a 50 m^2 wetland that is built to treat each runoff, which can be modeled as a steady-state first-order system using the following equations:

$$\frac{C_o}{C_i} = e^{-K_T t}$$

$$K_T = 1.15^{T-20}$$

$$t = \frac{A\epsilon h}{Q}$$

Here, C_o and C_i are the concentrations of nutrients in the outflow and inflow, respectively; K_T is the rate constant with the unit of day^{-1}, measured at T, which represents the water surface temperature; T can be assumed to be 20 °C. A (in m^2) is the area of wetland; ϵ is the bed porosity (assumed to be 0.42); h (in m) is the wetland bed depth (assumed to be 0.6 m); Q (in m^3/day) is the hydraulic loading rate. What are the net nitrate outflow concentrations from each farm? (Developed by Xinyu Liu)

16.10 A land use study at a refinery in Texas finds that 10 percent of the land under consideration is covered by trees, 20 percent has no trees, 34 percent is roads, 7 percent is impervious buildings, and 29 percent is water. The benefits provided by trees in physical and monetary units are shown in Table 16.6. Determine the total monetary value provided by the trees with the current tree cover, and also if the non-tree area was covered by identical trees. Do these monetary benefits contribute directly to the refinery's financial bottom line?

References

[1] A. Leopold. *A Sand County Almanac*. Oxford University Press, 1949.

[2] B. R. Bakshi, G. Ziv, and M. D. Lepech. Techno-ecological synergy: a framework for sustainable engineering. *Environmental Science & Technology*, 49(3):1752–1760, 2015.

[3] M. Wackernagel and W. E. Rees. *Our Ecological Footprint: Reducing Human Impact on the Earth*. New Society Publishers, 1996.

[4] M. Borucke, D .Moore, G. Cranston, et al. Accounting for demand and supply of the biosphere's regenerative capacity: the National Footprint Accounts' underlying methodology and framework. *Ecological Indicators*, 24:518–533, 2013.

[5] X. Liu and B. R. Bakshi. Ecosystem services in life cycle assessment while encouraging techno-ecological synergies. *Journal of Industrial Ecology*, 2018. DOI: 10.1111/jiec.12755.

[6] X. Liu, G. Ziv, and B. R. Bakshi. Ecosystem Services in Life Cycle Assessment, Part 1: A Computational Framework. *Journal of Cleaner Production*, 197:314–322, 2018.

PART IV
Solutions for Sustainability

••

Part IV focuses on solutions toward sustainable development. As we learned in Part II, virtually all disciplines contribute to the unsustainability of human activities. It should not be surprising that solutions also stem from multiple disciplines. Our focus will be on solutions that may be designed by engineers, but we will think of "design" in a broad sense. It will include the design of technological solutions, ecosystems, policies, and human behavior. We will also consider solutions that stem from economics and societal transformation.

17 Designing Sustainable Processes and Products

> The last word in ignorance is the man who says of an animal or
> plant: "What good is it?" If the land mechanism as a whole is good,
> then every part is good, whether we understand it or not. If the biota,
> in the course of eons, has built something we like but do not
> understand, then who but a fool would discard seemingly useless
> parts? To keep every cog and wheel is the first precaution of
> intelligent tinkering.
>
> Aldo Leopold [1]

The sustainability assessment methods described in previous chapters are useful for comparing alternatives based on their environmental impact. For choosing among alternatives, their footprint or life cycle environmental impact is usually one among multiple objectives. Rarely is an option selected just because it has the smallest environmental impact. Other factors such as economic feasibility and societal preference are also important and need to be considered in the decision. We learned about these and other requirements for sustainability in Section 3.4.

In this chapter we will learn about approaches that utilize environmental sustainability assessment methods, along with methods for assessing economic feasibility, in order to choose among alternatives. Such approaches are used for incorporating sustainability considerations in tasks such as engineering design, corporate strategy, government policy, and consumer choices. Our emphasis will be on the design of manufacturing processes and products.

We will first learn about methods for techno-economic analysis (TEA), in which the goal is mainly to assess and choose between technological options on the basis of conventional economic analysis. We will then focus on methods that combine such economic analysis with environmental sustainability assessment. These methods include eco-efficiency, in which metrics combine the quantification of environmental impact with measures of economic feasibility. We will also learn about how engineering design has evolved over the decades toward its current focus on sustainable development. This relies on formulating sustainable product and process design problems as multi-objective optimization in order to understand the trade-off between these objectives. Heuristic principles or rules of thumb have also been suggested for environmentally friendlier decisions in chemistry and engineering.

Table 17.1 Components of total capital investment. Adapted from [3].

1. Direct costs (Δ)
    1. Equipment etc.
        1. Purchased equipment
        2. Installation
        3. Instrumentation and controls
        4. Piping
        5. Electrical
    2. Buildings, process, and auxiliary
    3. Service facilities and yard improvements
    4. Land
2. Indirect costs
    1. Engineering and supervision
    2. Legal expenses
    3. Construction expense and contractor's fee
    4. Contingency
3. Fixed-capital investment = direct costs + indirect costs
4. Working capital
5. Total capital investment = fixed-capital investment + working capital

17.1   Techno-Economic Analysis and Design

Techno-economic analysis estimates the economic feasibility of technological alternatives. This popular and practical approach has been used for many decades for assessing economic feasibility and designing manufacturing processes and products. Our focus will be on common methods for engineering economic or cost analysis, followed by examples to illustrate their use. Such knowledge is essential for quantifying the economic bottom line, which is an essential component of sustainability.

17.1.1   Costs and Earnings

The cost of a manufacturing process consists of capital and operating expenses. The capital cost consists of equipment, buildings, construction, land, royalties, and other items that are required before the process can start operating. Among these, some items deteriorate or wear out over time, and are considered to depreciate. Others, like land and royalties, do not depreciate. Components of the total capital investment (T) are listed in Table 17.1.

> **Table 17.2** Constituents of total production cost and gross profit. Adapted from [4].
>
> 1. Manufacturing cost
>   1. Direct production costs
>   1. Raw materials
>   2. Utilities
>   3. Maintenance and repairs
>   4. Operating supplies
>   5. Operating labor
>   6. Direct supervision and clerical labor
>   7. Laboratory charges
>   8. Patents and royalties
>   2. Fixed charges
>   1. Depreciation
>   2. Local taxes
>   3. Insurance
>   4. Rent
>   5. Interest
>   3. Plant overhead
> 2. General expenses
>   1. Administrative costs
>   2. Distribution and selling costs
>   3. Research and development costs
> 3. Total product cost = manufacturing cost + general expenses
> 4. Gross profit = total income − total product cost

Operating costs are incurred by raw materials, utilities, labor, depreciation, maintenance, taxes, and other activities that take place throughout the life of the manufacturing process. The depreciation (D) is commonly calculated by a straight-line method that simply divides the capital cost of the depreciating assets by their life. Typical components in determining the total production cost (C) are given in Table 17.2.

Revenue (R) is earned from selling products in the marketplace. Additional revenue may be earned from credits such as those due to carbon trading or selling waste heat or byproducts. We will learn more about these options in Chapters 19 and 21. At the end of life, the equipment, buildings, and other capital may have some salvage value. Many methods have been developed for determining these costs at various stages of design. These methods rely on results of the engineering design, if they are available. At the early stages, approximate cost estimation methods for specific types of engineering systems are also available and are commonly used. These methods are described in detail in many textbooks on engineering design.

Profitability metrics are often used to determine economic feasibility. The gross profit is the difference between the revenue and the costs before depreciation:

$$P_g = R - C \tag{17.1}$$

The profit after taxes, or net profit, is

$$P_n = (1 - \phi)(P_g - D) \tag{17.2}$$

where ϕ is the tax rate and D is the depreciation allowance, which may be subtracted before taxes.

Example 17.1   The total capital investment needed for a new process is $20 million. Out of this, $8 million depreciates at the rate of 2 percent per year. The expected revenue per year is $1 million and the total cost before depreciation is $800,000. Calculate the gross and net profit for this process. The tax rate is 20 percent.

Solution

The gross profit is the difference between revenue and costs:

$$P_g = 1 \times 10^6 - 8 \times 10^5$$
$$= 2 \times 10^5 \ \$/yr$$

The net profit is

$$P_n = (1 - 0.2)(2 \times 10^5 - (0.02 \times 8 \times 10^6))$$
$$= 32{,}000 \ \$/yr$$

17.1.2   Time Value of Money

Combining costs and earnings incurred over different time points requires consideration of the time value of money. As we learned in Chapter 4, the value of money changes with time since people generally value the present more than the future. This difference is captured by means of the discount factor or interest rate. For a present value of P and an annual interest rate r, the future value after n years with simple interest is calculated as

$$F = P(1 + rn) \tag{17.3}$$

With compound interest, the future value is calculated as

$$F = P(1 + r)^n \tag{17.4}$$

If the interest is compounded at a rate other than annual, the future value is

$$F = P\left(1 + \frac{r}{m}\right)^{mn} \tag{17.5}$$

where m is the number of periods per year. Thus, the interest is compounded m times per year.

Example 17.2 What is the value after five years of $100 invested in an account that pays (a) 10 percent simple interest, (b) 10 percent compound interest, and (c) 10 percent compound interest every six months?

Solution

(a) For 10 percent simple interest, using Equation 17.3,

$$F = 100(1 + (0.1 \times 5))$$
$$= \$150$$

(b) For 10 percent compound interest, using Equation 17.4,

$$F = 100(1 + 0.1)^5$$
$$= \$161$$

(c) For 10 percent compound interest with two periods per year due to six-monthly compounding, using Equation 17.5 we obtain,

$$F = 100\left(1 + \frac{0.1}{2}\right)^{2 \times 5}$$
$$= \$163$$

The present value for a situation in which an operating cost of A is incurred at the end of each year for n years is

$$P = \frac{A}{(1+r)} + \frac{A}{(1+r)^2} + \cdots + \frac{A}{(1+r)^n}$$

$$= A \sum_{i=1}^{n} \frac{1}{(1+r)^i}$$

$$= \frac{1 - (1+r)^{-n}}{r} A \tag{17.6}$$

Notice that we used this equation in Example 4.2. If payments are made in m periods per year,

$$P = \frac{1 - (1 + (r/m))^{-mn}}{(r/m)} A \tag{17.7}$$

Example 17.3 A particular car costs $20,000. Fuel, insurance, and maintenance cost $1800 per year. Assuming that the operating costs are paid at the beginning of each year, what is the total cost of the car over a five-year period in terms of present dollars? You may consider the annual interest rate to be 8 percent.

Solution

Here, the capital cost is $20,000, while the operating cost is $1800 per year. Using Equation 17.6, the total cost over five years in terms of its present value is

$$P = 20,000 + \left(\frac{1 - 1.08^{-5}}{0.08}\right) 1800$$

$$= \$27,187$$

Example 17.4 If the payment in the previous example changes to $150 per month, what will be the present value of the expense?

Solution

Now there will be $m = 12$ periods per year. Therefore, using Equation 17.7,

$$P = 20,000 + \left(\frac{1 - (1 + (0.08/12))^{-5 \times 12}}{(0.08/12)}\right) 150$$

$$= \$27,398$$

17.1.3 Profitability Metrics

Determining and comparing the economic feasibility of alternatives requires metrics that utilize information about costs and revenue streams. Two categories of such metrics are based on whether they account for the time value of money. The first two metrics defined below do not take into account the time value of money, while the next two do.

Return on investment (ROI). This metric is the ratio of profit to total capital investment. Usually, the annual net profit or profit after tax is used in this calculation:

$$ROI = \frac{P_n}{T} \tag{17.8}$$

This metric indicates the annual rate of monetary return that may be obtained from investing the capital in the design. If profit or investment vary over time, average values are used to calculate the ROI. Corporations usually have a minimum annual ROI that is required for a project to be considered monetarily feasible. It is interesting to note the similarities between this monetary ROI and the energy ROI defined in Equation 12.4.

Payback period (PBP). This is the time required for the investment to pay off. It is calculated as the ratio of the direct capital investment (item 1 of the total capital investment in Table 17.1) to the annual cash flow, which consists of the net earnings and depreciation:

$$PBP = \frac{\Delta}{P_n + D} \tag{17.9}$$

Net present value (NPV). This is the present value of all the cashflows. In general, for an initial investment of C_0 and a recurring earning of C_i every year, the NPV in the nth year is

$$\text{NPV} = \sum_{i=1}^{n} \frac{C_i}{(1+r)^i} - C_0 \tag{17.10}$$

In typical design problems, C_0 is the total capital investment and C_i is the cashflow in the ith year. Feasibility requires a positive NPV.

Discounted cashflow rate of return (DCFR). This quantity is the interest rate i that results in a zero NPV. That is,

$$0 = \sum_{i=1}^{n} \frac{C_i}{(1+r)^i} - C_0 \tag{17.11}$$

This equation usually needs to be solved by trial and error to find the DCFR. For the project to be feasible, the DCFR should be greater than the minimum rate of return that is acceptable to the investor.

Example 17.5 A capital investment of one million dollars yields a steady earnings stream of $50,000 every year as net profit. Calculate the return on investment, and the payback period, NPV, and DCFR. You may use an interest rate of 8 percent, and assume depreciation to be 10 percent per year.

Solution
The ROI is given by

$$\text{ROI} = \frac{5 \times 10^4}{1 \times 10^6}$$
$$= 0.05 \text{ or } 5 \text{ percent}$$

For calculating the PBP we consider the capital investment to be the same as direct capital. The depreciation per year is 10 percent of one million, which is $100,000. Thus,

$$\text{PBP} = \frac{1 \times 10^6}{5 \times 10^4 + 1 \times 10^5}$$
$$= 6.67 \text{ years}$$

The NPV is calculated as follows. The period is ten years since with a depreciation of 10 percent the value depreciates to zero in ten years.

$$\text{NPV} = \sum_{i=1}^{10} \frac{5 \times 10^4}{1.08^i} - 1 \times 10^6$$
$$= \frac{1 - 1.08^{-10}}{0.08} \times 5 \times 10^4 - 1 \times 10^6$$
$$= -6.64 \times 10^5$$

The DCFR is calculated by solving the following equation for r:

$$\sum_{i=1}^{10} \frac{5 \times 10^4}{(1+r)^i} - 1 \times 10^6 = 0$$

This equation becomes

$$\frac{1-(1+r)^{-10}}{r} \times 5 \times 10^4 - 1 \times 10^6 = 0$$

By trial and error, we find that no value of r satisfies the equation.

In this section, we have learned about methods to analyze and design engineering systems based on their economic feasibility. However, as we know, sustainability requires the consideration of environmental and societal aspects that are outside the market and therefore are not captured by monetary values. The rest of this chapter will focus on addressing these multiple goals of sustainable engineering.

17.2   Eco-Efficiency

This approach, popularized by the World Business Council for Sustainable Development (WBCSD), is a way of reducing environmental impact while increasing economic value to a company [3]. It is meant to find "win–win" solutions in which environmental improvements also result in economic benefits, and has become popular among many corporations as an approach for moving their activities and business model toward sustainability. The WBCSD defines eco-efficiency as follows.

> Eco-efficiency is achieved by the delivery of competitively-priced goods and services that satisfy human needs and bring quality of life, while progressively reducing ecological impacts and resource intensity throughout the life-cycle to a level at least in line with the Earth's carrying capacity.

This definition goes on to say, "in short, it [eco-efficiency] is concerned with creating more value with less impact." Notice that this definition focuses on economic, environmental, and societal aspects, the demand and supply of ecosystem services, and the life cycle impact. Thus, it satisfies many of the requirements of sustainability that we covered in Section 3.4.

Seven elements have been identified for businesses to improve their eco-efficiency, and are listed in Table 17.3. These elements aim toward the following three main objectives:

Table 17.3 Seven elements that can be used to improve eco-efficiency [3].

No.	Item
1	Reduce material intensity
2	Reduce energy intensity
3	Reduce dispersion of toxic substances
4	Enhance recyclability
5	Maximize use of renewables
6	Extend product durability
7	Increase service intensity

1. *Reduce consumption of resources*. This focuses on reducing the use of resources such as materials, energy, water, and land, which could be achieved by enhancing recyclability and closing material loops.
2. *Reduce environmental impact*. This involves reducing emissions of pollutants and their impact.
3. *Increase product or service value*. This implies providing more benefits to consumers through increasing product functionality, flexibility, and modularity. It also encourages different business models such as those based on providing a service, like computing or transportation, instead of a product like a computer or car.

Efforts to become eco-efficient are meant to encourage companies toward four types of changes:

1. *Re-engineer their processes* by decreasing their resource intensity and pollution intensity. It would make the most sense to look for such opportunities in a company's own operations first, followed by focusing on their suppliers.
2. *Revalorize their byproducts* by efforts toward a zero-waste or 100 percent product. This would encourage finding markets or uses for a company's waste products, or modifying their processes to eliminate certain wastes. Such efforts often reduce resource use and pollution, while saving money.
3. *Redesign their products* to use more renewables, enhance recyclability, improve durability, etc.; these are some features that a consideration of eco-efficiency will encourage companies to include in their products.
4. *Rethink their markets*, which means that to improve their eco-efficiency companies will rethink their supply and demand. For example, a company may consider offering a service instead of a product since it will enhance eco-efficiency by greater control over the production of waste at the end of

the product's life. Since the product belongs to the company, it can ensure recyclability or enhance service life by remanufacturing.

An example of using eco-efficiency metrics is described in Box 17.1.

BOX 17.1 Eco-Efficiency in Industry [3]

Volkswagen introduced the Lupo 3L in 1999. It was developed by incorporating the principles of eco-efficiency. This car was extremely efficient, with a fuel consumption of only 3 L per 100 km, or 80 mpg. In addition, the car was designed to minimize its life cycle impact, which included a low-emissions engine and the use of recycled materials and highly efficient manufacturing methods. From the perspective of life cycle impact, the phase with the largest impact is the use phase. Therefore, efficiency of use is most important. To reduce impact at the end of life, this car was put together for easy disassembly and made from parts that could be easily recycled. These characteristics made this car among the most eco-efficient in terms of environmental impact per kilometer driven. This model was on the market until 2005, after which it was replaced by the Volkswagen Fox.

Eco-efficiency emphasizes a reduction of the intensity of various kinds of resources. Intensity is usually defined as the ratio of a flow to or from the environment to production from the manufacturing process, represented in physical or monetary units:

$$\text{Intensity} = \frac{\text{Interaction between the system and the environment}}{\text{Interaction between the system and the economy}} \quad (17.12)$$

Interaction between the system and the environment may be indicated by the mass of resources consumed or pollutants emitted. Common measures of economic value are the quantity produced, in physical or monetary units. Indicators of economic value are usually calculated over the company's boundary, while environmental indicators may be calculated over the company's boundary or its life cycle. Most companies measuring eco-efficiency focus on processes that they can influence most easily, including their own operations and some of their suppliers' operations that have a large impact. A narrow boundary runs the risk of shifting the impact outside the system, resulting in a false impression of enhancing eco-efficiency.

A typical example of eco-efficiency metrics reported by industry is shown Table 17.4. These results demonstrate increasing eco-efficiency over time since the reported intensities are decreasing. Such results are being produced by a multitude of companies through their annual sustainability reports and other outlets, and efforts are being made to modify existing designs to enhance their eco-efficiency.

Table 17.4 Results from typical industrial efforts toward eco-efficiency [4].

Parameter	Units	1995	2015	2016	2017	2017 performance: change compared to 2016	2017 performance: change compared to 1995
Energy	GJ/ton	2.92	1.35	1.34	1.30	−2.8%	−55.6%
CO_2 from energy	kg/ton	238.7	88.5	83.5	76.77	−8.1%	−67.8%
Total water	m^3/ton	7.95	1.88	1.85	1.80	−2.8%	−77.3%
Total COD	kg/ton	3.94	1.14	1.17	1.13	−3.4%	−71.4%
Disposed waste	kg/ton	24.27	0.26	0.35	0.18	−49.5%	−99.3%

Example 17.6 A factory emits 200 t/yr of CO_2 and uses 10,000 L/yr of water to produce 50,000 units/yr of a bearing, with an annual profit of $10,000. Determine its eco-efficiency. By using a more efficient furnace, CO_2 emissions can be decreased by 15 percent for a 5 percent decrease in the profit. Water use and production rate remain unchanged. What will be the eco-efficiency if this furnace is installed?

Solution

The eco-efficiency of the factory may be calculated by using Equation 17.12 with either the production or profit as the denominator. With production as the denominator,

$$CO_2 \text{ intensity} = \frac{200}{50,000}$$

$$= 4 \times 10^{-3} \text{ t } CO_2/\text{unit}$$

$$\text{Water intensity} = \frac{10,000}{50,000}$$

$$= 0.2 \text{ L } H_2O/\text{unit}$$

Results for both denominators are shown in Table 17.5.

With the higher-efficiency furnace, the CO_2 emissions become $0.85 \times 200 = 170$ t/yr and the profit becomes $0.95 \times 10,000 = \$9500$. Eco-efficiency values with the furnace installed are given in Table 17.5. From these results, we can see that with the new furnace, the CO_2 intensity improves for both normalization factors, while the water intensity remains unchanged with respect to production and becomes slightly worse with respect to profit. Thus, in this system, improving CO_2 intensity by the proposed technological change will result in a higher water intensity.

Table 17.5 Current and future eco-efficiency for Example 17.6.

Denominator	CO_2 intensity	Water intensity
Current		
Production	4×10^{-3} t/unit	0.2 L/unit
Profit	2×10^{-2} t/$	1 L/$
Future		
Production	3.4×10^{-3} t/unit	0.2 L/unit
Profit	1.79×10^{-2} t/$	1.05 L/$

Eco-efficiency measurements and environmental objectives often normalize environmental flows, with production measured in physical or monetary terms, as shown in Tables 17.4 and 17.5. Environmental impact occurs due to the total emissions and resource use, not to the normalized quantities, and sometimes these metrics can show improvement without any reduction in environmental impact. For example, the manufacturing process could be modified to reduce costs without reducing emissions, market prices may change to increase profit, and increasing production may increase emissions, but increases in profitability may be larger due to benefits such as economies of scale. To prevent misleading insight by the use of scaled metrics, it is important to report numbers with and without scaling, and many corporations have adopted this practice.

17.3   Process and Product Design

Techno-economic analysis, which we covered in Section 17.1, focuses on economic feasibility. This continues to be the primary goal of engineering design and decisions. As we have learned in previous chapters, such as Chapter 3, sustainability requires consideration of other criteria in addition to economic feasibility. In Part III we learned about various approaches for assessing sustainability by accounting for the direct and indirect environmental impacts of human activities. In this section, we will learn about evaluating economic and environmental aspects of engineering activities for making decisions toward sustainability.

17.3.1   Evolution of Engineering Design

The evolution of engineering design over the last several decades is depicted in Figure 17.1. Before the 1980s, industry did not consider the environment in its decisions: the primary focus was to maximize profit. As we learned in Chapter 5, during this time protecting the environment was considered to be a liability

Figure 17.1 Evolution of the approach for engineering design.

or a necessary evil. Increasing environmental impact resulted in environmental regulation and laws, and industry started accounting for environmental impact in design as a constraint that had to be satisfied. Imposing environmental considerations as constraints means that the resulting design is not likely to improve environmental performance beyond the imposed constraint. Thus, innovative solutions that can improve both economic and environmental performance may not be found by such an approach. We referred to such solutions as "win–win" in Chapter 5. This view evolved toward environmental protection as an opportunity. Gradually, design methods started including the reduction of local environmental impact as an objective along with profit maximization. This approach required decisions based on understanding and addressing the trade-off that often exists between economic and environmental goals. It also encouraged the identification of win–win designs. In the last 20 years, the environmental objective has evolved to consider the environmental impact beyond the direct emissions from the process being designed, to include the impact of emissions from the entire life cycle of the engineering activity. Now, it is increasingly common to develop designs based on understanding the trade-off between profit and life cycle environmental impact. This approach is evolving further toward working with nature and respecting its limits, which we will learn about in Chapter 20.

17.3.2 Decisions with Multiple Objectives

Decisions involving multiple objectives commonly encounter trade-offs between the objectives. For example, removing sulfur and nitrogen oxides from the emissions of coal-based power generation is expected to make the process more environmentally friendly, but to cost more. This is a "win–lose" situation, of the type that we first encountered in Chapter 5, and is illustrated in Figure 5.2a. "Win–win" situations may also be present, when a design results in simultaneous economic and environmental benefits. This may happen if removing SO_x and NO_x means avoiding fines due to violating regulations, or if the byproduct, such as gypsum, has a market value that offsets the cost of removal, or if society prefers less pollution and is willing to pay more for products from the more environmentally friendly corporation. This case is illustrated in Figure 5.2b.

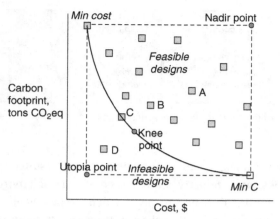

Figure 17.2 Pareto curve showing trade-off between multiple objectives.

Alternative design options may be plotted, as shown in Figure 17.2, where the objectives form the axes. In this figure, each point, except the one labeled D, represents a feasible design that will result in a corresponding cost and carbon footprint. The goal of this design exercise is to minimize cost and minimize carbon footprint. If we consider designs corresponding to points A and B, we can see that design B is better than A for both objectives. Thus, B represents a "win–win" design compared to A, or design B is said to *dominate* design A. Similarly, design C dominates designs A and B. Design D would dominate designs, A, B, and C, but it turns out that it is not possible to develop design D owing to physical, safety, and other constraints. Thus, the curve in Figure 17.2 represents all the solutions that cannot be dominated. That is, all the "win–lose" solutions lie on the curve. This is the Pareto or trade-off curve. At one extreme of the Pareto curve is the design with minimum carbon footprint, while at the other extreme is the design with minimum cost. Designs to the right and above the Pareto curve are feasible, while designs to its left and below cannot be obtained for the given system. The point that represents both optima is called the utopia point and the other extreme is called the nadir point, as shown in Figure 17.2.

Solutions that lie on the Pareto curve are called Pareto optimal, and all of them represent an "optimal" solution. In practice, since only one design can be implemented, we need to choose the "best" design. This solution is a subjective decision based on the relative importance to the decision maker of the two objectives. The knee point is sometimes chosen as the solution since it is the optimal point that is closest to the utopia point.

Example 17.7 Your friend would like to buy a compact car from the list in Table 17.6. Excluding hybrid cars, find the cars on the Pareto curve and suggest a car for her to buy. Now include hybrid models and answer the same questions.

Table 17.6 Price and fuel economy data for 2015 compact cars.

No.	Model	Price ($)	Fuel economy (mpg)
1	Honda Fit	19,257	37
2	Mazda 3	21,893	35
3	Chevrolet Cruze	21,631	28.5
4	Honda Civic	21,607	32
5	Kia Soul	17,700	27
6	Ford Focus	20,696	31
7	Kia Forte	19,576	31
8	Kia Rio	15,995	32
9	Subaru Impreza	20,322	29.5
10	Chevrolet Sonic	18,638	30.5
11	Ford Fiesta	18,600	32.5
12	Hyundai Accent	16,469	32.5
13	Hyundai Elantra	20,348	32
14	Toyota Corolla	19,050	31.5
15	Toyota Prius	21,059.5	49.5
16	Volkswagen Jetta	21,924	29
17	Volkswagen Jetta hybrid	28,727.5	45
18	Dodge Dart	20,623	30.5
19	Mazda 2	16,159	32
20	Fiat 500	21,928	35.5
21	Honda Civic hybrid	25,308	45.5
22	Scion xB	17,948.5	25
23	Honda CR-Z	22,051.5	34.5
24	Nissan Cube	18,651.5	27.5
25	Scion xD	17,116.5	30
26	Volkswagen Beetle	26,564	28.5
27	Hyundai Veloster	21,081.5	32
28	Nissan Sentra	18,540	31.5
29	Scion tC	20,495.5	27
30	Toyota Yaris	16,586.5	33.5
31	Scion iQ	17,670.5	36.5
32	Mitsubishi Lancer	27,591	30
33	Nissan Versa	15,652.5	31.5

What is the effect of hybrid technology on the Pareto curve and on your suggestion of a car for your friend?

Solution

The price and fuel economy of the cars listed in Table 17.6 are plotted in Figure 17.3. From the Pareto curve that excludes hybrid cars, we can see that cars on this Pareto curve are the Nissan Versa, Scion iQ, and Honda Fit. Your friend should choose between these cars depending on the relative importance of fuel economy

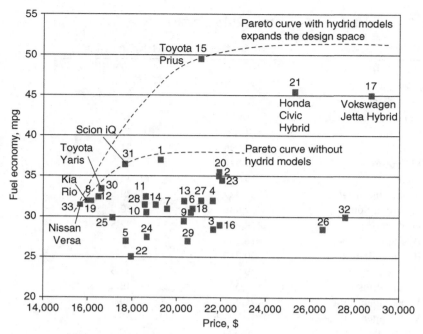

Figure 17.3 Fuel economy versus price for 2015 compact cars, based on Table 17.6.

versus price for her. Thus, if she is highly price-conscious, she should choose the Nissan Versa, while if she is highly conscious about fuel use, then the Honda Fit may be a good choice. The Toyota Yaris may be a good compromise between price and fuel economy.

The approximate Pareto curve after including hybrid cars is also shown in Figure 17.3. Owing to the small number of hybrid models in this table, it is not possible to obtain a more accurate curve. The Toyota Prius lies on this curve and may be a good one to suggest to your friend. Notice that with the inclusion of hybrid cars, the Pareto curve for conventional cars shifts to a region that is infeasible with conventional technology. This an example of innovation resulting in expansion of the design space and finding win–win solutions, which we first learned about in Chapter 5.

17.3.3   Heuristic Design

With experience, engineers usually identify heuristics or rules of thumb to guide the design process. Such rules do not guarantee an optimal design, and may even be incorrect in some situations. However, in most design situations, the heuristics help by eliminating many alternatives and expediting the design process. For example, some heuristics used by chemical engineers are as follows. In the design of a sequence of distillation columns, corrosive compounds should be removed first. This heuristic is based on the insight that since corrosive compounds are likely to require more expensive equipment, removing them first will mean a less expensive sequence of columns. Another heuristic is to remove close-boiling components last, because their separation is likely to be difficult and expensive, and is best done after all the other components have been removed.

Owing to their power and practicality, heuristics have also been developed for sustainable design. One example of such heuristics are the 12 principles of green chemistry [5] listed in Box 17.2. These principles are meant to guide chemists toward reducing the toxicity of their synthesis methods and products. Heuristics have also been developed for "green engineering," as shown in Box 17.3. Some similarities between these principles of green chemistry and green engineering are evident by their focus on hazardous waste and toxicity. However, the latter have a broader scope as they consider issues relevant to the end of life of products and their durability. Like any heuristic, these principles are also not always valid and can benefit from further analysis using methods like optimization and cost analysis. For example, some principles in Boxes 17.2 and 17.3 may result in products that are not economically feasible. For example, for many products such as plastics, Principle 7 of durability rather than immortality is usually not economically attractive. This is one reason for the formation of plastic trash islands across the world's oceans, as we learned in Chapter 2. Similarly, the lower cost and higher convenience of nonrenewable energy resources causes the violation of Principle 7 in Box 17.2 and Principle 12 in Box 17.3. Despite the relatively narrow focus of these principles of green chemistry and engineering as compared to the requirements for sustainability that we learned about in Chapter 3, they do provide sensible heuristics that could be used before applying more detailed methods.

BOX 17.2 Twelve Principles of Green Chemistry. Reproduced with permission from [5].

1. Prevention. It is better to prevent waste than to treat or clean up waste after it has been created.
2. Atom economy. Synthetic methods should be designed to maximize the incorporation of all materials used in the process into the final product.
3. Less hazardous chemical syntheses. Wherever practicable, synthetic methods should be designed to use and generate substances that pose little or no toxicity to human health and the environment.
4. Designing safer chemicals. Chemical products should be designed to effect their desired function while minimizing their toxicity.

5. Safer solvents and auxiliaries. The use of auxiliary substances (e.g., solvents, separation agents, etc.) should be made unnecessary wherever possible and innocuous when used.

6. Design for energy efficiency. The energy requirements of chemical processes should be recognized for their environmental and economic impacts and should be minimized. If possible, synthetic methods should be conducted at ambient temperature and pressure.

7. Use of renewable feedstocks. A raw material or feedstock should be renewable rather than depleting, whenever technically and economically practicable.

8. Reduce derivatives. Unnecessary derivatization (use of blocking groups, protection/deprotection, temporary modification of physical/chemical processes) should be minimized or avoided if possible, because such steps require additional reagents and can generate waste.

9. Catalysis. Catalytic reagents (as selective as possible) are superior to stoichiometric reagents.

10. Design for degradation. Chemical products should be designed so that at the end of their function they break down into innocuous degradation products and do not persist in the environment.

11. Real-time analysis for pollution prevention. Analytical methodologies need to be further developed to allow for real-time in-process monitoring and control prior to the formation of hazardous substances.

12. Inherently safer chemistry for accident prevention. Substances and the form of a substance used in a chemical process should be chosen to minimize the potential for chemical accidents, including releases, explosions, and fires.

BOX 17.3 Twelve Principles of Green Engineering. Reproduced with permission from [6].

1. Inherent rather than circumstantial. Designers need to strive to ensure that all material and energy inputs and outputs are as inherently non-hazardous as possible.

2. Prevention instead of treatment. It is better to prevent waste than to treat or clean up waste after it is formed.

3. Design for separation. Separation and purification operations should be designed to minimize energy consumption and materials use.

4. Maximize mass, energy, space, and time efficiency. Products, processes, and systems should be designed to maximize mass, energy, space, and time efficiency.

5. Output-pulled versus input-pushed. Products, processes, and systems should be "output pulled" rather than "input pushed" through the use of energy and materials.

6. Conserve complexity. Embedded entropy and complexity must be viewed as an investment when making design choices on recycle, reuse, or beneficial disposition.
7. Durability rather than immortality. Targeted durability, not immortality, should be a design goal.
8. Meet need, minimize excess. Design for unnecessary capacity or capability (e.g., "one size fits all") solutions should be considered a design flaw.
9. Minimize material diversity. Material diversity in multicomponent products should be minimized to promote disassembly and value retention.
10. Integrate local material and energy flows. Design of products, processes, and systems must include integration and interconnectivity with available energy and materials flows.
11. Design for commercial "afterlife." Products, processes, and systems should be designed for performance in a commercial "afterlife."
12. Renewable rather than depleting. Material and energy inputs should be renewable rather than depleting.

17.4   Shortcomings

The design methods in this chapter transform the sustainability assessment methods we covered in the chapters of Part III into methods for designing sustainable systems. These methods bring sustainability thinking to bear upon the design of engineering products and processes. Owing to the strong industrial interest and effort toward using these methods, one important side-effect is that data about the dependence and impact of a large number of business and industrial activities on the environment are now measured and have become available to the public. However, current efforts are no more than a step in the journey toward sustainability, since they suffer from many shortcomings. If we consider the requirements for sustainability that we covered in Chapter 3, the methods we have learned about so far focus mainly on the spatial shifting of impacts by incorporating life cycle methods. Dealing with the economic and environmental aspects together also reduces the chances of shifting impacts across disciplines. Many other requirements have not yet been considered, as discussed below.

Economic rebound effect. The expected benefits of enhancing eco-efficiency may not be realized due to the economic rebound effect or the Jevons paradox. This happens when increasing efficiency encourages consumption due to the lower cost of the more efficient item. As a result, even though consumption per unit decreases, total consumption does not decrease as much. We learned about this effect in Section 3.2.

Status quo of doing "less bad." Sustainable design methods may encourage continuous improvement by reducing harmful emissions and reducing life cycle impact. As we discussed in Chapter 16, while this is better than not reducing impact or ignoring life cycle impacts, such efforts often tend to justify the continued use of inherently unsustainable technologies and resources, and focus on doing "less bad" as opposed to "more good." For example, despite statements by many corporations acknowledging the risks of emitting greenhouse gases and using toxic chemicals, their use and emissions continue in parallel to claims about corporate sustainability. This acceptance of continuous improvement may discourage breakthroughs in innovation that enable a fundamental shift away from inherently unsustainable resources [7].

Ignoring ecological carrying capacity. Methods that we covered in this chapter do not account for the role of ecosystems and nature's capacity to supply goods and services. As we learned in Chapter 16, with regard to including the role of ecosystems in LCA, such extension is also possible for sustainable design. We will learn about this in Chapter 20.

Complexity, dynamics, and the "wicked" nature of sustainability. As discussed in Chapter 3, determining the sustainability of any system belongs to the class of wicked problems. One approach for solving such problems is to learn from other sustainable systems and try to emulate them. Since nature is a system that has sustained itself for millennia, learning from and emulating ecosystems would seem to be a pragmatic approach to sustainable development.

17.5   Summary

In this chapter we learned about methods that include sustainability considerations in engineering design. We learned about TEA, which is used extensively in conventional design, and then extended this approach to account for environmental impact at local and life cycle scales. The approach of eco-efficiency is popular in industry and aims to reduce impact per unit of production. To design sustainable systems, we need to consider multiple objectives: economic, environmental, and societal. Trade-offs between these goals are captured by the Pareto curve. Existing methods for sustainable design have inclusion of the environment as an objective, along with the conventional monetary goal. Some shortcomings of existing methods include ignoring the behavioral aspects and market response, the capacity of ecosystems to supply goods and services, and the wicked nature of sustainability. In the next few chapters, we will learn about approaches that address some of these shortcomings.

Key Ideas and Concepts

- Time value of money
- Payback period
- Net present value
- Multi-objective optimization
- Heuristic design

- Return on investment
- Discounted cashflow rate of return
- Eco-efficiency
- Pareto curve
- Green engineering

17.6 Review Questions

1. What is a desirable value of the following measures for an investment to be attractive: (a) return on investment, (b) payback period, (c) discounted cash flow rate of return?

2. Fill in the blanks. "Eco-efficiency aims to create more _____ with less _____."

3. How does a Pareto curve represent a "win–lose" situation?

4. Can the use of a heuristic for green engineering lead to an environmentally inferior design?

5. Why is doing "less bad" not good enough for sustainability?

Problems

17.1 You plan to take a loan of $500,000 and wish to pay it off each month over a 30-year period. If the annual interest rate is 5 percent, what will your monthly payment be?

17.2 Many corporations are working toward developing products for which they can claim a zero-carbon footprint. A large manufacturing site uses 25,000 MWh electricity, which is the largest contributor to the carbon footprint of its products. This electricity is obtained from the local grid, which emits 0.763 kg CO_2/kWh and costs $0.101 per kWh. The company is considering the installation of solar panels to generate this electricity. The life cycle CO_2 emission of the solar farm that is to generate the 25,000 MWh is estimated to be 40 g/kWh and the installation cost is $17.69 million. What will be the reduction in the manufacturing site's carbon footprint with this solar farm? Determine its payback period. (Based on a problem provided by Evan Boehlefield, Gabrielle Grigonis, Zakirah Mohd Fazil, and Andrew Nouanesengsy.)

17.3 A cellulosic ethanol process is estimated to have capital costs of $15 million two years before the present, $118 million one year before the present, and $63 million in the present year. Operation is expected to start next year, and is estimated to result in a net annual cash income including

Table 17.7 Typical data for beverage packaging.

Year	Packaging type	Packaging mass (million pounds)	Packaging recycled (million pounds)	Volume of beverage (billion gallons)
	PET	800	250	3
1990	Aluminum	2000	1000	5
	Glass	4000	1400	1
	PET	1200	550	5
1995	Aluminum	2000	1000	5.5
	Glass	1000	350	0.2
	PET	1700	600	6
2000	Aluminum	1900	1000	0.1
	Glass	400	100	1
	PET	1800	500	5.5
2005	Aluminum	1850	1	6
	Glass	300	90	0.06

sales, operating costs, depreciation, and taxes of $19 million in the first year, $34 million in each of the second to fifth years, and $25 million per year from the sixth to the tenth year. Calculate the net present value of this process for a 10 percent per annum interest rate, and its discounted cash flow rate of return.

17.4 Typical data about beverage packaging are given in Table 17.7. Answer the following questions based on these data.
1. Calculate the eco-efficiency of beverage packaging using the data the this table.
2. Using the carbon footprint information for the three types of packaging materials, calculate the eco-efficiency over the life cycle.
3. Compare the change in the carbon footprint for all packaging per unit of beverage sold with the carbon footprint for the entire quantity of beverage sold. Use these results to discuss the pros and cons of eco-efficiency.

17.5 A car rental company owns 100 Chevrolet Cruzes, 50 Toyota Yarises and 50 Ford Focuses. They plan to buy another 50 cars, and owing to new government regulations need to ensure that the average fuel efficiency of their fleet is at least 30 mpg. Using the data in Table 17.6, determine the model they should buy. They require all the new purchases to be the same model.

17.6 The definition of eco-efficiency in Section 17.2 refers to the Earth's carrying capacity. Do current eco-efficiency practices meet this goal of the original definition? Justify your response.

Table 17.8 Carbon and nitrogen footprints of transportation fuels [8].

	Nitrogen footprint	Carbon footprint	Type
Gasoline	2.18E–04	2.96E–01	Fossil
E10 corn	2.20E–04	2.84E–01	First gen.
E10 yellow poplar	2.09E–04	2.74E–01	Second gen.
E10 switchgrass	2.05E–04	2.72E–01	Second gen.
E10 LIHD	2.05E–04	2.72E–01	Second gen.
E10 MSW	2.00E–04	2.74E–01	Waste
E10 newsprint	2.01E–04	2.72E–01	Waste
E10 stover ethanol	2.03E–04	2.71E–01	Second gen.
E10 stover thermo	2.01E–04	2.71E–01	Second gen.
E85 corn	4.62E–04	2.30E–01	First gen.
E85 yellow poplar	3.26E–04	1.11E–01	Second gen.
E85 switchgrass	2.77E–04	8.55E–02	Second gen.
E85 LIHD	2.86E–04	9.46E–02	Second gen.
E85 MSW	2.24E–04	1.13E–01	Waste
E85 newsprint allocation	2.56E–04	6.69E–02	Waste
E85 newsprint as waste	2.38E–04	8.48E–02	Waste
E85 stover mass allocation	4.00E–04	1.28E–01	Second gen.
E85 stover monetary allocation	3.33E–04	1.07E–01	Second gen.
E85 stover as waste	2.55E–04	8.29E–02	Waste
E85 stover thermo mass allocation	3.88E–04	1.26E–01	Second gen.
E85 stover thermo monetary allocation	3.10E–04	1.04E–01	Second gen.
E85 stover thermochemical as waste	2.38E–04	7.89E–02	Waste
Butanol mass allocation	6.79E–04	1.59E–01	Second gen.
Butanol monetary allocation	5.41E–04	1.16E–01	Second gen.
Butanol as waste	4.35E–04	8.71E–02	Waste
Diesel	3.23E–04	2.57E–01	Fossil
B20 soybean	3.57E–04	2.38E–01	First gen.
B20 stover mass allocation	3.70E–04	2.33E–01	Second gen.
B20 stover monetary allocation	3.58E–04	2.29E–01	Second gen.
B20 stover	3.43E–04	2.24E–01	Second gen.
B100 soybean	5.12E–04	1.65E–01	First gen.
B100 stover mass allocation	6.04E–04	1.48E–01	Second gen.
B100 stover monetary allocation	5.30E–04	1.22E–01	Second gen.
B100 Stover as waste	4.36E–04	8.96E–02	Waste

17.7 For a manufacturing process, the dominant value of net present value (NPV) is given by $f_1 = 35\sqrt{x} + 22$, and the dominant value of the global warming potential (GWP) is $f_2 = 5x^2 + 10x + 100$. Here, x represents the energy used in the process. Determine the Pareto curve for this process. Determine the feasibility and optimality of designs with [GWP, NPV] values of [6000, 100], [260, 92], [12,000, 300]. Justify your answers.

17.8 Figure 5.2 provides perspectives on the relationship between business and the environment. Interpret these figures in terms of Pareto curves. Identify all the Pareto curves (lines) in these figures. Label each point from A through F as dominant or non-dominant with respect to each other.

17.9 Table 17.8 shows the carbon and nitrogen footprints for transportation fuels [8]. Answer the following questions based on these data.

1. Plot the footprints for fossil and first-generation biofuels and determine the Pareto curve. Identify the feasible and infeasible regions for this curve.
2. Now include the points for second-generation biofuels. Do they expand this design space? Justify your finding.
3. Discuss how you would use these results to choose the "best" transportation fuel(s).

17.10 Consider the design of a paper coffee cup, keeping in mind that it is made of layers of paper and plastic and that such cups are often not accepted by recycling companies. Which of the 12 Principles of Green Engineering does this cup satisfy or violate? Repeat the exercise for a ceramic coffee mug. Does applying these principles to the two products indicate greater environmental friendliness of the ceramic mug?

References

[1] Aldo Leopold. *Round River*. Oxford University Press, 1972.

[2] M. S. Peters, K. D. Timmerhaus, and R. E. West. *Plant Design and Economics for Chemical Engineers*. McGraw Hill, 2003.

[3] M. Lehni, S. Schmidheiny, and B. Stigson. Eco-efficiency: creating more value with less impact. World Business Council for Sustainable Development 2000.

[4] Unilever Global. Eco-efficiency performance overview. `www.unilever.com/ sustainable-living/reducing-environmental-impact/ eco-efficiency-in-manufacturing/ our-eco-efficiency-performance`, accessed January 31, 2019.

[5] P. T. Anastas and J. C. Warner. *Green Chemistry: Theory and Practice*. Oxford University Press, 2000.

[6] P. T. Anastas and J. B. Zimmerman. Design through the 12 principles of green engineering. *Environmental Science & Technology*, 37(5):94A–101A, 2003.

[7] M. Braungart, W. McDonough, and A. Bollinger. Cradle-to-cradle design: creating healthy emissions – a strategy for eco-effective product and system design. *Journal of Cleaner Production*, 15(13–14):1337–1348, 2007.

[8] X. Liu, S. Singh, E. L. Gibbemeyer, et al. The carbon–nitrogen nexus of transportation fuels. *Journal of Cleaner Production*, 180:790–803, 2018.

18 | Ecosystem Ecology

> The beauty of nature lies in detail; the message, in generality.
> Optimal appreciation demands both.
>
> Stephen Jay Gould [1]

Ecological systems are able to sustain themselves for long periods of time while providing diverse goods and services. These systems are certainly not static, and have the capacity to recover from or adapt to various perturbations while maintaining their basic structure and function. Of course, this adaptive capacity is not unlimited and can be exceeded if the perturbations are too large. In the case of such a disruption, the ecosystem may shift to a different state that may provide different, and less useful, goods and services. Over the last several decades, much insight has become available about the behavior and characteristics of ecological systems, and various efforts have been directed toward learning from this insight and incorporating it into human-designed systems. An important motivation for these efforts to learn from and emulate nature is that this may be one way of dealing with the wicked nature of sustainability that we learned about in Chapter 3. Before we cover methods based on learning from nature in subsequent chapters, we will focus in this chapter on some basic principles of ecosystem ecology. We will learn about the characteristics of ecosystems, such as materials and energy cycling, that determine their behavior. Such insight can provide the foundation for addressing many of the challenges of sustainable engineering.

The word ecology combines the Greek words *oikos*, which means home, with *logos*, which means study. Thus, ecology is the study of "life at home," with an emphasis on "the totality or pattern of relations between organisms and their environment" [2]. Ecology is a relatively recent science, and the word was coined in 1869. It is interesting to note that the word economics combines *oikos* with *nomos*, which means management. Thus, economics stands for management of the home, while ecology is its study. Given that both disciplines focus on the "home," which may be a specific region or the entire planet, it seems that study of the home (ecology) and management of the home (economics) should be closely integrated. However, as we learned in Chapter 4, this has not been the case, at least not in recent times.

18.1   Characteristics of Ecosystems

Ecology recognizes the inseparable connection between living (biotic) and non-living (abiotic) components. An ecological system or ecosystem is "any unit that includes all the organisms (the *biotic* community) in a given area interacting with the physical environment so that a flow of energy leads to clearly defined biotic structures and cycling of materials between living and nonliving components" [2]. An ecosystem is the smallest unit that is self-sustaining. It contains all the biotic and abiotic elements needed for survival. Examples of ecosystems include a forest, wetland, pond, farm, and garden. Abiotic components in such systems include soil, nutrients, water, sunlight, and air, while biotic components could be bacteria, fungi, plants, insects, birds, and animals. A typical ecosystem is depicted in Figure 18.1.

Ecosystems are part of a hierarchy that consists of small and large units, as shown in Figure 18.2. Ecology usually focuses on systems larger than an organism. Below the organism level, there is homeostasis, or control to a set point. For these levels, there are mechanisms to maintain each level at a fixed state. For example, cells have ways of maintaining their structure and composition, as do tissues and organs. Above the organism level, such maintenance to a fixed state does not happen. These levels exhibit homeorhesis, or pulsing within limits. Thus, populations of predators and prey show oscillations within limits as opposed to

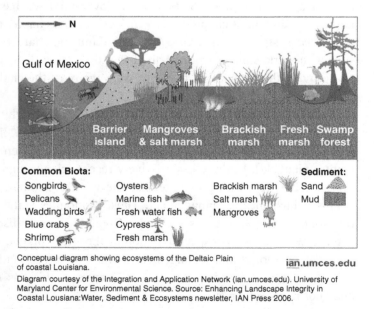

Conceptual diagram showing ecosystems of the Deltaic Plain of coastal Louisiana.

iān.umces.edu

Diagram courtesy of the Integration and Application Network (ian.umces.edu). University of Maryland Center for Environmental Science. Source: Enhancing Landscape Integrity in Coastal Lousiana:Water, Sediment & Ecosystems newsletter, IAN Press 2006.

Figure 18.1 Ecosystem on a marine delta. Courtesy of Jane Thomas, IAN Image Library (ian.umces.edu/imagelibrary).

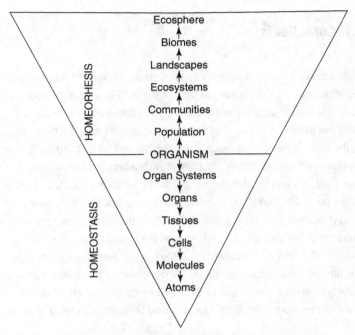

Figure 18.2 Hierarchy in ecosystems. Adapted with permission from [3].

maintenance at fixed points, and the quantity of water in a river tends to vary with seasons and rainfall.

An important characteristic of this hierarchical organization is that systems at larger scales often exhibit properties that were not present at the level below. For example, the water molecule has properties that are very different from its constituent hydrogen and oxygen atoms, an organism is more than the properties of constituent organ systems, the behavior of a flock of starlings includes murmuration, which cannot be predicted by knowing the behavior of a single bird, as shown in Figure 6.1. Examples of such emergence include the structure and vast diversity of coral reefs, which are a coevolution of algae and coelenterate animals, as described in more detail in Box 18.1. This property of emergence is a nonreducible property; that is, it cannot be reduced to the sum of the properties of its parts. As we learned in Chapter 6, the reductionist approach common in science and engineering is inadequate for understanding such systems, and holistic approaches are needed. Such approaches are particularly important for understanding ecosystems.

18.2 Material Cycles and Energetics

Ecosystems involve strong interactions and networking between biotic and abiotic components. The nature of these interactions plays a key role in the ability of

BOX 18.1 Coral Reefs

Coral reefs are among the most diverse ecosystems on the planet, but they prosper in a nutrient-poor environment. This paradox of high productivity in a resource-poor environment is explained by the mutualistic relationship that exists between algae and corals. Algae called zooxanthellate live inside the corals. They carry out photosynthesis to produce resources that are used by the coral. These include oxygen, glucose, glycerol, and amino acids, which are produced by photosynthesis. In return, the coral provide a protected environment to the algae and resources such as nitrogen, phosphorus, and sulfur, which are secreted as wastes by the coral polyp. The polyp also builds a calcium carbonate skeleton. This tight cycling creates a self-sustaining or autocatalytic system that allows corals to be independent of the surrounding water for nutrients. A large number of other species live in and around the coral reef, including colorful parrotfish that feed on the algae within the coral, scavengers such as cleaner shrimp, crabs, sea cucumbers, and many other species. When subjected to stress, the coral polyp ejects the algae, causing the coral reef to lose its color and become bleached. After the stress is gone, the coral can recolonize the algae, but extended bleaching can result in death of the reef.

ecosystems to be self-sustaining. Insight into these interactions can help in developing human-designed systems with similar properties. In this section we will learn about material cycling and energy transformation in ecosystems.

18.2.1   Food Web

The three main categories of organisms in an ecosystem are producers, consumers, and decomposers.

- *Producers* are typically green plants that can produce their own food through photosynthesis. The following overall reaction converts carbon dioxide and water into glucose and oxygen in the presence of sunlight:

$$6CO_2 + 12H_2O \xrightarrow{\text{sunlight}} C_6H_{12}O_6 + 6O_2 + 6H_2O \tag{18.1}$$

  The glucose can then be converted into other molecules needed by plants, such as proteins, carbohydrates, and cellulose. This conversion requires other nutrients such as nitrogen, phosphorus, and sulfur, which are typically obtained from the soil. It also requires energy, which is obtained by breaking down some of the glucose through respiration:

$$C_6H_{12}O_6 + 6O_2 \rightarrow 6CO_2 + 6H_2O \tag{18.2}$$

  For a plant to grow and maintain itself, the quantity of glucose formed by photosynthesis has to be more than the glucose decomposed by respiration.

Thus, plants absorb more CO_2 during photosynthesis than they emit during respiration. This makes them capable of sequestering atmospheric CO_2.

- *Consumers* include many types of organisms. Primary consumers are herbivores, since they rely directly on consuming producers. Secondary consumers feed on primary consumers. Multiple higher levels of consumers also exist, as shown in the food web in Figure 18.3. Some consumers occupy multiple levels since they feed on plants and other levels of consumers. Humans are one such consumer.
- *Decomposers* feed on dead plants and animals, and include bacteria, fungi, earthworms, millipedes, termites, dung beetles, vultures, hyenas, etc. As with consumers, decomposers may also be categorized as primary, secondary, and so on. These organisms release nutrients contained in biotic components of ecosystems and make them available for reuse.

Despite the complex interactions and diversity of organisms in food webs, they may all be organized in terms of multiple feeding or trophic levels. Producers may

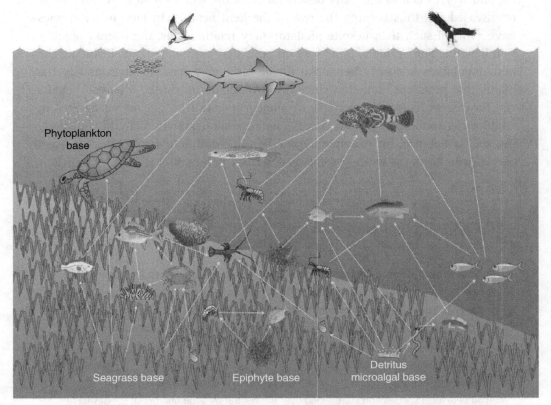

Figure 18.3 Food web in a South Florida seagrass meadow. Courtesy of Kris Beckert, IAN Image Library (ian.umces.edu/imagelibrary).

be considered to form the bottom level and have the largest collective mass. The next level is primary consumers or herbivores, followed by secondary consumers, and so on, as depicted in Figure 18.3.

Ecological food webs convey a predator–prey relationship between organisms. However, other types of relationships also exist that do not involve feeding. These relationships are mutually beneficial and are called mutualism. One such example is the role of birds in dispersing seeds of the fruits they eat. This relationship between fruit-bearing trees and birds is mutually beneficial since the birds get food, while the tree is able to grow in new locations to which the birds transport the seeds. Another example is that of algae and fungi, where algae provide food and fungi provide protection. Such situations in which species are living together is called symbiosis, and it could involve mutualism where both species benefit, or parasitism where only one benefits. In fact, parasitism is a predator–prey type of relationship. The coral reef example in Box 18.1 is an example of symbiotic mutualism.

Most relationships in ecosystems cannot be cleanly categorized since they often combine multiple features. For example, the predator–prey relationship between elk and wolves is also mutually beneficial since the wolves tend to remove weak or diseased elk, thus keeping the rest of the herd healthy. In fact, many species have evolved such that, despite predator–prey relationships, the overall population benefits. For example, birds feed on pests that consume plants. Reducing the pest population is clearly beneficial to the birds who feed on them and to the plants that pests consume, but since it allows plants to thrive, it is also beneficial for insects. This is an example of homeorhesis or the "balance of nature." In engineering terminology, these are examples of autocatalytic activities. Disrupting such a balance, for example by removing predators or introducing invasive species, can harm the entire system, as described in Box 18.2.

BOX 18.2 Ecological Disruption

Deer overpopulation in the eastern USA. Removal of predators such as cougar, bear, and wolf in the eastern USA has resulted in an overabundance of white-tailed deer. Their population is estimated to have grown from about 500,000 to 20 million over the last 100 years. This overpopulation has many negative impacts, including the degradation of forests due to excessive grazing, the spread of diseases such as Lyme disease, and increases in fatalities due to deer–automobile collisions.

Invasive species. A foreign species in an ecosystem becomes invasive if it can outcompete local species and start taking over the habitat. A large number of species have become invasive across the world and are causing ecological disruption and loss of ecosystem goods and services. Some examples of invasive species and their impacts are as follows.

- The kudzu vine was introduced into the USA from southern and eastern Asia for controlling soil erosion and as a decorative plant. It has become invasive, grows very fast, and deprives native plants of sunlight.
- The brown tree snake, a native of Australia and Indonesia, was introduced on the island of Guam and has been responsible for the elimination of local species of birds, bats, amphibians, and lizards. This has caused further ecological deterioration since these species were responsible for pollination of many local flowers and dispersal of seeds of some of the most important fruit trees.
- Water hyacinth is an aquatic plant that has taken over many freshwater bodies in tropical countries. It spreads very quickly, crowds out local species, and blocks sunlight from reaching under the water. Its effects are not all negative since it is now harvested and used as a feedstock for making biogas.

18.2.2 Biogeochemical Cycles

The intense recycling of materials and the efficient use of energy in the over-all system are essential characteristics for sustaining ecosystems. Material cycles include those of carbon, nitrogen, water, and minerals. These cycles of abiotic materials involve biotic and abiotic components, and are essential for various planetary activities and services such as climate regulation, maintenance of soil fertility, availability of fresh water, concentration of mineral deposits, etc. They are described in brief in this subsection, with indications about the influence of human activities.

The carbon cycle, depicted in Figure 18.4, involves the cycling of carbon that is released from processes such as respiration, weathering, and combustion. Carbon dioxide is taken up by plants during photosynthesis, as we saw in Reaction 18.1, to form biomass, which then supports the food web. Some carbon dioxide is fixed by conversion to calcium carbonate and in the form of biomass sediments that are converted into fossil fuels over millions of years. As we learned in Chapter 2, human activities, such as fossil combustion and changes in land cover, have been increasing the emissions of CO_2, resulting in atmospheric accumulation.

The nitrogen cycle, shown in Figure 18.5, is essential for sustaining life owing to the critical role of nitrogen in proteins and other molecules. The atmosphere is made up mostly of nitrogen, but this is relatively inert and cannot be used directly by primary plants. Natural processes that convert this nonreactive nitrogen to its reactive form include lightning and nitrogen-fixing bacteria in the roots of legu-minous plants. All other life gets nitrogen from plant matter. Decomposers return

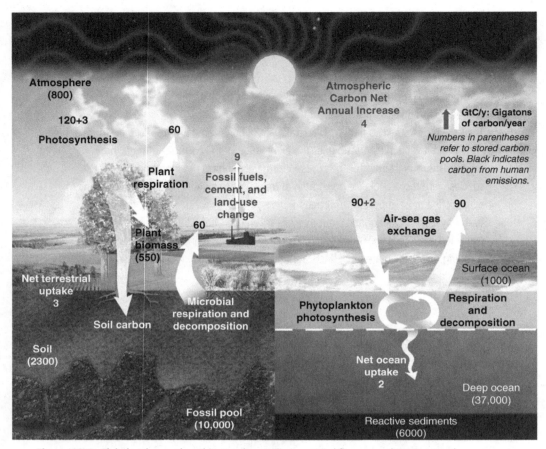

Figure 18.4 Global carbon cycle. White numbers indicate natural fluxes, numbers in parentheses are stored values, and black numbers are anthropogenic flows [4].

nitrogen from animal wastes and at the end of life back to the soil. The technological activities of humans now convert atmospheric nitrogen into its reactive form through the Haber–Bosch process to produce ammonia, and through high temperature combustion that produces nitrogen oxides. The ammonia is converted into various kinds of fertilizers and chemicals, while the nitrogen oxides get deposited on land, often through acid rain. As we saw in Figure 2.22, the anthropogenic flow of reactive nitrogen is now comparable to the natural flow, causing severe disruption of this cycle and resulting in impacts such as the creation of aquatic dead zones, shown in Figure 2.23.

18.2.3 Energy Transformation

Ecologists commonly assess and model ecosystems as networks of exergy flow. Primary producers transform solar exergy into biomass exergy, which is then consumed by heterotrophs at higher trophic levels. All biotic components convert

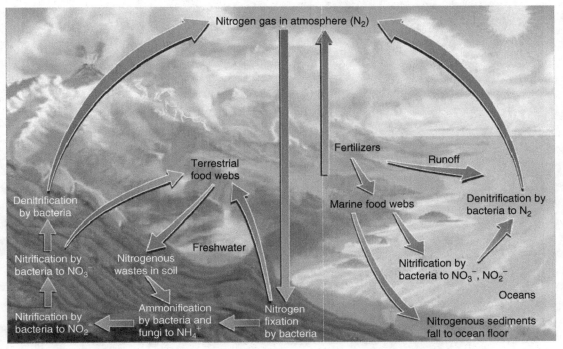

Figure 18.5 Global nitrogen cycle.

consumed exergy to do work such as respiration, hunting, reproduction, etc. The flow and transformation of solar radiation is depicted in Figure 18.6. Greenhouse gases in the atmosphere affect the "back radiation" flux. As little as 1 percent of the solar input absorbed by the surface is converted into biomass through photosynthesis. The rest is not wasted as it plays an essential role in keeping Earth livable, contributing to the water cycle by causing evaporation, keeping the planet warm, and creating wind and waves.

The quantity of exergy flow in ecosystems decreases with increasing level, as depicted in Figure 18.7. Correspondingly, the overall efficiency decreases, but the quality of the available energy increases. Thus, species that are higher up the food-chain tend to have greater influence on their surroundings. For example, tigers influence the ecosystem to a greater extent than deer, and bees more than the flowers they feed on. This higher-quality energy is fed back to lower levels, as shown in Figure 18.7. This feedback creates an autocatalytic or self-sustaining loop since the higher-quality energy feedback maintains the resources at lower levels.

These transformations in nature are governed by the laws of thermodynamics, in the same way as thermodynamics governs the behavior of industrial systems. Biotic components utilize resources to maintain a distance from equilibrium with the surroundings, or a non-zero exergy. The resource transformation to maintain this non-zero exergy or low-entropy state means that, as per the second law, all

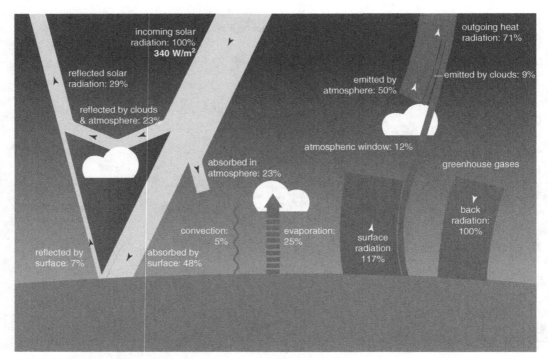

Figure 18.6 Earth's energy budget. Image from NASA's Earth Observatory, https://earthobservatory.nasa.gov/Features/EnergyBalance.

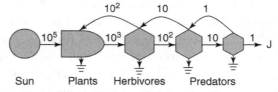

Figure 18.7 Exergy transformation in foodchains. Adapted with permission from [5].

biotic components cause an increase of entropy in the surroundings. Thus, when entropy is decreased (exergy is increased) in an ecological process or component such as by the growth of a tree or the maintenance of other living beings, it also causes an increase in the entropy of the surroundings. Examples of such an entropy increase include that due to the shedding of leaves in the fall, the droppings of sea birds that collect to form guano, and the large quantity of oxygen "excreted" by plants during photosynthesis. Such waste could result in a large and negative environmental impact. After all, ecological and industrial systems are both governed by the same physical laws. As per the second law of thermodynamics, waste is inevitable in any activity that reduces entropy (disorder). Then, why is it that ecosystems do not encounter unsustainability in the way industrial

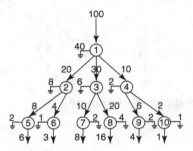

Figure 18.8 Energy network in a typical ecosystem.

systems do? As we will learn in Section 18.3, unsustainable situations also appear in ecological succession, but the overall system develops toward greater sustainability by increased networking: in a mature ecosystem, waste from one process is utilized as a resource in another. Even though the efficiency of an individual ecological component can be very small, owing to intense cycling the overall efficiency of ecosystems can be very high. Waste is still produced in the overall system, however, usually in the form of low-quality heat. If the ecosystem is disrupted by changes such as removal of predators or introduction of invasive species, as described in Box 18.2, the overall efficiency can decrease.

Example 18.1 An ecosystem is represented by the energy network shown in Figure 18.8. In Node 1, out of the 100 J input, 40 J is wasted, while 60 J is used by Nodes 2, 3, and 4. Calculate the overall energy efficiency of this network. If Node 4 is not available any more owing to the removal of a species, what is the new network efficiency?

Solution
The total input to the network is 100 J and the total useful output is $6 + 3 + 8 + 16 + 4 + 1 = 38$ J. Thus, the network efficiency, which is the ratio of the useful output to the total inputs (Equation 12.3) is

$$\eta = \frac{38}{100}$$
$$= 38 \text{ percent}$$

If Node 4 is removed, the 10 J that was used by it will now be wasted. The network output will be $6 + 3 + 8 + 16 = 33$ J. The new network efficiency will be 33 percent.

Example 18.2 A single node in an ecological network has an efficiency of 10 percent, as depicted in Figure 18.9a. If a new node develops to use the waste from Node 1, the resulting network is shown in Figure 18.9b. Determine the efficiency of a network with 2, 20, and 200 such nodes.

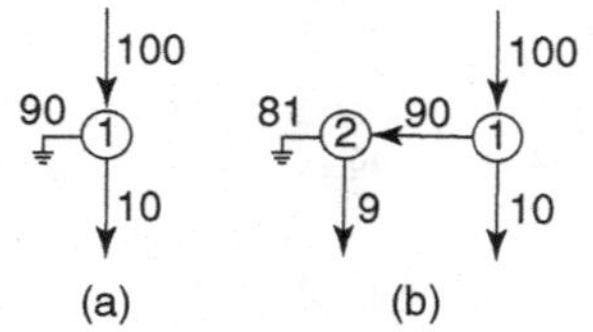

Figure 18.9 Energy network development for Example 18.2.

Solution

For the two-node network shown in Figure 18.9b, the efficiency is

$$\eta = \frac{10 + 9}{100}$$
$$= 19 \text{ percent}$$

For one node, the waste is $(0.9)(100) = 90$, for two nodes in series it is $(0.9^2)(100) = 81$, for three nodes it will be $(0.9^3)(100) = 72.1$, and so on; for n nodes it will be $(0.9^n)(100)$. Thus, the efficiency for n nodes in series may be written as

$$\eta_n = 1 - 0.9^n$$

Using this equation, the efficiency for 20 nodes will be $\eta_{20} = 87.8$ percent and for 200 nodes it will be $\eta_{200} = 99.9$ percent.

These simple illustrations demonstrate an important characteristic of ecosystems: even though each ecological activity may have low efficiency, the overall ecological network can be extremely efficient. Conversely, the removal of nodes from an ecosystem by activities such as habitat destruction and biodiversity loss can make the overall system less efficient.

18.3   Dynamics of Ecosystems

18.3.1   Nature of Ecosystem Dynamics

The development of ecosystems over time has been studied extensively, and the basic insights are summarized in this section. Even though it is popular to think about the "balance of nature," in reality ecosystems are not at a steady state or equilibrium and are constantly changing. Despite this dynamic character, ecosystems are resilient to many disturbances and are able to maintain their structure and function. However, there are thresholds beyond which if a system is perturbed then it may collapse to a different state that lacks the structure and function of the original state.

Changes in ecosystems over time are represented by the phenomenon of ecological succession. This involves changes in the species that constitute an ecosystem, and is usually of two types: primary or secondary. Primary succession is the development of a biotic community in an area for the first time, while secondary succession involves the development of a biotic community in an area that had a community which was disturbed. An example of *primary* succession is the gradual covering of bare rocks with lichens, followed by moss and hardy species that create soil particles and hold water, to further encourage the growth of grasses, shrubs, and finally trees. *Secondary* succession relies on the substrate that was created by the ecosystem which occupied the land before it was disrupted. Secondary succession happens when an agricultural field is abandoned and left to nature, as has happened in the eastern USA. As shown in Figure 18.10, among the early species to occupy the land are grasses such as crabgrass, owing to their ability to grow quickly on bare soil and their drought tolerance. However, they can be shaded out by taller plants, which colonize the land more slowly than crabgrass, but end up dominating in 1–3 years. These species gradually build detrital matter and soil quality. In 3–10 years, pine cones are able to establish themselves, resulting in a pine forest, which gradually gets replaced by hardwood trees in about 30 years. This happens because pine cones cannot germinate in the shade of pine trees, while the seeds of deciduous trees can do so. The resulting forest in about 70 years can sustain itself since the seeds of deciduous trees can thrive beneath the parent trees.

For secondary succession to happen as described above, the basic substrate needs to be such that it can support various species, starting with crabgrass. The soil also needs to have the seeds of successive species in it from previous ecosystems, or these seeds need a mechanism to be dispersed to the ecosystem. This makes the system resilient to perturbations. However, if the substrate itself is lost or if the disturbance continues, then the secondary succession may be of a very different type compared to the original ecosystem, and the system may lose its resilience and cannot retain its original structure and function. Examples of such ecosystems include the conversion of previously forested lands into grasslands by the continual use of fire and grazing. Another example is water bodies with

Figure 18.10 Secondary succession of oak–hickory forest. Left panel: pioneer species; annual plants are succeeded by perennials; middle panel: intermediate species; shrubs and pines grow, followed by oak and hickory; right panel: climax community; mature oak and hickory forest. By CNX OpenStax [CC BY 4.0 (http://creativecommons.org/licenses/by/4.0)], via Wikimedia Commons.

Table 18.1 Exergy degraded and stored for various ecosystem. With permission from [6].

Ecosystem	Percent of incoming solar radiation degraded in nonradiative processes	Exergy storage (MJ/m^2)
Quarry	6	0
Semi-arid	2	0.07
Clear-cut	49	0.59
Grassland	59	0.94
25-year-old Douglas fir plantation	70	12.7
23-year-old natural growing forest	71	26
400-year-old fir forest	72	38
Tropical rain forest	70	64

excessive input of nutrients, which results in harmful algal blooms and a loss of their original productivity.

As succession progresses, ecosystem efficiency improves, as is apparent from Table 18.1. At one extreme, the quarry is able to utilize only 6 percent of the incoming solar radiation and no stored exergy, while at the other extreme a tropical rain forest utilizes 70 percent of the incoming solar radiation and has 64 MJ/m^2 of stored exergy. This depicts the increased use and storage of the incident exergy in more developed ecosystems.

18.3.2   Understanding Ecosystem Dynamics

Ecosystem succession may be understood in terms of its gross primary production (GPP) and autotrophic respiration (R_a). The GPP is measured as the total of the carbon flux generated by photosynthesis. It is represented by Equation 18.1. The autotrophic respiration is the carbon flux used for respiration, and is represented by Equation 18.2. The difference between these two quantities is the net primary productivity (NPP). As shown in Figure 18.11, young or developing ecosystems at an early stage of succession involve species that are able to use most resources for growth (production), and less for maintenance (respiration). This includes species like crabgrass and other annuals that grow quickly and produce a lot of biomass but do not last for very long as they get crowded out by longer-lasting species, as illustrated in Figure 18.10. At this early stage, the GPP/R_a ratio is greater than 1. In this situation, biomass accumulates, as indicated by the expanding gap between GPP and R_a shown in Figure 18.11. However, as the structure of the ecosystem builds up and more resources are needed to maintain it in a low-entropy or organized state, so fewer resources are available for production. Gradually, the system shifts to a state in which the GPP/R_a ratio approaches 1, that is, production is just enough to maintain the system. This occurs when the system is mature

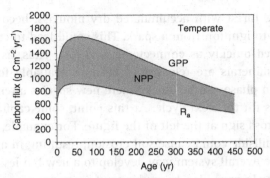

Figure 18.11 Development of a temperate forest: gross primary production, respiration, and net primary production. Reproduced with permission from [7].

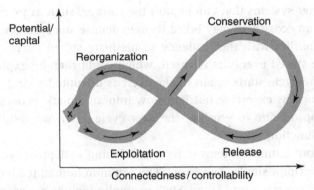

Figure 18.12 Adaptive cycle in ecosystems. Based on [8].

or reaches the climax stage. Typically, a system stays in this state until hit by another perturbation.

Another insightful way to understand the development of ecosystems is by Holling's adaptive cycle, shown in Figure 18.12 [8]. In this figure, the x-axis is connectedness or controllability, and the y-axis is potential or capital or number of options available for the future. When potential and connectedness are both low, ecosystems tend to exploit whatever resources are available. This could be the state after a disruption such as a forest fire or volcanic eruption. Species that grow under such conditions tend to be those that can exploit the available resources most quickly, like crabgrass. Under such conditions, material cycles are not closed and overall efficiency is not high. This exploitation phase can last for a long time, and the ecosystem develops by becoming more connected and accumulating potential in the form of biomass. This is the gradual transformation of the burned forest to a mature forest, and is also depicted in the early stages of Figures 18.10 and 18.11. During this phase, the overall system becomes more efficient by becoming increasingly tightly bound within the existing system and preventing competition. This also makes the system more rigid and vulnerable. Such a system could be a mature forest with accumulated dry biomass, becoming a tinderbox that is susceptible to ignition from a spark. This could lead to collapse, where the potential is released quickly as connectedness is lost. Thus, as the forest burns, the accumulated minerals are released to become available for reorganization. This reorganization phase can be quick where new species attempt to exploit the available resources for the next cycle. At this point, some capital may be lost, as indicated by the cross sign at the left of the figure. For example, intense rain after a forest fire on a hillside could cause soil erosion, resulting in a loss of nutrients. Because of this, the overall system may develop to a new but less productive state.

This adaptive cycle may also be understood in terms of exergy flow: the x-axis may be considered to be the stored exergy. Thus, the forward loop involves increasing the storage of exergy. As this storage increases, it also increases opportunities for other systems that can exploit the storage, such as pests or fire. Thus, the success of an ecosystem may breed its own demise since the tendency to store exergy is in conflict with the tendency to dissipate stored exergy. In the backward loop, the stored exergy is released, which could then be exploited by other processes as the cycle starts again. If the cycle is disrupted owing to an external perturbation such as excessive nutrient flow into an aquatic ecosystem or excessive loss of topsoil due to erosion, the new cycle need not result in the same structure and function.

Thinking about human systems in terms of Holling's adaptive cycle, we see that many systems exhibit similar behaviors. The common human tendency is to reach and stay at the "conservation" stage. Such examples include government intervention to prevent large disruption of specific industries or corporations. For example, a few years ago the US government bailed out the automobile industry. Understanding about the adaptive cycle indicates that rather than sustaining a mature and rigid corporation, it might have been better to use some of the resources to enable "creative destruction" by encouraging the development of new transportation ideas and technologies. This could have overcome the inefficient status quo of current transportation options while allowing a "soft landing" for the economy.

18.4   Summary

The introduction to ecosystem ecology in this chapter only scratches the surface of this discipline, but even this basic insight may be used to develop nature-inspired human-designed systems, as discussed in this section and described in detail in the next two chapters.

Perhaps the most common observation about ecosystems is their near complete absence of waste. This high efficiency is due to intense networking, which can have several important implications for human-designed systems as well.

- Ecosystems are able to utilize more of their resources as they become more networked. This happens because the networking allows waste from one activity to be used as a resource in another activity. Thus, even though an individual ecosystem component operates with low efficiency (the efficiency of plants is only about 4 percent), the overall system can be highly efficient.
- Reliance on networking for efficient resource use can make the ecosystem more resilient to perturbations, since the networking can provide more alternative pathways, with limited dependence on any single species or path.
- If the network is disrupted by the removal of a process or species, it will result in less efficient operation and more waste. Such disruption beyond a limit could cause a loss of resilience and a shift to another state that might be less productive and useful.

Such efficient networking is not always present, since in the early stages of development the ecosystem produces waste that may not have any use, at least not in the near term. Examples of such waste include coal, crude oil, and coral reefs. Over time, ecosystems develop to adapt to the presence of such products. Sustainable development requires human activities to undergo similar adaptation.

Key Ideas and Concepts

- Homeostasis
- Mutualism
- Trophic level
- Adaptive cycle

- Homeorhesis
- Parasitism
- Succession
- Creative destruction

18.5 Review Questions

1. Fill in the blank. "An ecosystem is the smallest unit that is _____."
2. Why is a reductionist approach inadequate for understanding ecosystems?
3. State two benefits of networking in ecosystems.
4. As an ecosystem grows, how does the ratio of growth to maintenance change?
5. What is the significance of leakage in Holling's adaptive cycle?

Problems

18.1 Study the graphs of the atmospheric concentration of various gases shown in Figure 2.18. What type of behavior do you see? Does this system exhibit homeostasis or homeorhesis? On the basis of your knowledge about ecosystem behavior, what do you expect will happen to these atmospheric concentrations in the future?

18.2 Identify some species in various trophic levels of the following ecosystems: (a) a local garden, (b) intensive agriculture, (c) a fresh water lake, (d) a forest. You may consider such ecosystems near your location.

18.3 Trees in England were cut down in the early years of the Industrial Revolution to produce charcoal for making steel. Even today, centuries after the logging, many of these logged areas lack trees. From the point of view of Holling's cycle, explain why these areas did not recover to become forested again.

18.4 Maintaining a lawn requires regular mowing. Which stage of ecological succession does the mowing maintain? Explain with the help of Figures 18.11 and 18.12.

18.5 To understand the benefits of ecological networking, consider all ecosystem activities to have an energy efficiency of 5 percent. The first activity or node in a certain network converts sunlight, while subsequent activities utilize waste from another node in this network.

1. Compare the efficiency of networks consisting of 2, 5, and 10 nodes each.
2. Derive a general formula to determine the efficiency of a network with n nodes. What is the efficiency as $n \to \infty$?

18.6 Consider an ecological network consisting of activities as nodes, with each node having an efficiency of 4 percent. The entire network is run by 1000 W of sunlight, which is converted to useful products by one node. The waste from this node is used as a resource by another node, and so on. The mature ecological network consists of 500 nodes.

1. Calculate the energy efficiency of this network.
2. If 50 nodes are eliminated owing to a disruption, what will be the network efficiency?
3. Calculate the efficiency if 200 nodes are eliminated.
4. What do these results imply about the removal of species from an ecosystem due to local extinction?

18.7 Migration is a popular strategy among many species. Monarch butterflies migrate between the Sierra Madre mountains in Mexico and many locations across the USA and Canada. Answer the following questions about this amazing phenomenon. You may benefit from learning more with the help of sources such as this article `http://nyti.ms/ 2iiT4yV`.

1. What type of control strategy do the wings of monarch butterflies adopt? Is it homeostasis or homeorhesis? Justify your answer.
2. Do migrating groups of monarch butterflies exhibit homeostasis or homeorhesis? You may answer this question with respect to some

characteristics such as their migration path or number of butterflies. Do they have a leader that they can follow during their migration?

3. If you are in North America, some of these migrating monarch butterflies may make it to your location. Look for them during the summer. Inquire about how their location population has changed over the decades.

18.8 The relationship between deer and wolves is considered to be of a predatory nature. Describe how the interaction between grass, deer, and wolves can result in an autocatalytic or self-sustaining system.

18.9 Explain with examples why the release phase of Holling's adaptive cycle is often referred to as creative destruction. It is often claimed that the development of digital photography resulted in creative destruction. Explain this claim based on Holling's cycle as applied to the evolution of photography from analog to digital.

18.10 Some of the most diverse ecosystems are in regions that lack a rich supply of resources. Coral reefs and tropical rain forests are two such examples. Explain how these systems are able to thrive in resource-poor areas. What does this insight imply for developing human-designed systems?

References

[1] S. J. Gould. *Wonderful Life*. Norton & Company, 1989.

[2] E. P. Odum and G. W. Barrett. *Fundamentals of Ecology*, 5th edition. Thomson Brooks/Cole, 2005.

[3] G. W. Barrett, J. D. Peles, and E. P. Odum. Transcending processes and the levels-of-organization concept. *BioScience*, 47(8):531–535, 1997.

[4] US DOE. Carbon cycling and biosequestration: integrating biology and climate through systems science. Report from the March 2008 Worshop. US Department of Energy, Office of Science, 2008.

[5] M. T. Brown, M. J. Cohen, E. Bardi, and W. W. Ingwersen. Species diversity in the Florida everglades, USA: a systems approach to calculating biodiversity. *Aquatic Sciences*, 68(3):254–277, 2006.

[6] B. D. Fath, S. E. Jørgensen, B. C. Patten, and M. Straškraba. Ecosystem growth and development. *Biosystems*, 77:213–228, 2004.

[7] J. Tang, S. Luyssaert, A. D. Richardson, W. Kutsch, and I. A. Janssens. Steeper declines in forest photosynthesis than respiration explain age-driven decreases in forest growth. *Proceedings of the National Academy of Sciences*, 111(24):8856–8860, 2014.

[8] C. S. Holling. Understanding the complexity of economic, ecological and social systems. *Ecosystems*, 4:390–405, 2001.

19 Industrial Symbiosis and the Circular Economy

> It is not the strongest of the species that survive, nor the most intelligent, but the most responsive to change.
>
> Charles Darwin

Many human activities and products are inspired by nature. For example, early man learned about edible species and food sources by observing other animals and ecological cycles, the technology to fly is inspired by birds and insects, and enhancing the strength of materials benefits from an understanding of the structure of strong naturally occurring materials such as spider silk and abalone shells. With increasing realization of the unsustainability of modern activities, there is greater emphasis on mimicking nature in developing new products, processes, and industrial systems. In this chapter, we will learn about several approaches and activities that rely on learning from and emulating nature, and their implications for sustainable engineering. Such efforts for mimicking nature in systems designed by people involve considering nature as a model, mentor, and measure [1].

- *Nature as model* means imitating or being inspired by the way nature does things to meet human needs.
- *Nature as mentor* considers nature as a way of judging the appropriateness of human activities, using the argument that since nature has been around for so long, it has figured out what is right and what works.
- *Nature as measure* implies that our focus should be on learning from nature and not on just extracting from it.

In this chapter, we will see the relevance of each of these goals in the approaches of biomimicry, industrial symbiosis, and circular economy.

19.1 Biomimetic Product Innovation

Developing new products based on mimicking nature is a successful and active area of research and innovation. Successful biomimicry applications may be broadly classified into the following four categories [2]:

Figure 19.1 Biomimicry of a kingfisher's beak to design superfast trains. (a) Photo by author, (b) By tansaisuketti / Wikimedia Commons CC-BY-SA-3.0.

- *Materials*. Examples of materials based on biomimicry include Velcro, inspired by barbs on seeds that attach to fur for dispersal; bioadhesives based on bivalve mollusks; and ceramic designs based on sea shell structures.
- *Movement*. Examples of movement inspired by biomimicry include design of fast trains based on the design of a kingfisher's beak, which can penetrate water with little turbulence and high efficiency, as depicted in Figure 19.1; a flying robot inspired by the common swift; and needles based on a mosquito's injector.
- *Function and behavior*. Biomimetic function and behavior include self-cleaning surfaces based on understanding the structure of lotus leaves; more efficient solar cells based on the internal reflection mechanism used by butterfly wings; processes such as genetic algorithms based on natural selection; and optimization methods based on ant foraging.
- *Sensors*. Sensors developed by mimicking nature include radar based on bat navigation and infrared sensors that mimic a jewel beetle's sensors to detect burnt wood.

Such innovations have resulted in many new products and approaches for meeting various human needs. A natural question about such biomimetic or biologically inspired products is whether they take us closer to sustainability. Unfortunately, the answer is often negative, since the products from biomimicry can have a large environmental impact owing to their reliance on toxic materials and nonrenewable sources of energy. For example, Velcro is made of plastic that relies on nonrenewable resources and will last in our landfills for much longer than the seeds they mimic. Similarly, unlike the ephemeral butterfly wings that they mimic, solar panels rely on nonrenewable resources such as minerals and often involve hazardous waste materials in their life cycle. In fact, evaluation of many biomimicry products by the sustainability assessment methods discussed in previous chapters indicate their high life cycle impact. Thus, developing products by mimicking nature need not be more sustainable, particularly if they are developed by a reductionist biomimicry. However, as we will learn in the next few sections, mimicking nature at a larger, more holistic scale is a promising approach for sustainable engineering.

19.2   Industrial Symbiosis

The goal of industrial symbiosis, also called byproduct synergy, is to establish a network of industrial processes in which waste from one process is used as a resource in another. The aim is to mimic ecological systems, and their self-sustaining nature. As we learned in Chapter 18, the sustainability of ecosystems is achieved by ensuring that waste from one activity is used as a resource in another, resulting in very little net waste from the overall ecosystem. Industrial symbiosis aims to establish such networks of industrial systems. This emphasis of industrial symbiosis on a network of industrial systems as opposed to a single product means that this approach aims toward a more "holistic biomimicry" or "eco-mimicry."

The expected evolution of the industrial network with increasing symbiosis is shown in Figure 19.2. Here, the Type I industrial ecosystem has little symbiosis and large input of resources and large waste streams. As symbiosis increases, the system evolves to Type II, with smaller input and output streams. Finally, the system could develop to Type III with a minimum input of resources and minimum waste. We learned about such development of natural ecosystems in the previous chapter and through Figure 18.11. Some sites with industrial symbiosis are described in Box 18.1.

BOX 19.1 Examples of industrial symbiosis.

Kalundborg, Denmark. Perhaps the best-known example of industrial symbiosis is the network in Kalundborg, Denmark. This network developed by itself over a few decades. As shown in Figure 19.3, it involves the exchange of waste materials such as: hot water, heat, ash, and gypsum from the power station; sulfur from the refinery; and sludge from the pharmaceutical company. Development of the network was facilitated by the executives of various companies knowing each other socially owing to organizations such as a local Lions Club.

Uimaharji, Finland. Figure 19.4 shows the evolution of an industrial ecosystem in eastern Finland that is associated with forestry [4]. The effect of this evolution on various flows is summarized in Table 19.2. It shows an improvement in eco-efficiency for all flows, although the absolute value of the emission of carbon dioxide has increased.

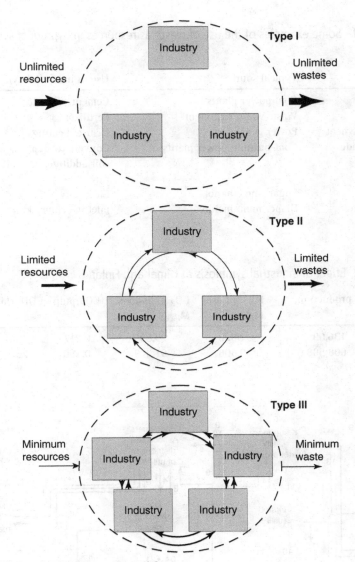

Figure 19.2 Types of industrial ecosystems [3].

Exchange of goods between industrial sectors is common since industries buy each other's products. What is unique in industrial symbiosis is that, like natural ecosystems, the industrial network involves the exchange of waste materials that would otherwise be discarded. Such interaction between industrial systems is much less common. Table 19.1 lists several examples of how waste streams are put to use by means of industrial symbiosis.

Industrial symbiosis can have many benefits from both the economic and environmental points of view. These include significant reductions in energy

Table 19.1 Some examples of the use of waste as resources in byproduct synergy networks.

Waste	Typical source	Use in industrial symbiosis
Ash	Coal power plants	Cement, asphalt
Biosolids	Waste water treatment	Fertilizer
Low-quality heat	Power generation	District heating
Sulfur dioxide	Coal-burning power plants	Convert to gypsum and use as soil additive, drywall manufacture
Bagasse	Sugarcane ethanol	Electricity
Plastic waste	Diaper manufacture	Fuel in cement kiln

Table 19.2 Effect of industrial symbiosis at Uimaharji, Finland.

Year	Pulp production, Mg	Fuel oil use, Mg	CO_2 emissions, Mg	CO_2/pulp	COD, t/d	BOD, t/d
1991	136,095	9828	29,599	0.217	30	5
2002	608,588	30,308	91,279	0.150	20	0.5

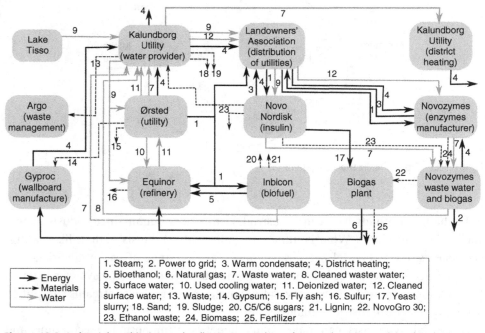

1. Steam; 2. Power to grid; 3. Warm condensate; 4. District heating; 5. Bioethanol; 6. Natural gas; 7. Waste water; 8. Cleaned waster water; 9. Surface water; 10. Used cooling water; 11. Deionized water; 12. Cleaned surface water; 13. Waste; 14. Gypsum; 15. Fly ash; 16. Sulfur; 17. Yeast slurry; 18. Sand; 19. Sludge; 20. C5/C6 sugars; 21. Lignin; 22. NovoGro 30; 23. Ethanol waste; 24. Biomass; 25. Fertilizer

Figure 19.3 Industrial symbiosis at Kalundborg, Denmark as of September 2018. Adapted with permission from the Kalundborg Symbiosis website.

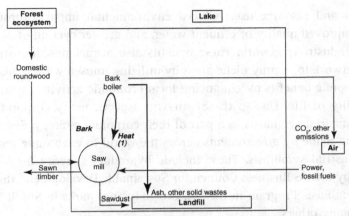

(a) Type I industrial ecosystem, 1955–1966.

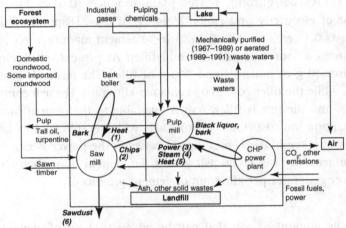

(b) Type II industrial ecosystem, 1967–1992.

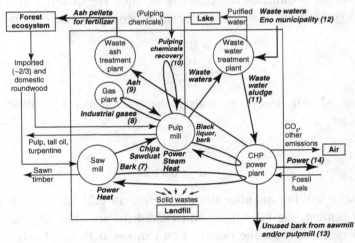

(c) Type III industrial ecosystem, 1992–2003.

Figure 19.4 Evolution of industrial symbiosis in Uimaharji, Finland. Reproduced with permission from [4].

consumption and resource use, reduced environmental impacts through fewer emissions, improved quality of effluent water, and greater cycling of many materials in the industrial network. These benefits also attract new industries to the symbiotic network to occupy niche areas by utilizing unused waste. This results in broader economic benefits by enhancing local economic activities while improving the quality of life. Due to these win–win aspects, many corporations have embraced industrial symbiosis as a part of their corporate strategy. Several efforts are also being made by governments across the world to encourage regional and national industrial symbiosis. These include byproduct synergy networks being encouraged by the US Business Council for Sustainable Development, the National Industrial Symbiosis Program in the UK, eco-industrial parks in South Korea and China, and many others.

Example 19.1   A coal-burning power plant produces 0.0135 kg of ash per kilowatt-hour of electricity and emits 0.994 kg of CO_2. Disposal of this ash by landfilling costs 0.1 cents per kilogram. A local cement manufacturer can add the ash to its kiln as a source of minerals and filler. At present, producing 1 kg of cement requires 10 g of minerals and 300 g of filler. The minerals cost 0.5 cents per kilogram, while the filler costs 0.05 cents per kilogram. The ash contains 2 percent minerals, and the rest is filler. Ash is available to the cement plant without cost, but transportation costs 0.02 cents/kg ash, which is split equally between the two processes. If a symbiosis is established between the power generator and the cement manufacturer, how much ash would the cement manufacturer purchase from the power generator per kilogram of cement produced?

Solution

Let x kg be the amount of ash that can be added to 1 kg of cement. This ash contains $0.02x$ kg of minerals and $0.98x$ kg of filler. The quantity of ash that will satisfy the 10 g (0.01 kg) of minerals/kg cement requirement is calculated as

$$0.02x = 0.01$$

$$x = \frac{0.01}{0.02} = 0.5 \text{ kg}$$

The quantity of ash needed to satisfy the 300 g (0.3 kg) filler/kg cement requirement is

$$0.98x = 0.3$$

$$x = 0.31 \text{ kg}$$

The filler is the limiting quantity since if more than 0.31 kg of ash is added per kilogram of cement, the filler input will exceed the specified limit of 300 g per kilogram of cement. Thus, the cement kiln can use 0.31 kg of ash per 1 kg of cement produced.

Example 19.2   Will the symbiosis discussed above be mutually beneficial from a cost point of view?

Solution

Considering 0.31 kg ash, for the cement producer, minerals and filler are saved, but transportation is an additional cost.

$$\text{Cement cost savings} = (0.02 \times 0.31 \times 0.5) + (0.98 \times 0.31 \times 0.05) - (0.01 \times 0.31)$$

$$= 0.015 \text{ cents}$$

For the power generator, the symbiosis will mean that, per 1 kg of cement, some landfilling cost is saved, but an additional cost of transportation is incurred. Thus,

$$\text{Power generation cost savings} = (0.32 \times 0.1) - (0.01 \times 0.31)$$

$$= 0.029 \text{ cents}$$

Thus, this symbiosis is mutually beneficial.

Despite the economic and environmental benefits of industrial symbiosis, such networks rarely develop by themselves owing to obstacles such as the following.

- Industrial symbiosis requires trust between businesses, which is often missing and takes time to develop. Organizations where people can meet, government incentives, and legal systems are often needed to help develop the needed trust between firms.
- Waste is usually less consistent in quality as compared to virgin resources. This results in reluctance among potential producers and consumers of waste. Potential consumers are less confident of the ability of waste to meet their needs in place of virgin materials. Potential suppliers, whose waste could be a resource for other businesses, may be reluctant to sell it owing to the fear of litigation due to such variability and the low predictability of the quality of their waste.
- Information about the supply and demand of wastes that could be resources is often not readily available. An intermediary such as a non-governmental organization or other mechanism for sharing information between industries is often necessary.
- Symbiosis results in greater interdependence between businesses. This could increase the vulnerability of the overall network since disruption in one business could affect other businesses that rely on its waste products, and lack other alternate sources.

Because of such challenges, the role of facilitating organizations is essential for uncovering industrial symbiosis. One such effort, the Materials Marketplace, is described in Box 19.2.

BOX 19.2 Materials Marketplace

This is a project led by the US Business Council for Sustainable Development (USBCSD) to bring together suppliers and users of waste materials and enable byproduct synergy networks. It relies on advanced cloud-based software into which companies with waste streams or opportunities for using waste enter their information, allowing other companies and project staff to identify symbiosis opportunities. Several regional projects have also been established. Examples of opportunities identified by this effort include the following. The life cycle environmental impact is calculated by Eco-LCA (see Table 10.7).

- Silica sand and ground silica are byproducts that can be used as substitutes for virgin sand; 4200 tons/yr of byproduct is available, which if diverted from landfills is estimated to result in the avoidance of emissions of 1110 MT CO_2eq/yr, with monetary savings of $172,000 per year due to less disposal and value creation.
- Residue from bauxite mining may be added to cement for its fuel value and mineral content. Two million tons/yr of this residue is available. Diverting all of it from storage or disposal to cement production could avoid 14,400 MT CO_2eq/yr, and result in $40 million/yr of savings and value creation.
- Off-spec polymers from a chemical company are used to construct impermeable barrier floors. This reduces the use of fossil and other resources for manufacturing the floors. It is estimated to avoid 313 MT CO_2eq and save $322,000 per year.
- Acetic acid byproduct from chemical manufacturing is used to manufacture acetates for descaling purposes. This displaces the use of petrochemical feedstocks and disposal of the acid. It is estimated to avoid 1408 MT CO_2eq and to save $1,430,000 per year.

For industrial symbiosis to truly contribute to sustainable development, it needs to satisfy the requirements we covered in Chapter 3, such as preventing the shifting of impacts outside the system boundary, and keeping the system within nature's capacity. The sustainability assessment methods from Part III can be used for this purpose, as illustrated in the following example.

Example 19.3 An industrial ecosystem has developed spontaneously around paper manufacturing in the town of Kouvola in southeastern Finland. The life cycle emission of greenhouse gases from the industrial ecosystem and two reference scenarios that could have developed in the absence of industrial symbiosis are shown in Table 19.3. The industrial ecosystem includes a power plant that runs on wood residues and sludge from the pulp and paper mill. This plant also provides electricity and district heat to the town. In reference scenario 1, instead of using electricity from the power plant in the pulp and paper mill, the town would

Table 19.3 Selected emissions (kg) from industrial ecosystem in Kouvola, Finland and two alternate scenarios [5].

Emission	Industrial ecosystem	Reference scenario 1	Reference scenario 2
CO_2	6.20E+08	6.96E+08	7.48E+08
CH_4	1.30E+06	1.42E+06	1.24E+06
N_2O	4.68E+04	5.33E+04	5.12E+04
SO_2	1.31E+06	1.50E+06	2.28E+06
NO_x	3.03E+06	2.99E+06	3.45E+06
NH_3	1.06E+05	1.05E+05	1.04E+05

have had to rely on electricity and heat from average Finnish markets. In reference scenario 2, the town would use heat from peat that is available locally. Compare the life cycles of the three industrial systems in Kouvola in terms of their global warming and acidification potentials.

Solution
Using the characterization factors in Table 11.1, the GWPs for the three systems are

$$\text{GWP}_{\text{IE}} = (6.20 \times 10^8) + (25 \times 1.30 \times 10^6) + (298 \times 4.68 \times 10^4)$$
$$= 6.66 \times 10^8 \text{ kg } CO_2\text{eq}$$

Similarly, $\text{GWP}_{\text{R1}} = 7.5 \times 10^8$ kg CO_2eq and $\text{GWP}_{\text{R2}} = 7.94 \times 10^8$ kg CO_2eq.

For the acidification potential, the characterization factors are 0.7 kg SO_2eq/kg NO_x and 1.88 kg SO_2eq/kg NH_3. Then,

$$\text{AP}_{\text{IE}} = (1.31 \times 10^6) + (0.7 \times 3.03 \times 10^6) + (1.88 \times 1.06 \times 10^5)$$
$$= 3.63 \times 10^6 \text{ kg } SO_2\text{eq}$$

Similarly, $\text{AP}_{\text{R1}} = 3.79 \times 10^6$ kg SO_2eq and $\text{AP}_{\text{R2}} = 4.89 \times 10^6$ kg SO_2eq. This shows that the industrial ecosystem has the smallest GWP and AP, followed by reference scenarios 1 and 2 in that order.

Example 19.4 In the state of Pennsylvania, waste products such as ash, foundry sand, and baghouse dust are being used instead of virgin sand in many tasks such as building and road construction. The quantity of each waste product in thousands of tons is 1499 for coal-derived bottom ash, 279 for other ash, 153 for foundry sand, and 180 for baghouse dust. The fractions of each that are reused are 74, 95, 45, and 92 percent, respectively. The environmental impact of one ton of virgin sand is 0.03 GJ of primary energy, 2.4 kg of CO_2eq, 0.02 kg of SO_2 and 0.02

kg of NO_x. Determine the total reduction in environmental impact in Pennsylvania due to this symbiosis. You may assume a 1 : 1 replacement between the recycled and virgin materials (adapted from [6]).

Solution
Given these data, the amount of virgin sand replaced by waste products is

$$(1499 \times 0.74) + (279 \times 0.95) + (153 \times 0.45) + (180 \times 0.92) = 1609 \text{ thousand tons}$$
$$= 1.609 \times 10^6 \text{ tons}$$

The reduction in environmental impact due to the use of this material instead of virgin sand is

- primary energy: $0.03 \times 1.609 \times 10^6 = 4.827 \times 10^4$ GJ
- carbon dioxide: $2.4 \times 1.609 \times 10^6 = 3.86 \times 10^6$ kg CO_2eq
- sulfur dioxide: $0.02 \times 1.609 \times 10^6 = 3.218 \times 10^4$ kg SO_2
- nitrogen oxides: $0.02 \times 1.609 \times 10^6 = 3.218 \times 10^4$ kg NO_x

This calculation ignores some steps such as waste collection and transportation. It also assumes that the production of the waste streams themselves has no emissions or impact, which implies that no emissions or resources are allocated to the waste products. Given the additional details in [6], the total savings in Pennsylvania are shown in Figure 19.5. Positive values are savings, while negative values are additional impacts.

19.3   The Circular Economy

The concepts of industrial symbiosis and byproduct synergy encourage resource efficiency by closing loops between multiple industrial processes. The concept of a circular economy is similar, but focuses on closing loops for products and materials. As shown in Figure 19.6, the circular economy encourages reuse and extended use by remanufacturing and repair, and keeping materials in the economy by recycling and recovery. Resources that are lost during manufacturing, shown by the arrow leaving the cycle, may be kept in the economy by using them in other industries by industrial symbiosis.

The circular economy goes against conventional thinking that treats materials as resources to be consumed and products as things to be bought continually. Circular economy treats materials as resources to be preserved and products as things whose life needs to be extended for as long as possible. Such thinking has already resulted in changes in manufacturing so that products may be refurbished, disassembled, and remanufactured with greater ease. This may require efforts to avoid the mixing of diverse materials that are difficult to separate for recycling.

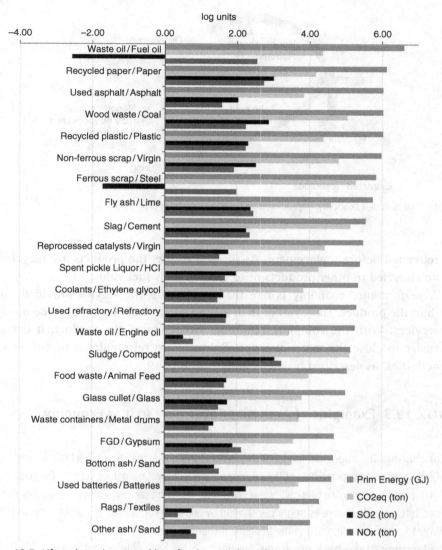

Figure 19.5 Life cycle environmental benefits due to industrial symbiosis in Pennsylvania. Reproduced with permission from [6].

A change in thinking is required from a linear to a circular to a performance economy, as described below.

- A *linear* economy is one in which materials are used once and then discarded. Examples of products in this economy are single-use beverage cups, plastic grocery bags, and electronic parts that cannot be recycled. A linear economy is a Type I industrial ecosystem, shown in Figure 19.2.
- A *circular* economy is one in which materials are reprocessed. Thus, single-use containers are not used, glass bottles are cleaned for reuse, and tires are

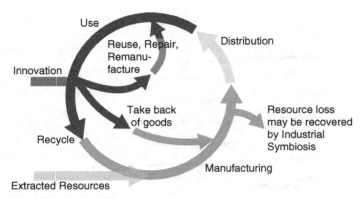

Figure 19.6 Circular economy. Reproduced with permission from [7].

retreaded before replacement. At the end of life, the products are recycled or downcycled to other products or to recover their fuel value.

- A *performance* economy is one that emphasizes the service provided rather than the product. Thus, rather than selling a car, firms would provide mobility services. With the company retaining ownership of the product, it becomes easier to close the material loops. Products are reformulated to enable such activities, as described by the examples in Box 19.3.

BOX 19.3 Examples of Activities Toward a Circular Economy

Car sharing. The traditional model of car ownership is giving way to a model of car sharing, which has many characteristics of the performance economy. Companies such as Car2Go and Zipcar provide the service of mobility by car without requiring car ownership. Companies like Uber, Lyft, and Ola enable even greater sharing by allowing car owners to have the benefits of ownership and providing others the benefits of sharing.

   Electronics take-back programs. Several countries have policies that require manufacturers to take back their products at their end of life. A prominent example is the Waste Electrical and Electronics Equipment (WEEE) Directive of the European Union. Many companies have voluntary product take-back programs. For example, Best Buy stores accept products made by many electronics manufacturers. Companies such as Xerox have redesigned their products to enable disassembly, reuse, or remanufacturing after take-back. This includes changes such as reducing the diversity of materials, in particular those that are difficult to recycle; different mechanisms for putting parts together in such a way that they can be disassembled; and choosing recycled and recyclable materials. This bears similarities to the case of the Volkswagen Lupo 3L that we learned about in Box 17.1.

The circularity of a system may be quantified by means of information about its material and energy flows such as those obtained from methods like energy analysis, material flow analysis, and life cycle assessment. For example, for the material flow in an economy depicted in Figure 12.10, we could calculate quantities such as the following [8]:

- The domestically processed material (DPM) is the quantity of domestic consumption and recycling. Domestic consumption includes imports but not exports.
- The domestically processed output (DPO) is all the material output produced in the system, including gaseous emissions and solid wastes.

These quantities could be used to quantify circularity by calculating the following metrics:

- The stock growth is the fraction of DPM that contributes to the net addition of stock (addition to stocks − demolition).
- The degree of circularity is the fraction of DPM that is recycled.
- The biodegradable flow is the fraction of DPM that is biodegradable.
- The throughput is the fraction of DPM that forms DPO.

These quantities are used to quantify the global economy in the following example.

Example 19.5 Quantify the circularity of the global and European economies, depicted in Figure 12.13.

Solution
For the global economy, the quantities and metrics listed above are calculated as follows.

The DPM include domestic extraction and recycled material. This system has no imports or exports. Thus,

$$DPM = 58 + 4$$
$$= 62 \text{ Gt/yr}$$

The DPO is 41 Gt/yr.

$$\text{Stock growth} = \frac{26 - 9}{62}$$
$$= 0.27$$
$$\text{Degree of circularity} = \frac{4}{62}$$
$$= 0.06$$

Table 19.4 Circularity metrics for the global and European economies in 2005 based on Figure 12.13. [8]

Indicator	Unit	Global	EU
Domestically processed material (DPM)	Gt	62	7.7
	t/cap	9.6	15.8
Stock growth	%	28	22
Degree of circularity	%	6	13
Biodegradable fraction	%	32	28
Throughput	%	66	66

$$\text{Biodegradable fraction} = \frac{20}{62}$$
$$= 0.32$$
$$\text{Throughput} = \frac{41}{62}$$
$$= 0.66$$

A similar approach yields the values for the European economy that are listed in Table 19.4. From these results we can see that the European Union has a higher degree of circularity compared to the global average. However, Europe's per capita consumption is also much higher.

19.4   Summary

Learning from nature is attractive as a source of potential solutions for sustainable engineering. With this motivation, methods such as biomimicry, industrial symbiosis, and circular economy have been developed. Reductionist biomimicry helps in innovating new types of products such as adhesives and self-cleaning surfaces. However, simply mimicking nature in a reductionist manner may not contribute to sustainable development. The concepts of industrial symbiosis, byproduct synergy, and circular economy attempt to mimic ecosystems at a larger scale and constitute holistic biomimicry. The assumption underlying the methods covered in this chapter is that by mimicking ecosystems we will be able to develop human-designed systems that, like natural ecosystems, are self-sustaining. To determine whether this is indeed likely to be true, we considered the six requirements of sustainability assessment covered in Box 3.3. The effort to mimic ecosystems by closing material loops can be combined with life cycle assessment methods to consider multiple spatial scales and multiple flows. This can overcome the tendency for eco-industrial parks to outsource activities that have a low economic bene-fit and high environmental impact. Other requirements for sustainability, such as accounting for multiple disciplines and the supply of ecosystem services, are not considered in the approaches covered in this chapter.

Key Ideas and Concepts

- Reductionist biomimicry
- Type I, II, and III industrial ecosystems
- Linear economy
- Domestically processed materials

- Holistic biomimicry
- Byproduct synergy
- Performance economy
- Degree of circularity

19.5 Review Questions

1. Explain the meaning of using nature as a model, mentor, and measure.
2. Give two examples of materials developed by biomimicry. Why is the approach used for developing them called reductionist biomimicry?
3. What is a Type III industrial ecosystem?
4. What is the benefit of the materials marketplace for bauxite mining?
5. What is a performance economy?

Problems

19.1 Suggest an industrial symbiosis for establishing an eco-industrial park (EIP) based on the industrial processes and their inputs and outputs shown in Table 19.5. The current site only has the pulp and paper plant. This example is based on an EIP in Kymenlaasko, Finland.

19.2 Some results from a study of an EIP in Ulsan, Korea are summarized in Table 19.6. For each flow, determine the time required to recover the investment. What is the total reduction in CO_2 emissions due to this EIP? Is this a "win–win" situation?

19.3 A start-up company is based on the development of a material that can be used to make recyclable carbon fiber composites. This allows snowboards and scraps from snowboard production to be recycled to be used again. The recycled materials are sent back into the snowboard industry and other industries that use the same material in production. The recycling process produces water containing sodium acetate, which is considered waste. (Provided by Emily Ambuehl, Sean Finnessy, Joshua Froats, and Yangyang Zou.)

1. What level of industrial symbiosis is presented in the example?

Table 19.5 Industrial processes in an eco-industrial park in Kymenlaasko, Finland [9].

Process	Inputs	Outputs
Pulp and paper plant	Chlorine dioxide, sodium hydroxide, chlorine, steam, electricity, heat, water (fresh or waste), hydrogen peroxide, calcium carbonate	Steam, water, sodium hydroxide, biomaterials (fuel), sewage sludge, miscellaneous inert waste, electricity, carbon dioxide
Chlorine dioxide plant	Steam, water, external inputs	Chlorine, sodium hydroxide, chlorine dioxide, waste water, waste
Power plant	Sewage sludge, sodium hydroxide, biomaterials as fuel	District heat and electricity, waste water, steam, electricity, heat
Local energy plant	District heat and electricity	Electricity
Calcium carbonate plant	Carbon dioxide, steam, electricity, water	Waste water, calcium carbonate, miscellaneous waste
Municipal waste water treatment plant	Electricity	Sewage sludge
Hydrogen peroxide plant	None	Hydrogen peroxide, waste water
Landfill	Miscellaneous waste, miscellaneous inert waste, waste, ash	None

2. Propose a way in which the sodium acetate water could be used again instead of being waste.

3. Why might another business be hesitant to use the wastewater and recycled materials?

19.4 Selected material and energy flows from a bituminous coal-burning power plant are shown in Table 19.7. The ash and flue gas desulfurization (FGD) residue can be used to produce cement by replacing sand. The environmental impact per ton of virgin sand includes 0.03 GJ of primary energy use, 2.4 kg CO_2eq of greenhouse gas emissions, 0.02 kg of SO_2 emissions, and 0.02 ton of NO_x emissions. Assuming a 1 : 1 substitution by mass, determine the environmental impact of producing 1 kWh of electricity and 1 kg of Portland cement with and without symbiosis.

19.5 A coal-based power plant consumes 500 tons of coal per hour to produce 700 MW of electricity. Sixty percent of the heat content in the coal is lost as heat. The coal contains 20 percent ash. The fuel value of coal is 30,000 kJ/kg. By selling the electricity, the power plant earns a profit of 2×10^{-4}

Table 19.6 Benefits of industrial symbiosis in Ulsan eco-industrial park, South Korea [10].

Material	From	To	Economic benefit		Environmental benefit	
			Investment (million US$)	Profit (million US$/yr)	CO_2 reduction (ton/yr)	Air pollutant reduction (ton/yr)
Steam	Yoosung Corp.	Hankuk Paper	0.85	2.32	19,058	135
Steam	Sung-am MWIF	Hyosung Yongyeon (II)	5	7.1	55,500	176.8
Steam	KP Chemical Hansol EME	SKC	14	6.4	44,468	314
Aldehyde waste water	SK energy	Noksan MWWTF	0.13	1.98	NA	NA
Nutrient for micro-organisms	Dau Metal	Teakwang industry (I)	0.1	3.69	NA	NA
Neutralizing agent	POSCO	LS-Nikko	0.05	1.14	NA	NA
Aluminum chip	Dongnam Fine Hanjoo Metal	Dongnam Fine Hanjoo Metal	0.1	4.44	NA	1325

Table 19.7 Selected material and energy flows for electricity generation from bituminous coal [11].

Flow	Quantity	Unit	Category
	Coal to electricity		
Carbon dioxide	9.94E–01	kg	Air
Electricity	1.00E+00	kWh	Product
Nitrogen oxides	2.71E–03	kg	Air
Particulates	4.15E–05	kg	Air
Sulfur dioxide	6.77E–03	kg	Air
Ash and FGD sludge	1.41E–02	kg	Solid
Bituminous coal	4.42E–01	kg	Raw material
	Portland cement		
Carbon dioxide	3.7E–01	kg	Air
Nitrogen oxides	2.50E–03	kg	Air
Sulfur dioxide	1.66E–03	kg	Air
Bituminous coal	1.07E–01	kg	Fuel
Portland cement	1.00E+00	kg	Product
FGD sludge	6.15E–02	kg	Input
Ash	1.35E–02	kg	Input

per kilowatt-hour. The ash is temporarily stored behind large dams. The power plant is looking for alternative ways of disposing of the ash. Landfilling costs 10^{-6} per kilogram but prices are expected to increase by 4 percent per annum. Another option is to sell the ash to a local cement manufacturer. This option will involve transportation of the ash to the cement plant, which will cost 7.5×10^{-6} per kilogram. (a) When will it be profitable for the power plant to start transporting the ash to the cement plant? (b) By using ash, the cement plant will replace minerals that it buys for 10^{-5} per kilogram. Each kilogram of mineral is equivalent to 1.4 kg of ash. If landfilling remains almost free, how much should the cement manufacturer pay the power plant so that both have no net additional cost?

19.6 An EIP consists of three processes: leather processing, protein hydrolysates manufacturing, and plastofill production. Leather processing produces 65×10^6 m^2/yr of tanned leather, 85,000 ton/yr of chrome liquors, 80,000 ton/yr of fleshing, and 4.6×10^6 m^3/yr of waste water. Chrome liquors are processed to recover the chromium sulfate that is used in the tanning process. The recovery rate is 1 ton chromium sulfate per 17 tons of chrome liquor. The yield of protein hydrolysates is one ton per 16.7 tons of fleshing. This product is used as fertilizer. The waste water is treated in a facility that produces one ton of sludge per 70 m^3 of waste water. The plastofill process converts 1 ton of sludge into 0.157 ton of plastofill, which is a filler for asphalt. (a) On the basis of this information, is it possible to have an EIP with zero net waste? What is the minimum amount of waste that must be produced? (b) Each of the three processes relies on fossil fuel-based energy and emits 0.2 ton of CO_2 per ton of tanned leather, 0.02 ton of CO_2 per ton of chromium sulfate, 0.01 ton of CO_2 per ton of protein hydrolysate, and 0.05 ton of CO_2 per ton of plastofill. Determine the CO_2 emission of the EIP. (c) Establishment of the EIP replaces conventional processes for the production of chromium sulfate, protein hydrolysate, and plastofill. The tons of CO_2 emitted per ton of each product from these processes are 0.015, 0.008, and 0.045, respectively. Determine the change in CO_2 emissions due to the symbiosis. Does the quantity of the CO_2 emission justify the EIP? (Based on a study in [12].)

19.7 Using the last three rows of the data in Table 19.3, compare the impact of the industrial ecosystem with the two reference scenarios. For all the emissions in the table and results in Example 19.3, identify trade-offs, if any, between the three cases.

19.8 Calculate circularity metrics for the paper flow system depicted in Figure 12.12. Compare this system with the global and European economies. What can you conclude about the relative circularity of the paper system? Identify opportunities for making this system more circular.

19.9 We learned in Chapter 2 about the islands of trash in global oceans (Figure 2.7). Will the concept of a circular economy be able to address this problem? Describe the changes that will be needed in this product and its use to make its economy circular. You may assume the product to be a polyethylene grocery sack.

19.10 Two common examples of activities needed to develop a performance economy are car sharing and electronics take-back (see Box 19.3). Sketch an approximate life cycle system boundary of these activities with and without sharing. Under what conditions would a sharing economy result in a smaller life cycle impact? Consider various practical scenarios and state your assumptions.

References

[1] J. M. Benyus. *Biomimicry: Innovation Inspired by Nature*. Harper Collins, 1997.

[2] E. Lurie-Luke. Product and technology innovation: what can biomimicry inspire? *Biotechnology Advances*, 32(8):1494–1505, 2014.

[3] T. E. Graedel and B. R. Allenby. *Industrial Ecology and Sustainable Engineering*, 2nd edition. Prentice Hall, 2010.

[4] J. Korhonen and J.-P. Snäkin. Analysing the evolution of industrial ecosystems: concepts and application. *Ecological Economics*, 52(2):169–186, 2005.

[5] L. Sokka, S. Lehtoranta, A. Nissinen, and M. Melanen. Analyzing the environmental benefits of industrial symbiosis. *Journal of Industrial Ecology*, 15(1):137–155, 2011.

[6] M. J. Eckelman and M. R. Chertow. Quantifying life cycle environmental benefits from the reuse of industrial materials in Pennsylvania. *Environmental Science & Technology*, 43(7):2550–2556, 2009.

[7] W. Stahel. The circular economy. *Nature*, 531(7595): 435–438, 2016.

[8] W. Haas, F. Krausmann, D. Wiedenhofer, and M. Heinz. How circular is the global economy? An assessment of material flows, waste production, and recycling in the European Union and the world in 2005. *Journal of Industrial Ecology*, 19(5):765–777, 2015.

[9] L. Sokka, S. Pakarinen, and M. Melanen. Industrial symbiosis contributing to more sustainable energy use: an example from the forest industry in Kymenlaakso, Finland. *Journal of Cleaner Production*, 19(4):285–293, 2011.

[10] S. K. Behera, J.-H. Kim, S.-Y. Lee, S. Suh, and H.-S. Park. Evolution of 'designed' industrial symbiosis networks in the Ulsan eco-industrial park: 'research and development into business' as the enabling framework. *Journal of Cleaner Production*, 29–30:103–112, 2012.

[11] National Renewable Energy Laboratory. U.S. life cycle inventory database. www.lcacommons.gov/nrel/search, accessed November 3, 2014.

[12] D. M. Yazan, V. A. Romano, and V. Albino. The design of industrial symbiosis: an input–output approach. *Journal of Cleaner Production*, 129(Supplement C):537–547, 2016.

20 Ecosystems in Engineering

> If Judeo-Christian monotheism took nature out of religion,
> Anglo-American economists (after about 1880) took nature out of
> economics.
>
> J. R. McNeill [1]

If Judeo-Christian monotheism and Anglo-American economists took nature out of their respective fields, as stated in the above quote, technology since the Industrial Revolution has done the same for engineering. As we learned in Chapter 19, many efforts toward the sustainability of engineering activities are inspired by ecological systems, and attempt to learn from nature. If we apply the requirements for sustainability that are summarized in Box 3.3 to these methods, we realize that approaches such as industrial symbiosis and circular economy do not take into account the ecological constraints imposed by the limited capacity of nature to provide resources and absorb emissions. In fact, ecosystems are kept outside the boundary of these methods, making them techno-centric in nature. In this chapter, we will learn about methods that explicitly include ecosystems in engineering decisions and could enable environmental sustainability by staying within ecological constraints. As we will see, many such approaches are very old. These practices became much less popular in the last few centuries, for reasons such as the dominance of high-quality and cheap fossil energy, the spread of Western values, and the availability of other technological solutions. However, due to their potential for greater environmental friendliness, they need to be rediscovered and adapted to help address modern challenges. We will learn about such methods in this chapter, from traditional knowledge to emerging solutions based on including nature in engineering.

20.1 Traditional Ecological Knowledge

Traditional ecological knowledge has been defined as "the cumulative body of knowledge, practice, and belief, evolving by adaptive processes and handed down through generations by cultural transmission, about the relationship of living beings (including humans) with one another and with their environment" [2]. It includes knowledge about the ecosystems available among indigenous

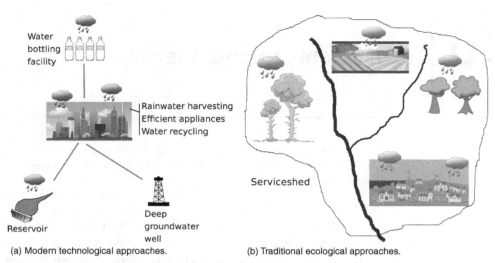

(a) Modern technological approaches.   (b) Traditional ecological approaches.

Figure 20.1 Approaches for obtaining fresh water.

populations that has been used for sustaining human societies and enhancing human well-being. Many such knowledge systems and practices have been lost, but some continue to be used in traditional societies even today. This knowledge is based on a deep understanding of local landscapes and ecosystems, and has been used to manage many systems in a sustainable manner. This is in contrast to more recent ways of satisfying human needs, which can cause much environmental harm and are often unsustainable.

As an example, consider methods for obtaining fresh water for human activities. As depicted in Figure 20.1a, the modern, technology-based approach relies on building large dams to create reservoirs, and canals to transport the water over long distances. Bottled water is also popular, and rainwater harvesting is used in some regions. This approach has brought water and prosperity to many areas of the world, but at the cost of pollution and ecological destruction, and the forced migration and resettlement of many communities. In addition, the building and operation of such systems relies on large inputs of nonrenewable resources, and many systems are plagued by problems such as a reduction in reservoir capacity due to silting and the salinization of agricultural soils due to too much water use in previously arid regions and due to increase in the water table. Many such modern practices have resulted in a tragedy of the commons, as in the examples of the Aral Sea and the Colorado River discussed in Chapter 2. In addition, the approaches used across the world, be it for the Hoover Dam in the USA, Aswan in Egypt, Sardar Sarovar in India, Three Gorges in China, or Belo Monte in Brazil, are similar, with limited adaptation to the local landscape, ecology, and society.

In contrast, as depicted in Figure 20.1b, the traditional approach practiced for centuries before the advent of modern dams and canals relied on the protection of rivers and forests and low-impact farming practices. Small-scale technologies such as small earthen dams and weirs, terraced land, and gullies for transporting water by gravity flow were also used. These were combined with land management practices such as the rotation of grazing lands based on tracking local ecological conditions, and with simple rules of thumb or traditions based on wisdom collected over the generations for adaptive management [3].

Often, modern methods of managing natural resources tend to be directed toward maintaining an equilibrium, such as obtaining a maximum sustainable yield of fish or timber. This yield is often calculated using detailed scientific models of the ecosystem along with various economic aspects, and such policies are implemented in a top-down manner by people who may not be part of the managed system. Such approaches attempt to impose homeostasis on a system that is naturally inclined toward homeorhesis. We learned about these concepts in Chapter 18. Thus, these modern approaches implicitly assume ecological stability, and may cause a gradual decay in the resilience of the overall system, moving the system toward thresholds and surprises such as the collapse of fisheries that we learned about in Chapter 2. The gradual loss of resilience is often addressed by increasing reliance on fossil-based alternatives such as artificial fertilizers and bigger fishing boats.

In contrast, the traditional approach to resource management may be based on rules developed locally and enforced by the resource users themselves, often in the form of societal norms and taboos. Such rules typically focus on maintaining system resilience as opposed to maximizing the yield of a single species. We will learn more about such societal approaches toward sustainability in Chapter 22. This approach relies on accumulated ecological wisdom passed down the generations, and includes feedback mechanisms that ensure overall system resilience. Unlike relatively rigid rules prohibiting catching species smaller than a specified size, or continuing to farm on the same land, as done in conventional management, traditional management tends to be more flexible and allows rotation in farming location and in the size of species caught at a given time. Such practices are often captured by holistic worldviews that consider humans to be part of nature, where all beings interact.

Some specific examples of traditional ways of meeting human needs that are practiced even today or have been rediscovered are described in Boxes 20.1 and 20.2. The development of such approaches requires many years of accumulated local knowledge that is usually distilled into simple rules and conventions, often through religious practices. It requires closeness with nature, quite in contrast to the modern trend of decreasing ecological knowledge with increasing prosperity, which we learned about in Chapter 7. Many such traditional approaches have been forgotten or are not suitable for today's more populated, consumption-oriented, and globalized world. However, opportunities exist for learning from traditional approaches, and rediscovering and adapting them to today's conditions. We will learn about some such approaches in the rest of this chapter.

BOX 20.1 Traditional Ecological Knowledge for Water Management

Many water acquisition and management methods have been developed in the arid and semi-arid parts of the world. In the eastern part of Rajasthan state in India, the water table had been depleting owing to excessive withdrawal for agriculture and opencast mining activities. The mines would punch through the aquifer, thus contributing to its depletion and degradation. Efforts to build dams and reservoirs on the local rivers and bring water by canals were not very successful because of the low rainfall and high evaporation rates in the desert sun. These hardships prompted a local non-governmental organization, *Tarun Bharat Sangh* (TBS), to take the mining companies to court, which resulted in curtailment of their most destructive activities. With support from local villages, TBS rejuvenated their local traditional technique of building small earthen dams to collect rainwater, not for storage in an above-ground reservoir as done in modern engineering but to allow the accumulated water to seep underground and recharge the aquifer. Sites were selected on the basis of local knowledge about underground fissures in the rocks and the location of aquifers. This approach has resulted in the revival of several local rivers and improved agricultural productivity and local well-being. It has enhanced many ecosystem services and satisfies human needs with virtually no reliance on nonrenewable resources. It is an example of how traditional ecological knowledge can be adapted to modern times.

BOX 20.2 Water Temples of Bali, Indonesia

An irrigation system practiced in Bali, Indonesia that has survived for over a thousand years relies on a complex interplay between the farmers, priests, and rainfall patterns. The following quote is from a statement describing the inclusion of this site among the UNESCO World Heritage Sites [4]:

A line of volcanoes dominate the landscape of Bali and have provided it with fertile soil which, combined with a wet tropical climate, make it an ideal place for crop cultivation. Water from the rivers has been channelled into canals to irrigate the land, allowing the cultivation of rice on both flat land and mountain terraces.

Rice, the water that sustains it, and subak, the cooperative social system that controls the water, have together shaped the landscape over the past thousand years and are an integral part of religious life. Rice is seen as the gift of god, and the subak system is part of temple culture.

Water from springs and canals flows through the temples and out onto the rice paddy fields. Water temples are the focus of a cooperative management of water resource by a group of subaks. Since the 11th century the water temple networks have managed the ecology of rice terraces at the scale of whole watersheds. They provide a unique response to the challenge of supporting a dense population on a rugged volcanic island.

The overall subak system exemplifies the Balinese philosophical principle of Tri Hita Karana that draws together the realms of the spirit, the human world and nature. Water temple rituals promote a harmonious relationship between people and their environment through the active engagement of people with ritual concepts that emphasise dependence on the life-sustaining forces of the natural world.

This system is able to have high productivity without the use of artificial fertilizers, pesticides, or mechanization. In fact, efforts were made in the 1970s to force farmers to shift to the "modern" approach of the green revolution, but it quickly turned out to be unsustainable owing to pest outbreaks and pollution. Fortunately, they were allowed to revert to their traditional subak system, which continues to be practiced today.

20.2 Nature-Based Solutions

Efforts to rely on nature for meeting human needs have originated in many disciplines and will be the focus of this section. These include ecological engineering, which originated in ecology and environmental engineering, and green infrastructure, which has roots in landscape design and urban planning.

20.2.1 Ecological Engineering

Ecological engineering involves the development and use of various types of ecological systems to satisfy human needs. It has been defined as "the design, construction, operation and management (that is, engineering) of landscape/aquatic structures and associated plant and animal communities (that is, ecosystems) to benefit humanity and, often, nature" [5]. The definition continues: "other terms with equivalent or similar meanings include ecotechnology and two terms most often used in the erosion control field: soil bioengineering and biotechnical engineering. However, ecoengineering should not be confused with 'biotechnology' when describing genetic engineering at the cellular level, or 'bioengineering' meaning construction of artificial body parts." For thousands of years, human beings have practiced some form of ecological engineering in the form of traditional farming and water management practices. With increasing concern about the degradation of ecosystems, ecological engineering was proposed in the 1960s as a way to "fashion synthetic systems partly under old energy budgets of nature and partly with special power take-off from civilizations" [6]. In this section we will learn some basics of ecological engineering and its applications, such as wetlands for water treatment and "green infrastructure" for meeting urban needs.

Concepts and principles. Some basic concepts of ecological engineering are discussed here, with emphasis on those that are different from conventional engineering practice. More details are available in sources such as [7].

Ecological engineering exemplifies *systems thinking*, that is, it takes a holistic viewpoint that considers operation of an entire ecosystem instead of just its component parts. Thus, even though ecological engineering projects have specific goals, they try to balance meeting those goals with maintaining ecological benefits.

A properly designed engineered ecosystem should rely only on *natural renewable energy*, and should be self-sustaining, with only limited human intervention. Unlike conventionally engineered technologies, ecologically engineered technologies may require an input of nonrenewable energy at the beginning, but then should be sustainable with solar energy.

Ecosystem conservation is an essential aspect of ecological engineering. This goal must be satisfied while satisfying human needs. Consequently, ecological engineering will not indulge in activities that destroy or disturb nature, unless this is absolutely essential. Since ecosystems, communities, species, and organisms are the tools of ecological engineers, disrupting any of them does not make sense. This attitude represents a conservation ethic like that stated by Aldo Leopold and summarized in the quote at the start of Chapter 17.

The design of engineered ecosystems relies on the approach of *self-design*, which refers to the ability of ecosystems to design themselves or self-organize in a bottom-up manner without any top-down control. Left to their own devices and with access to resources such as nutrients, water, and various species, ecosystems can develop without any human intervention or imposed top-down organization. We learned about ecological development during primary and secondary succession in Chapter 18. The result is a system that is highly flexible and resilient, and capable of providing a variety of goods and services. Such systems can adapt to external influences and sustain themselves for long periods of time, which is a desirable property for sustainability.

However, such self-designed systems often lack the predictability and performance guarantee of conventionally engineered systems developed by the *imposed design* approach. For example, the ability of ecosystems to treat pollutants can depend on factors like seasons and cloud cover, while the ability of conventional waste treatment is much more predictable. Systems developed by conventional design are best at doing the single task or the few tasks they are designed to do, with high predictability. However, such systems usually require high-quality inputs

Table 20.1 Ecological engineering versus conventional engineering: typical differences [5].

Category/characteristic	Conventional engineering (imposed design)	Ecological engineering (self-design)
Project goal	Single purpose	Multiple benefits
Benefits to the ecosystem	Low priority	High priority
Structures	Concrete and steel, human-made	Landscape/aquatic features, naturalized
Energy source	Fossil-fuel combustion, electricity	Solar, gravity, plants, animals
Material movement mechanisms	Pumps, blowers, conveyors	Convection, gravity, plant/microbial processes
Processes	Human-driven, human-regulated	Natural, self-regulated
Climate and landscape setting	Relatively unimportant	Critical
Useful lifespan	Relatively short	Relatively long
Performance	Controlled	More variable
Robustness	Often low	Usually high
Operation and maintenance costs	High	Low
Land requirements	Low	High

of resources such as fossil energy and purified minerals (metals). The trade-off between imposed and self-design is summarized in Table 20.1. Since conventional engineers are trained to do imposed or top-down design, they have mostly ignored or considered inferior the ability or use of ecosystems. Perhaps this is also due to the attitude of dominating nature that we learned about in Section 6.1. Given the complementary nature of conventional engineering and ecological engineering, one way of meeting some of the challenges of sustainability is to find the appropriate balance between the use of these approaches. This would require greater interaction between engineers and ecologists. We will learn about such approaches in Section 20.3.

Wetlands for water treatment. One of the most successful applications of ecological engineering is for treating water in wetland ecosystems. Wetlands are areas of land that are saturated with water for at least part of the year. These areas are traditionally referred to by names such as swamps, marshes, and bogs. Their ability to improve water quality is described in Section 2.8, and has been known for many centuries. Relying on this ability was common in societies all over the world. Box 20.3 describes one such wetland that is still operating, near Kolkata, India.

With increasing reliance on modern technology, and greater pressure from urbanization, population growth, and consumerism, the role of wetlands has been gradually forgotten in many parts of the world, and a very large number of wetlands have been drained or have become dumping grounds for waste water from domestic and industrial sources, and for other types of waste. Such changes invariably perturb wetlands beyond their natural carrying capacity, resulting in their severe degradation. In addition, many wetlands have been filled to obtain land for farming and urban development, as we learned in Box 2.2 for the Black Swamp of Ohio. The extent of decline in wetlands all over the world is shown in Figure 2.20.

In the last few decades, the natural ability of ecosystems to treat waste has been being rediscovered. Wetlands and other ecological systems are making a comeback [8] and societies are once again beginning to appreciate the ability of wetlands to treat many pollutants. Types of water treatment problems that wetlands have been effective for include: the removal of suspended solids; reducing chemical and biological oxygen demand, mainly due to microbial growth; removing nitrogen and phosphorus due to take-up by plants and soils; capturing organic chemicals including chlorinated hydrocarbons, phenols, surfactants, pesticides, and cyanides, which can be removed through processes such as volatilization, plant uptake, and photochemical oxidation; and removing halogens, sulfur, metals, and metalloids to varying extents.

Commonly used treatment wetlands may be categorized as surface flow and subsurface flow wetlands. Free-water surface flow wetlands look like natural wetlands and attract fish, birds, and other wildlife. Such wetlands are popular for treating urban stormwater and agricultural runoff and require little input, making them very cost-effective. The wetlands described in Boxes 20.3 and 20.4 are of this type. Subsurface flow wetlands may have horizontal or vertical flow. They keep the water below the soil surface, which may be important for suppressing odors and human or wildlife contact. These wetlands are commonly used for the treatment of household waste water and industrial wastes. They do require some maintenance but tend to be less expensive than traditional solutions, as we will see in Section 20.3. A typical schematic of such a wetland is shown in Figure 20.2. Vertical flow wetlands may be used in a batch manner by putting water on top and allowing it to percolate through the bed. This approach can enhance oxygen transfer and nitrification. Water on the surface can also be used to block oxygen transfer and create a reducing environment at the bottom, which can immobilize metals.

BOX 20.3 Traditional Treatment Wetlands in Kolkata, India

The East Kolkata Wetlands [9] treat urban sewage, provide fertilizer, grow rice and fish, support biodiversity, and enable the livelihood of thousands. All of this is accomplished in a cost-effective manner, and the entire system functions with little supervision or input of nonrenewable resources. It is the world's largest aquaculture system based on sewage. It was established in the late 1800s and has gone through phases of degradation and loss. Of the approximately 8000 hectares in 1945 only about 5000 hectares remain today. It is now protected by the Ramsar Convention.

BOX 20.4 Industrial Treatment Wetlands in Victoria, Texas

DuPont built a wetland to treat waste water from its Nylon Intermediates manufacturing plant in Victoria, Texas. The water entering this wetland is high in carbonaceous and nitrate content. It occupies about 22 hectares and receives about 12,000 m^3/day of water after some pretreatment in a biological plant (an anoxic to oxic reactor). The first stage of wetland treatment is provided by five cells in parallel, followed by two cells with deep zones, and a final stage that is mainly wildlife habitat. This facility has worked well, with COD removal above 99 percent, the complete removal of nitrite and nitrate, and 100 percent permit-compliance. Another successful experiment at this plant was with the use of biosolids from the waste water treatment to produce biomass that is used as animal feed in the local area. In addition, this site has become very popular with visitors, including school groups and bird watchers. Before implementation of this wetland, this plant used deep-well disposal to get rid of its waste. Success of the wetland strategy required process changes to permit the biological treatment of its wastes.

Treatment wetlands do have some *limitations*, particularly for many modern waste streams. Wetlands usually require a larger land area than a conventional treatment process. Given their self-designed nature, their performance may vary seasonally, and, if strict emissions standards are to be met, wetlands may not be able to satisfy the requirements by themselves. Furthermore, if a wetland is not managed properly it can be a source of mosquitoes or odors owing to the anaerobic environment. Various techniques are available for preventing such problems. Despite the many attractive environmental characteristics and economic feasibility of wetlands, most current uses of wetlands are as end-of-pipe or add-on treatment systems that are not fully integrated with the technological system they support.

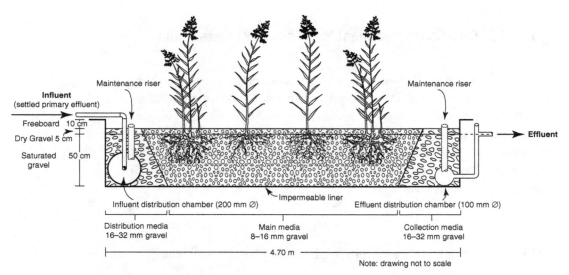

Figure 20.2 Schematic of typical subsurface flow treatment wetland. Reproduced with permission from [10].

Example 20.1 A wetland may be modeled as reactors in series. Using such an approach, the capacity of a wetland to remove nitrates may be represented by the following equation:

$$\frac{C_o - C^*}{C_i - C^*} = \left(1 + \frac{k\tau_n}{Nh}\right)^{-N}$$

$$\overset{N \to \infty}{=} \exp(-k\tau_n/h)$$

Here, C_i, C_o, and C^* are the inlet, outlet, and background nitrate concentrations, respectively in mg/L; h is the water depth in m; k is the apparent rate coefficient in m/day; N is the number of tanks in series; and τ_n is the nominal detention time in days. The nominal temperature is 20 °C and the background concentration of nitrate in natural wetlands is known to be zero. The input stream has a concentration of 0.68 mgN/L, the wetland depth is 0.5 m, the rate coefficient is 25 m/year, and the detention time is two days. As a result of nonidealities such as shortcircuiting in the flow, the wetland is equivalent to 7.2 tanks in series. If the flowrate of polluted water is 10 m³/day, determine the outlet concentration and area of the wetland.

Solution
For this example, $C_i = 0.68$ mgN/L; $C^* = 0$ mg/L; $k = 25$ m/yr; $N = 7.2$; $h = 0.5$ m; $\tau_n = 2$ days. Then we get

$$\frac{C_o}{0.68} = \left(1 + \frac{(25/365) \times 2}{7.2 \times 0.5}\right)^{-7.2}$$

$$= 0.764$$

Thus, the outlet concentration is $C_o = 0.52$ mg/L.

The volume of wetland is

$$V = F_i \tau_n$$
$$= 10 \times 2$$
$$= 20 \text{ m}^3$$

Given that the wetland depth is 0.5 m, its land area is

$$A = V/h$$
$$= 40 \text{ m}^2$$

20.2.2 Green Infrastructure

Green infrastructure refers to natural systems that provide goods and services for meeting human needs. Conventional technological solutions are known as gray infrastructure. These terms and approaches originated in landscape design and urban planning, and are used mainly to address issues such as treating stormwater runoff, maintaining air quality, and harvesting rainwater. Given the definition of ecological engineering that is provided in Section 20.2.1, green infrastructure may be considered to be a type of ecological engineering directed toward urban systems and land use.

In the urban context, categories of green infrastructure include the following:

- *Tree canopy* includes street trees, urban forestry, shrubs, and forests.
- *Green open spaces* includes ground covers, open land, wetlands, and greenways.
- *Green roofs* of many types have been developed to provide a vegetative cover on building roofs.
- *Vertical greenery systems* involve vines and plants growing vertically.

Various ecosystem service benefits from trees and wetlands regarded as green infrastructure are illustrated in Figure 20.3. Wetlands can remove water pollutants, while trees can remove air pollutants. In addition, both types of ecosystems provide goods and services such as oxygen, pure water, and biomass to industrial systems. They also provide many other co-benefits to society, such as aesthetic value, recreational opportunities, flood regulation by wetlands, etc. For many tasks, grey and green alternatives are available, as shown in Table 20.2. As we can see in this table, these alternatives cover a spectrum of options between gray and green. Some practical uses of green infrastructure are described in Box 20.5.

Example 20.2 In C_3 plants (trees and temperate grasses), the water lost by evapotranspiration during the summer is 0.4 to 0.66 tons H_2O per day. This process

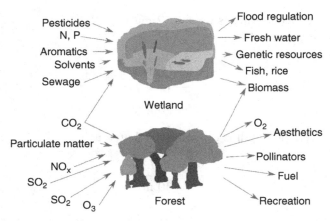

Figure 20.3 Ecosystem services from wetlands and forest.

Table 20.2 Gray and green infrastructure options for various tasks.

Task	Gray	Intermediate	Green
Water quality regulation	Anaerobic baffle reactor	Constructed subsurface wetland	Natural wetland
Air quality regulation	Scrubber	Tree plantation	Natural forest
Flood regulation	Storm sewer	Blue or green roof	Bioswale

BOX 20.5 Green versus Gray Infrastructure

We describe some practical examples in which green infrastructure has been adopted and found to be more attractive than the conventional gray alternatives.

   Drinking water supply. New York City draws its drinking water from the Catskill Mountains. With deterioration of the existing infrastructure, in the 2000s the city needed to determine an approach for maintaining the quality of its drinking water. It was determined that if the city spent money on protecting the Catskill watershed by conserving land and its cover, it would be much cheaper than building conventional water treatment facilities. The green infrastructure was estimated to require setting aside $300 million over ten years to acquire land and restrain development, while the gray alternative would have cost $8 billion.

   Air quality. A collaboration between the Nature Conservancy and Dow Chemical has been exploring the benefits of green infrastructure to industry. A pilot study at Dow's site in Freeport, Texas considered the role of trees in mitigating NO_x. The following quote summarizes some of its findings [11]:

   Analysis of a test case in southeast Texas found that reforestation can be a cost-competitive method to generate NO_x emissions credits. The hypothetical 1,000 acre project, estimated to cost

$470,000 to implement, would be expected to remove quantities of ozone and NO_2 equivalent to a range between 122 and 202 tons of NO_x total over the next 30 years, at a cost of $2,400 to $4,000 per ton of NO_x. The cost per ton compares to the estimated industry NO_x control costs range, from about $2,500–$5,000 per ton of NO_x for traditional NO_x abatement strategies. The analysis assumes that reforestation could occur on land owned by Dow, the State of Texas or the Conservancy. The analysis also assumes the planting of tree seedlings.

takes heat from the surroundings, resulting in a cooling effect. Knowing that the latent heat of evaporation of water is 540 kcal/kg, calculate the cooling effect of trees. If this cooling effect translates directly into less air-conditioning for the houses in the neighborhood, calculate the monetary savings per day given that electricity costs 15 cents/kWh.

Solution
For the evaporation of 0.4 tons of water,

$$\text{Cooling provided} = 400 \text{ kg } H_2O/\text{day} \times 540 \text{ kcal cooling/kg } H_2O$$
$$= 2.16 \times 10^5 \text{ kcal of cooling/day}$$

Similarly, the evapotranspiration of 0.66 tons of water results in 3.564×10^5 kcal of cooling/day.

In units of kWh, this amount of cooling is equal to from 251 kWh/day to 414 kWh/day. Therefore, the cost saving due to the lower value is given by

$$\text{Cost saving} = 251 \text{ kWh/day} \times 0.15 \text{ \$/kWh}$$
$$= 37.65 \text{ \$/day}$$

The range of savings in the neighborhood is 37.65–62.1 $/day.

20.3   Techno-Ecological Synergy

20.3.1   Motivation

As we learned in Chapter 1, sustaining human well-being is critically dependent on the availability of goods and services from ecosystems. Despite ecological degradation, which we learned about in Chapter 2, and the need to stay within ecological capacity, as discussed in Chapter 3, most methods for assessing and designing sustainable systems ignore the role of ecosystems, particularly their capacity to supply goods and services. The approaches based on biomimicry that we covered in Chapter 19, such as industrial symbiosis and circular economy, take inspiration from nature, but ignore its role in supporting human systems. These methods are techno-centric owing to their primary focus on technology and its impact on ecosystems. In contrast, approaches based on traditional ecological knowledge and ecological engineering that we covered in Section 20.2 are more eco-centric because of their focus on ecosystems and ways of using their natural capacity to support conventional technological activities. However, using them to meet the needs of modern society can be challenging owing to factors such as their self-designed nature.

As an illustration of the differences between techno-centric and eco-centric methods, let us revisit ways of satisfying human water requirements. As we learned in Section 20.1, eco-centric methods focus on the role of rain, land use, rivers, and the hydrological cycle in satisfying water requirements, as illustrated in Figure 20.1b. This reflects the ecologist's inclination toward avoiding ecological disruption and discovering how nature works. While such an approach may provide greater ecological protection, the system may not be adequate for meeting human needs, particularly in a consistent, continuous and predictable manner, as indicated by the differences summarized in Table 20.1. However, eco-centric approaches can be inherently sustainable and have little or no dependence on nonrenewable resources.

In contrast, to satisfy water requirements, techno-centric methods consider the use of reservoirs and dams, pipes, bottling facilities, and tubewells to withdraw groundwater, as illustrated in Figure 20.1a. Such technological, gray infrastructure approaches reflect the engineering inclination to design things with maximum human control and predictability. This tendency of engineering to control and dominate nature often ignores eco-centric options and nature's natural ability to meet many human needs, particularly if it is not very well understood. As ecological literacy decreases in society, and particularly among engineers (see Section 7.2), the link between engineering and ecology weakens and the gap between technological solutions and ecological systems increases, as has been happening for several decades.

As depicted in Figure 20.4, the techno- and eco-centric approaches lie at opposite ends of the techno-ecological spectrum. Techno-centric sustainable engineering analysis methods such as life cycle assessment and various footprints; solutions such as industrial symbiosis and circular economy encourage products and technologies with a smaller life cycle impact and greater circularity of products, but without considering whether the supporting ecosystems can sustain them or whether ecosystems can provide an alternate solution. In contrast, the eco-centric approaches of traditional ecological knowledge and ecological engineering are usually not integrated with technological methods and alternatives. For example, treatment wetlands are usually developed for end-of-pipe treatment instead of being fully integrated with conventional equipment during process design. Thus, eco-centric approaches tend to ignore the needs of the technological systems that

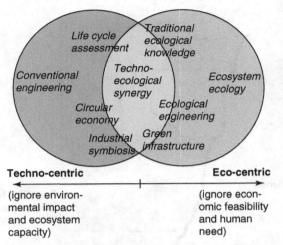

Figure 20.4 Bridging techno- and eco-centric methods by techno-ecological synergy.

they may be supporting, while techno-centric approaches tend to ignore nature's ability to satisfy many human needs.

Techno-ecological synergy (TES) is an approach that encourages combination of the features of techno- and eco-centric design methods. It recognizes that synergies could exist between technological or gray infrastructure and ecological or green infrastructure systems, as shown in Figure 20.5: healthy forests purify and sustain the supply of water, which permits the use of less expensive and environmentally friendlier technological infrastructure for further water purification and transport, which in turn encourages protection of the forest ecosystem. Thus, TES lies at the intersection of techno- and eco-centric methods, as shown in Figure 20.4.

20.3.2 Approach

A general representation of techno-centric approaches such as industrial symbiosis and the circular economy is shown in Figure 20.6. These methods consider a single technological system, such as a manufacturing process, or a network of technological systems, such as an industrial ecosystem, in order to reduce their environmental impact, which is certainly an essential requirement for sustainable development. However, these approaches ignore the contribution from ecosystems and their capacity to provide the goods and services that can satisfy the needs of industrial and other human activities. Keeping nature outside the system boundary has at least two disadvantages:

1. Unintended harm due to decisions that unknowingly rely on degraded ecosystem services.

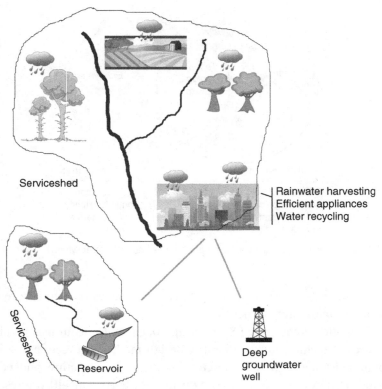

Figure 20.5 Techno-ecological synergy combines technological and ecological approaches for obtaining fresh water.

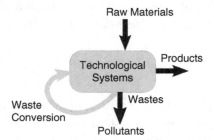

Figure 20.6 Techno-centric methods such as industrial symbiosis and the circular economy, consider networks of technological systems.

2. Lost opportunities for developing innovative win–win solutions that benefit from the ability of ecosystems to meet human needs in an ecologically and economically superior manner.

The TES approach includes ecosystems in the system boundary, as depicted in Figure 20.7. We can see that in addition to using wastes as resources in other industries, as is done in industrial symbiosis and the circular economy, TES also

Table 20.3 Technological and ecological approaches for various tasks.

Task	Technology (gray infrastructure)	Ecosystem (green infrastructure)
Carbon capture, utilization, and sequestration	Scrubbing with monoethanol amine; use for enhanced oil recovery; convert to chemicals	Primary production (photosynthesis); soil sequestration
Water purification	Anaerobic baffled reactor; sludge treatment	Surface and subsurface flow wetlands
Sulfur dioxide removal	Flue gas desulfurization; alkali scrubber	Vegetation
Nitrogen oxide removal	Selective catalytic reduction	Vegetation
Fresh water provision	Dams, reservoirs, desalination; cloud seeding	Native vegetation to sustain hydrological cycle, aquifer recharge
Pollination	Hand pollination, transported bees, genetic engineering	Local habitat for insect, bird, bat pollinators
Soil fertility	Artificial fertilizers: urea, ammonia, phosphates	Microbes, natural fertilizers: compost, manure
Shoreline protection	Seawalls	Mangroves, oyster reefs
Reliable power and flood control	Sediment dredging	Forests on slopes upstream of dam

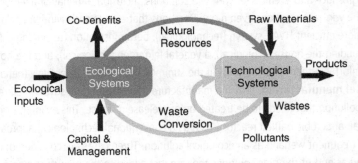

Figure 20.7 Techno-ecological synergy includes networks of technological and ecological systems.

includes the ability of ecosystems to capture and utilize wastes. For example, as we learned in Section 20.2.1, wetlands can remove water pollutants, while trees can remove air pollutants. In addition, ecosystems can also provide goods and services such as oxygen, pure water, and biomass to industrial systems. Ecosystems also provide many other co-benefits to society, such as aesthetic value, recreational opportunities, flood regulation by wetlands, etc. These interactions are illustrated in Figures 1.10 and 20.3.

The technological and ecological options for treating or utilizing many types of wastes and providing various goods and services are summarized in Table 20.3.

The TES approach considers the technological and ecological alternatives in terms of their physical, economic, and ecological characteristics, and chooses the best combination of approaches to meet the desired goals. The environmental and economic feasibility of TES for a biodiesel manufacturing site is described in Box 20.6.

BOX 20.6 Examples of Designs by Techno-Ecological Synergy

Residential system. [12, 13]A residential system such as a single-family home consists of technological and ecological systems. The building has various associated technologies such as the heating, ventilation, and air-conditioning (HVAC) system, appliances, wall insulation, faucets, and toilet. The surrounding yard constitutes the ecosystem, which may include a lawn, trees, a vegetable garden, and pond. Usually, the focus of environmentally friendly design is on the technological components. Figure 20.8 shows that by developing synergies between the technological and ecological variables, as in TES, we can obtain solutions that are win–win compared to the solutions obtained from conventional techno-centric design. This figure plots the cost versus the Scope 2 (direct emissions + electricity) carbon footprint. The base case (shown at the right of the figure) has zero cost and a carbon footprint of about 52 tons. The Pareto curve for techno-centric design shows that a smaller footprint is possible at a slightly lower cost, which is a win–win solution. The Pareto curve for the TES design indicates expansion of the design space and even better win–win solutions. Inclusion of behavioral variables improves the win–win solutions even more. Decisions that enable such win–win solutions include a more efficient HVAC system (technological), better insulation (technological), trees surrounding the house (ecological), a vegetable garden (ecological), and the hottest and coldest comfortable thermostat settings in the summer and winter, respectively (behavioral).

    Biodiesel manufacturing. [14] The manufacture of biodiesel from soybean oil results in a stream of pollution that needs to be treated before release or reuse. This example compares the use of an anaerobic baffled reactor (ABR) as a conventional technological approach with a subsurface treatment wetland as an ecological solution. Three cases are considered: (1) ABR and wetland as end-of-the-pipe solutions that are just added to the manufacturing process, and the water is released to the environment; (2) integration of water treatment solutions with the rest of the process design so that the process can adapt to the treatment approach and vice versa, and the water is released to the environment; (3) the same as case 2 but the water is reused in the process. The Pareto curves for these alternatives are shown in Figure 20.9. The x-axis variable is the net present value, while the y-axis variable is the net water supply (supply–demand). This figure identifies an integrated wetland with water recycling as the best option. In this TES design the process is operated in a different manner from the non-TES design. Since the use of wetland is a relatively inexpensive way of treating the water, the process relies less on distillation to purify the water and more on the wetland, resulting in a win–win TES design.

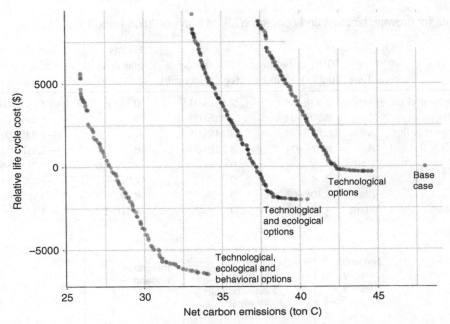

Figure 20.8 Pareto curves for the residential system, for the base case, for only technological options, for technological + ecological options, and for technological + ecological + behavioral options. Adapted from [13].

Designing a TES system involves determining the best combination of technological and ecological systems and its parameters that satisfy specified objectives, such as the net present value and sustainability metrics. The monetary objective may be quantified by metrics that we learned about in Chapter 17. The sustainability metric based on TES is defined as the gap between the supply and demand of a specified ecosystem service, as defined by Equation 16.2.

Example 20.3 An ethanol producer has been buying corn grown in the conventional manner, which is with tillage of the soil before planting the seeds. Recently, some farmers have started planting corn without tillage, and are claiming that this no-till corn is better for the environment, since the soil sequesters more carbon. From the data in Table 20.4, determine the validity of this claim with and without accounting for the ecosystem services provided by the farming.

Solution
Using the data in Table 20.4, greenhouse gas emissions from farming without considering the carbon sequestration ecosystem service are 6.83E+04 kg CO_2eq/acre with tillage and 7.30E+04 kg CO_2eq/acre without tillage. Thus, on the basis of the GHG emissions, corn grown with tillage is better owing to its smaller emissions.

If we include the carbon sequestration ecosystem service provided by agriculture, this quantity should be subtracted from the GHG emissions to calculate

Table 20.4 Data for growing biomass and converting it into transportation fuels [15].

Activity	Unit	GHG emission (kg CO_2eq/unit)	GHG uptake (kg CO_2eq/unit)	Energy (gasoline gallon eq/unit)	Net present value ($/unit)
Corn with conventional till.	Acre	6.83E+04	2.54E+03	0	6.67E+02
Corn without tillage	Acre	7.30E+04	2.48E+03	0	5.87E+02
Switchgrass without N fert.	Acre	5.67E+04	3.24E+03	0	−4.88E+02
Switchgrass with N fert.	Acre	6.07E+04	3.23E+03	0	−5.31E+02
Wind turbine	Unit	0	0	4.27E+06	−3.01E+05
Solar panel	Unit	0	0	7.9E+01	−7.23E+02
Corn ethanol	Plant	4.30E+07	0	8.02E+07	−6.88E+06
Switchgrass ethanol	Plant	5.00E+06	0	8.11E+08	−9.17E+06

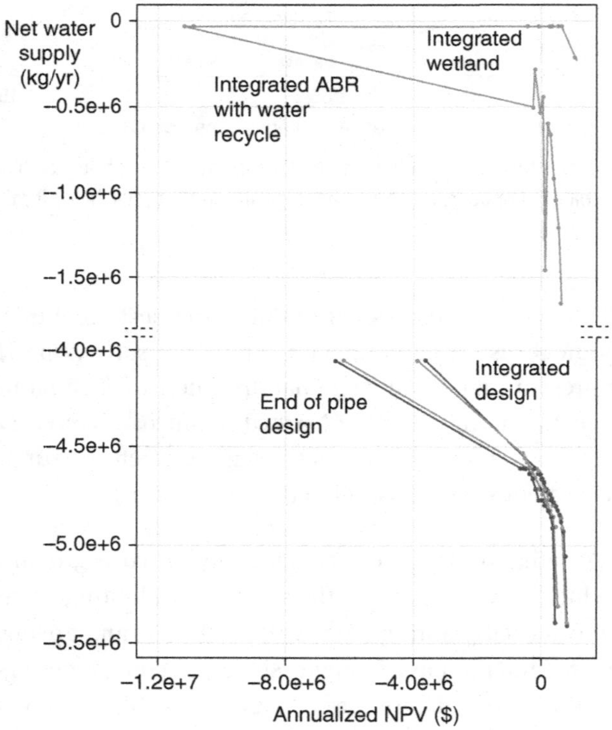

Figure 20.9 Pareto curves for techno-centric and TES design of biodiesel process and water treatment. Adapted from [14].

the net emission. Thus, the net emission from farming with tillage is 6.83E+04 − 2.54E+03 = 6.576E+04 kg CO_2eq/acre, and from farming without tillage is 7.30E+04 − 2.48E+03 = 7.052E+04 kg CO_2eq/acre. Therefore, even when we account for soil carbon sequestration, corn with tillage is better.

20.4 Summary

Modern engineering has kept ecosystems outside its decision boundary. This is so despite the critical role that nature plays in supporting all industrial and human activities. In this chapter we learned about the importance of shifting the engineering paradigm toward including the role of ecosystems. We learned about practices based on traditional ecological knowledge that have been sustained for centuries. We then focused on more modern approaches that aim to benefit from the ability of ecosystems to satisfy human needs. These include ecological engineering and green infrastructure. Techno-ecological synergy is an approach that integrates these nature-based solutions with conventional technological solutions to develop integrated networks of technological and ecological systems.

Key Ideas and Concepts

- Traditional ecological knowledge
- Treatment wetlands
- Self-design
- Green infrastructure
- Ecological engineering
- Imposed design
- Gray infrastructure
- Techno-ecological synergy

20.5 Review Questions

1. Define traditional ecological knowledge and provide two examples.
2. What is self-design? How is it different from imposed design?
3. What is a vertical flow subsurface wetland?
4. What is the key difference between techno-ecological synergy and byproduct synergy?
5. What are the technological and ecological options to maintain soil fertility?

Problems

20.1 Ethanol is a required additive in gasoline in the USA and many other countries. The main steps in producing ethanol from corn include farming to grow corn and the manufacturing step in which corn sugar is converted to ethanol by fermentation. For these two steps, discuss the ecosystem services that each activity depends on and how each activity impacts ecosystem services. Think of the direct and indirect inputs from nature and the impact of each activity. Use the CICES listing in Table 1.1 to identify the ecosystem services.

Table 20.5 Emergy metrics for green infrastructure solutions for managing stormwater.

Type of green infrastructure	Emergy yield ratio	Environmental loading ratio	Emergy sustainability index	Transformity
Rain garden	1.00262	450.56	0.0026	2.36E+07
Porous pavement	1.00092	1804.75	0.0009	3.96E+08
Green roof	1.01780	58.39	0.0181	2.01E+07
Tree planting	1.04611	69.39	0.0492	7.67E+06

20.2 Choose an indigenous community in any part of the world and determine two ways in which they use traditional ecological knowledge. Is their approach ecologically friendlier than the modern approach for accomplishing the same task? Does it seem likely that their knowledge will be forgotten in the next two to three decades?

20.3 Some communities in East Africa seem to have established a mutualistic relationship with a local bird. The bird, called a honeyguide, helps people find bee hives. People collect the honey, and then the honeyguide feeds on the wax combs left behind. Communities have developed different vocalizations to attract honeyguides, and the birds have unique sounds and behaviors to communicate with humans [16]. Is this an example of traditional ecological knowledge? Justify your response by determining whether the characteristics of this system match those of traditional knowledge systems.

20.4 The effect of temperature on the ability of wetlands to remove nitrates is captured through the relationship of the time constant with temperature, which is given by $k = k_{20}\theta^{(T-20)}$. Here, k_{20} is the removal rate coefficient at 20 °C, θ is the temperature factor whose value in such systems is 1.15, and T is in °C. How will the performance of the wetland in Example 20.1 change if the temperature drops to 4 °C?

20.5 The results of an emergy analysis of various green infrastructure options for treating stormwater in Syracuse, New York are summarized in Table 20.5. Using these results, discuss the environmental sustainability of each option.

20.6 A short-rotation coppice in the UK sequesters biomass at the rate of 1 kg/(m² yr²). Calculate the mean cooling over a typical 16-hour day in the summer. How much will the temperature of the surrounding air decrease due to this cooling? You may assume that the cooling affects air up to a height of 10 m and the heat capacity of air is 1 kJ/(kg K). Calculate the carbon sequestered by this coppice using the fact that 2 to 3×10^{-3} moles of CO_2 are assimilated per mole H_2O lost.

Table 20.6 Cost and net CO_2 emissions for biosolids management in central Ohio.

Case	Cost (million US)	Net CO_2 emissions (tons)
Base case	9.80	42,818
Techno-centric	6.23	36,696
	8.09	23,134
	7.05	27,000
TES without feedback	8.75	101
	6.34	36,594
	6.77	11,756
TES with feedback	7.61	−4500
	5.94	36,012
	6.04	24,370

Table 20.7 Carbon sequestration, land use, and cost per tree over 20 years.

Species	C sequestration (kg CO_2)	Land used (acre)	NPV ($)
White oak	1.61E+04	1.66E−03	−3.65E−01
Scots pine	8.65E+04	8.15E−04	−1.79E−01
American elm	1.23E+04	1.12E−03	2.47E−01
Spruce	1.31E+04	1.12E−03	2.47E−01
Birch	1.19E+04	7.92E−04	−1.74E−01
Eastern hemlock	5.24E+04	1.12E−03	2.47E−01

20.7 The treatment of urban waste water results in the production of biosolids. This byproduct may be landfilled, incinerated, composted to use as fertilizer, or applied directly to land. Each alternative involves different costs and greenhouse gas emissions. A study of biosolids management options in central Ohio focused on ways of making this activity carbon neutral. Four cases were considered: a base case that represented current operation: a techno-centric design that considers various technological options; a techno-ecological synergy design that considers ecological alternatives such as lengthening the timber cycle and planting new forests; and a TES design with feedback where wood is used to replace some of the fossil resources. Typical costs and net CO_2 emissions are listed in Table 20.6. Sketch the approximate Pareto curves for the three cases and compare them with the base case. Are there win–win opportunities in developing synergies with nature?

20.8 From the data in Table 20.4, evaluate the trade-off between GHG emissions and NPV, with and without including the agroecosystem service of carbon sequestration. Discuss the feasibility of producing ethanol from corn versus switchgrass. Note that the ethanol plant processes corn or switchgrass from 10,000 acres of cultivated land.

20.9 You wish to utilize 10,000 acres of land to produce energy with no net emission of greenhouse gases. Using the data in Tables 20.4 and 20.7, determine a combination of available land-use options to satisfy this goal. Now determine a combination that also provides the maximum NPV.

20.10 Box 20.5 suggests the possibility of relying on vegetation to mitigate industrial air pollution. Suppose that a manufacturing site wishes to get credit for the air quality regulation ecosystem service provided by the vegetation on its own site. Answer the following questions for this TES system.

1. Since there is no way to guarantee that the molecules of air emissions taken up by the trees are the same as those emitted by the process, should the manufacturing process be allowed to claim credit for the pollutant uptake by the trees on its site?
2. If the site is in a temperate region, deciduous trees lose their leaves during the winter, along with their capacity to take up emissions. Suggest ways of operating the manufacturing process in the presence of this intermittency of nature.

References

[1] J. R. McNeill. *Something New Under the Sun*. Norton, 2000.

[2] F. Berkes, J. Colding, and C. Folke. Rediscovery of traditional ecological knowledge as adaptive management. *Ecological Applications*, 10(5):1251–1262, 2000.

[3] F. Berkes. *Sacred Ecology*, 3rd edition. Routledge, 2012.

[4] UNESCO. Cultural landscape of Bali province: the subak system as a manifestation of the Tri Hita Karana philosophy. http://whc.unesco.org/en/list/1194, no date, accessed November 9, 2013.

[5] K. R. Barrett. Ecological engineering in water resources. *Water International*, 24(3):182–188, 1999.

[6] H. T. Odum. Ecological tools and their use: man and ecosystems. In P. E. Waggoner and J. D. Ovington, editors, *Proceedings of the Lockwood Conference on Suburban Forest and Ecology*, pages 57–75. The Connecticut Agricultural Experiment Station, 1962.

[7] W. J. Mitsch and S. E. Jørgensen. *Ecological Engineering and Ecosystem Restoration*. Wiley, 2004.

[8] J. Talberth, E. Gray, L. Yonavjak, and T. Gartner. Green versus gray: nature's solutions to infrastructure demands. *Solutions*, 1(4):40–47, 2013.

[9] N. Kundu, M. Pal, and S. Saha. East Kolkata wetlands: a resource recovery system through productive activities. In M. Sengupta and R. Dalwani, editors, *Proceedings of Taal 2007: The 12th World Lake Conference*, pages 868–881. International Lake Environment Committee Foundation, 2008.

[10] J. Nivala, T. Headley, S. Wallace, et al. Comparative analysis of constructed wetlands: the design and construction of the ecotechnology research facility in Langenreichenbach, Germany. *Ecological Engineering*, 61:527–543, 2013.

[11] The Nature Conservancy and Dow Chemical Company. *The Nature Conservancy–Dow Collaboration: 2012 Progress Report*. Nature Conservancy and Dow Chemical Company, 2013.

[12] R. A. Urban and B. R. Bakshi. Techno-ecological synergy as a path toward sustainability of a North American residential system. *Environmental Science & Technology*, 47(4):1985–1993, 2013.

[13] X. Liu and B. R. Bakshi. Extracting heuristics for designing sustainable built environments by coupling multiobjective evolutionary optimization and machine learning. *Computer Aided Chemical Engineering*, 44:2539–2544, 2018.

[14] V. Gopalakrishnan and B. R. Bakshi. Ecosystems as unit operations for local techno-ecological synergy: integrated process design with treatment wetlands. *AIChE Journal*, 64(7):2390–2407, 2018.

[15] R. J. Hanes, V. Gopalakrishnan, and B. R. Bakshi. Synergies and trade-offs in renewable energy landscapes: balancing energy production with economics and ecosystem services. *Applied Energy*, 199:25–44, 2017.

[16] C. N. Spottiswoode, K. S. Begg, and C. M. Begg. Reciprocal signaling in honeyguide–human mutualism. *Science*, 353(6297):387–389, 2016.

21 Economic Policies

> Our economies have been trading one form of capital, Earth's riches, for another – human riches. Without accounting accurately for this trade-off, we will continue to have a false impression of economic progress and growth. That is as dangerous as flying an aeroplane into the night without navigation tools or instruments.
>
> Edward Barbier [1]

In Chapter 4 we learned some basic principles of free-market economics and how their practice can contribute to unsustainability. We identified the following four reasons.

1. The external social cost of economic activities is not included in the price. This can result in excessive resource use and tragedies of the commons.
2. Discounting gives less importance to the future. This makes it difficult to justify decisions with long-term benefits.
3. Perfect substitutability is assumed between natural and economic capital. This could downplay ecological degradation by assuming that economic capital can act as a substitute for lost natural capital.
4. The physical basis of the economy is not fully considered. This means that the role of ecosystem services in supporting the economy is undervalued or ignored.

In this chapter we will focus on how these issues can be addressed by devising economic policies that encourage sustainable development. We will learn about ways of internalizing externalities, including non-market-based and market-based policies such as environmental taxes, emissions trading, and payment for ecosystem services. We will also learn about the approach of inclusive wealth, which is a more comprehensive way of accounting for economic, human, and natural capital.

21.1 Internalizing Externalities

In a large number of cases of ecological degradation, the environment is kept outside the market boundary, resulting in negative environmental externalities. Some such examples mentioned in Chapter 4 include the shrinking of the Aral sea and the appearance of harmful algal blooms in Lake Erie. In this section, we will learn about approaches for internalizing externalities. We will see how market forces could be harnessed to reach a societal optimum that balances human needs with environmental protection.

21.1.1 Non-Market-Based Policies

For many decades, the most popular approach for dealing with environmental problems has been government regulation. Figure 21.1 shows the significant increase in environmental regulations in the USA. These policies have evolved from banning pollutants, to command and control policies, to the trading of individual pollutants, to the trading of ecosystem services. When environmental problems first attracted significant public attention, for example, when the impact of DDT and other pesticides became known, the approach used was to completely ban the substances. With greater realization of the costs and benefits of various pollutants, "command and control" policies became more popular. These policies often specified the use of a "best available technology" and an acceptable pollution level. Examples of such policies include requiring catalytic converters for reducing tailpipe pollution from vehicles and requiring scrubbers for flue gas desulfurization to reduce sulfur dioxide emissions from coal-burning power plants.

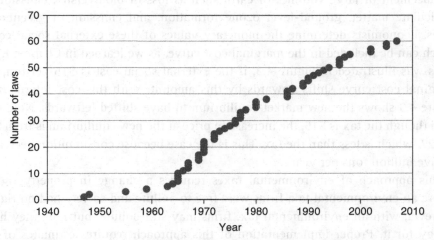

Figure 21.1 Major environmental regulations in the USA since 1945. Some important laws are: 1948, Federal Water Pollution Control Act; 1963, Clean Air Act; 1964, Wilderness Act; 1969, National Environmental Policy Act; 1970, Clean Air Act extensions including National Ambient Air Quality Standards; 1970, Occupational Health and Safety Act; 1972, Federal Insecticide, Fungicide, and Rodenticide Act; 1973, Endangered Species Act; 1976, Resource Conservation and Recovery Act; 1977, Clean Water Act; 1980, Comprehensive Environmental Response, Compensation, and Liability Act; 1989, Montreal Protocol; 1993, North American Free Trade Agreement; 1997, Kyoto Protocol; 2007, Energy Independence and Security Act.

The shortcomings of such an approach include the following:

1. Specifying the technology can stifle innovation toward the development of better technologies and other solutions.
2. Determining the acceptable pollution level to include in the regulations was not easy.
3. Regulations tend to be a source of conflict between industry and government.

Nonetheless, such methods continue to be popular for environmental protection all over the world since they are relatively easy to formulate, implement, and monitor.

21.1.2   Market-Based Policies

In recent decades, with advances in the field of environmental and resource economics, market-based policies have become increasingly popular for internalizing environmental externalities. These methods rely on taxation and/or trading.

Principles of Environmental Taxes

The economic impact of an externality can be internalized in the market by means of an environmental tax. This idea is attributed to the economist Arthur Pigou, and is also called a Pigouvian tax. As an example, consider the externalities due to the consumption of a metal ore. These include damage due to displacement of large volumes of earth such as loss of biodiversity, emission of particulate matter, ground-level ozone formation, and emission of greenhouse gases. Economists determine the monetary values of these external social costs, which can be included in the marginal cost curve, as we learned in Chapter 4.

As was illustrated in Figure 4.3, if the external social cost is $15 per ton, the marginal cost curve shifts upwards by this amount. With this cost internalized, Figure 4.3 shows the new market equilibrium to have shifted leftwards. Note that even though the tax is $15, the increase in price at the new equilibrium is from $16 to $27, which is less than the tax. This is the case because consumption decreases by five million tons per year.

This approach of environmental taxes requires a change in property rights: before an environmental tax, firms were free to pollute and society had no rights. However, with an environmental tax, firms may still pollute, but now they have to pay for it. Proper implementation of this approach requires estimates of the external social cost and intervention by a central entity such as a government to set up the system. Once the system is set up, it can run by itself but may need to be monitored.

Example 21.1   The use of coal as a source of energy results in the emission of mercury into the atmosphere. This heavy metal then gets deposited in the surrounding landscape, where it can accumulate in species such as fish, birds, and even humans. The relationship between the cost of electricity C in dollars per terawatt-hour and its demand D in terawatts is given by $C = -10D + 525$. The relationship between the marginal cost M and power generation D is given by $M = 20D$. Determine the market equilibrium price and production of electricity. If a tax of $10 per megawatt is introduced to internalize the impact of mercury emissions, determine the new equilibrium. How will the cost of electricity change for a consumer with and without considering the new market equilibrium?

Solution

The market equilibrium is the intersection point of the demand curve and marginal cost curve, so that at equilibrium, $M_e = C_e$. Therefore,

$$20D_e = -10D_e + 525$$

$$D_e = \frac{525}{30}$$

$$= 17.5 \text{ TWh}$$

and $C_e = M_e = 350$ $/TWh.

With 10 $/TWh tax, the marginal cost equation may be written as $M = 20D + 10$. At equilibrium,

$$20D_e + 10 = -10D_e + 525$$

$$30D_e = 515$$

$$D_e = 17.2 \text{ TWh}$$

and $C_e = M_e = 354$ $/TWh

If the new market equilibrium is not taken into account, the cost of electricity for the consumer would increase by $10/TWh due to the tax. However, at the new market equilibrium, the cost of electricity increases by only $4/TWh.

Tradeable Emissions Permits

This approach combines regulation, environmental taxes, and property rights. The basic idea is that society sets a quota for the maximum allowable amount of pollution or resource use. This quota is then distributed among stakeholders in the form of permits or individual quotas. These constitute the "right to pollute." Firms or any other interested parties may then buy or sell permits. Permits may be bought if a stakeholder wishing to emit the traded pollutant finds buying permits to be cheaper than reducing emissions by approaches such as more efficient operation or technology. Companies that have older technology that is expensive to upgrade are likely to find that buying permits is more cost-effective. Firms with newer or cleaner processes are likely to have permits to sell. To reduce the total quantity of pollutant or resource use, quotas are gradually decreased over time. This will cause the cost of permits to increase, and ultimately, even the firms that prefer buying

BOX 21.1 Sulfur Dioxide Trading in the USA [2]

The trading of SO_2 emissions was launched in 1995 to achieve the goal of the 1990 Clean Air Act Amendments of reducing emissions to ten million tons below 1980 levels. Phase I affected mainly coal-burning units, while Phase II in 2000 tightened the annual emissions limits and included most of the other emitting units. The permits or allowances are sold in annual auctions held by the Environment Protection Agency (EPA), and emissions are monitored by smokestack monitors wired directly to EPA computers.

As shown in Figure 21.2, in 2002, SO_2 emissions were 41 percent lower than those in 1980. This reduction has been achieved more quickly and at a lower cost than anticipated when the program was initiated. The market mechanism also encouraged tremendous improvements in SO_2 scrubber technology. The program is hailed as a success of market-based systems, as conveyed by the estimated costs and benefits in Table 21.1. The Office of Management and Budget in a 2003 study estimated this program to have benefits exceeding costs by more than 40 : 1 in the previous ten years.

Despite its overall success, this trading program did have some negative side-effects. Many old plants continued to operate, since they were "grandfathered" into the regulation. The high pollution from these plants resulted in hotspots of SO_2 emissions, such as near the Gulf Coast. The environmental harm due to acid rain included the lower productivity of affected lakes and forests. Their recovery has been very slow. The reduction in emissions happened less aggressively than originally intended since changes in emission quotas were controlled by Congress and not the EPA. Finally, the technologies for reducing SO_2 and NO_x emissions themselves require some energy. This parasitic energy consumption is mainly from fossil resources, resulting in an increase in CO_2 emissions per unit of electricity produced.

permits instead of reducing emissions may find it to be economically more attractive to upgrade their processes to emit less. This approach has been successfully used for regulating fishing in Europe, and reducing acid rain in the USA. The latter case is discussed in more detail in Box 21.1.

Probably the most ambitious use of tradeable permits so far has been for reducing global greenhouse gas (GHG) emissions. This scheme was set up as part of the 1997 Kyoto Protocol, and was meant to create a global market for GHG emissions. It also included the Clean Development Mechanism (CDM), which was meant to be a way to help developing countries adopt cleaner technologies by giving credit to developed nations supporting such developments. Thus, a developed country could offset its GHG emissions by supporting the development of wind farms in a developing country. However, this effort has not been particularly successful owing to bureaucratic snafus and lack of cooperation. Countries with large emissions such

Table 21.1 Summary of estimated costs and benefits of the acid rain program in 2010 (year 2000 US dollars (millions)). [3]

Quantified benefits	
PM2.5 mortality (USA and southern Canada)	$107,000
PM2.5 morbidity (USA and southern Canada)	$8000
Ozone mortality (eastern USA)	$4000
Ozone morbidity (eastern USA)	$300
Visibility at parks (three US regions)	$2000
Recreational fishing in New York	$65
Ecosystem improvements in Adirondacks (New York residents)	$500
Total annual quantified benefits	$122,000
Quantified costs	
SO_2 controls	$2000
NO_x controls	$1000
Total annual quantified costs	$3000

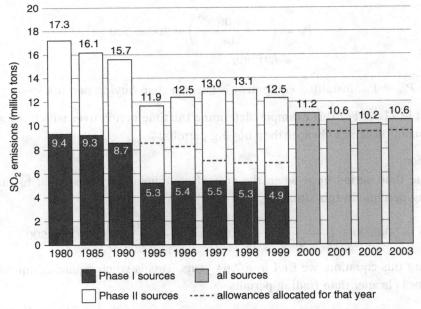

Figure 21.2 Reductions in sulfur dioxide emissions due to trading [2]. Maps showing changes in acid deposition are available at `http://nadp.slh.wisc.edu`.

as the USA did not ratify the protocol, and Canada withdrew from it. Other countries with large and rapidly increasing emissions such as China and India were exempt owing to their developing status and low historical emissions.

Example 21.2   Efficiency improvements to reduce emissions cost a company $500,000 in capital investment and $10,000 per year in operating cost. This investment will reduce emissions by 200 tons per year. Instead of this investment, the company could buy emissions permits that are expected to cost $500 per ton of annual emissions each year. Considering a plant life of ten years and a discount rate of 8 percent, is it better for the company to buy emission permits or install new equipment?

Solution
Over a ten-year period, the present value of installing new equipment may be calculated by Equation 4.3 as

$$P_{eq} = 500,000 + \left(\frac{1 - 1.08^{-10}}{0.08}\right) 10,000$$

$$= 567,100$$

The cost of buying permits for the same period, while assuming that prices remain unchanged is

$$P_{tr} = \left(\frac{1 - 1.08^{-10}}{0.08}\right) \times 200 \times 500$$

$$= 671,008$$

Since $P_{eq} < P_{tr}$, installing equipment is cheaper than buying permits.

Example 21.3   For this example, determine the time period over which installing equipment becomes cheaper than buying permits.

Solution
Let the time period be n when the cost of equipment becomes equal to that of buying permits. In this situation,

$$500,000 + \left(\frac{1 - 1.08^{-n}}{0.08}\right) 10,000 = \left(\frac{1 - 1.08^{-n}}{0.08}\right) \times 200 \times 500$$

Solving this equation, we find $n = 7.64$ years. This is when replacing equipment becomes cheaper than trading permits.

Payment for Ecosystem Services
We learned about the importance, types, and status of ecosystem services in Chapter 2. In Chapter 3, we learned about the necessity of staying within nature's capacity to supply the goods and services demanded by human activities. Since ecosystem services are outside the market in conventional economics, many schemes have been developed to internalize them. These methods involve monetizing ecosystem services and developing trading schemes to create markets for

them. Some schemes based on payment for ecosystem services (PES) already exist, as described in Box 21.2. These include the trading schemes for sulfur dioxide and carbon dioxide described in the previous subsection and Box 21.1.

BOX 21.2 Examples of Schemes Based on Payment for Ecosystem Services [4]

Self-organized deals. The water bottling company Perrier Vittel takes water from aquifers in the Rhine–Meuse watershed in northeastern France. Some years ago, it found that the water quality in the natural streams was deteriorating. Upon further exploration, the reason was found to be changing farming practices on land upstream of their bottling facility. Rather than building a filtration plant, Perrier Vittel found it less expensive to conserve the farmland surrounding their aquifers. They signed long-term contracts with local farmers and paid them for their use of conservation practices such as less intensive pasture-based dairy farming, reforestation of relevant zones, and improving the management of animal waste. They also bought 600 acres of sensitive habitat, which provides the water quality regulation and water provisioning ecosystem services for their bottling operation.

Public payments. Some governments allocate funds for PES. For example, in Paraná, Brazil the state provides funds to municipalities for the protection of watersheds and the restoration of degraded ecosystems. Part of the revenue from the Circulation of Goods and Services tax is distributed to municipalities according to the area under their protection and whether the municipality supplies water to adjacent municipalities.

Regulation-driven open trading. The acid rain program described in Box 21.1 is a PES scheme for the air quality regulation ecosystem service. The Kyoto Protocol had developed a PES scheme for climate regulation. Here, countries could purchase offsets for their GHG emissions in the form of forestry and other GHG-reduction activities in other countries.

Water quality trading. A trading scheme established in central Ohio involves exchange between farmers, fossil fuel users such as power plants, and water treatment facilities. Each entity is a potential contributor of nutrients such as nitrogen and phosphorus to local water bodies. Coal combustion requires the removal of sulfur dioxide, which involves the use of ammonia. The ammonia often volatilizes and gets deposited on local streams and lakes. Farm runoff also adds N and P nutrients to water, and so does municipal water treatment. This water quality trading scheme means that the power generation company pays the farmer to reduce nutrient runoff by changing farming practices, provided this payment is less than the cost of reducing emissions from the power plant. The farmer accepts the payment and changes farming practices provided that any loss due to the change is less than the payment from the power company. This scheme has been an effective way of reducing nutrient flow to water bodies in a "win–win" manner.

Table 21.2 Portfolio of a farm of the future with PES schemes. The conventional portfolio consists of only corn, alfalfa, and milk.

Commodity	Percentage of income	Customer
Biodiversity credits	5	Conservation trust
CO_2 offset credits	10	Steelmaker
Renewable electricity	15	Power market
Conservation tillage	10	Coal-burning power plant
Certified sustainable timber	5	Specialty market
Water credits	20	Watershed market
Pollinator credits	10	Local pollination market
Corn	15	World market
Alfalfa	5	World market
Milk	10	Local market

In addition to its environmental benefits, PES schemes are also likely to help in alleviating poverty since many poor people work close to the land in occupations such as farming, fishing, and forestry, and can influence local ecosystems. A PES scheme would pay such people for protecting and restoring ecosystems, thus creating incentives for their sustainable management. Potential benefits to the rural poor include increased cash income, gaining experience with business activities, and greater knowledge about sustainable resource use. In the long run these could enhance the resilience of local ecosystems and enhance their productivity. However, many risks also exist in such schemes, such as loss of the right to harvest products, increased competition for land and loss of land rights, and incompatibility with cultural values. In addition, global trading schemes such as those for CO_2 are sometimes opposed due to their being ways of transferring wealth to poorer nations. The successful development and implementation of each PES scheme needs careful thought and knowledge of local conditions, along with strong institutions. It is envisioned that with increasing development of PES schemes, the portfolio of the farm of the future will look like Table 21.2, where sources of income such as corn, alfalfa, and milk that dominate today will constitute only a small fraction of farm income. Depending on how the land is used, revenue may be earned from the production of other resources such as wind electricity and timber, and from activities that enhance ecosystem services, such as reducing nutrient runoff due to conservation tillage, biodiversity, and water credits due to multiple and ecologically friendly uses of the land.

Example 21.4 A farmer owns 1000 acres of land. If he plants corn with conventional tilling of the soil he can get a yield of 160 bushels/acre, which may be sold at $3.25 per bushel. One bushel of corn is 25 kg. If he switches to conservation farming practices, including the use of cover crops and no tilling, the soil would sequester more carbon but the yield of corn would decrease by 10 percent. If the value of sequestered carbon is \$40 per ton of CO_2, how much CO_2 will need to be sequestered to compensate for the reduction in yield?

Solution
With conservation farming, the yield will decrease by $0.1 \times 160 = 16$ bushels/acre. This will result in a monetary loss of $3.25 \times 16 = \$52$ per acre. Since the farmer can earn \$40 for each ton of sequestered CO_2, the mass of CO_2 that will need to be sequestered to compensate for the \$52 reduction in revenue is $\frac{52}{40} = 1.3$ tons.

21.2   Inclusive Wealth

In Part I of this book we learned that human well-being, as measured by many metrics such as gross domestic product (GDP), life expectancy, and the Human Development Index (HDI) has been improving across the world. We also learned about the degradation of ecosystem services across the world, much of it to enable improvement in human well-being. Among the reasons for this dichotomy is the lack of a physical basis to neoclassical economics, as we learned in Section 4.5, and because metrics such as GDP and the HDI do not account for factors such as ecological degradation and rising inequality, which need to be considered in any sustainability assessment. The approach of inclusive wealth aims to overcome these limitations.

Inclusive wealth is an approach that measures the wealth of nations by carrying out a comprehensive analysis of a country's productive base [5]. It measures all of the assets from which human well-being is derived and also a nation's capacity to create and maintain human well-being over time. Assets required for human well-being are classified into the following three categories:

1. *manufactured capital*, which includes items such as industrial products, machinery, and civil infrastructure like roads and buildings;
2. *human capital*, which includes education, employment, and health; and
3. *natural capital*, which includes fossil and mineral resources, forests, agricultural land, and fisheries.

These are analogous to the triple values that we learned about in Chapter 3. The "inclusive wealth" determines the social value of all these capital assets by converting them into the common unit of money, which reflects human preferences. Determining these monetary values is challenging, but economists have developed various approaches for their estimation, some of which are summarized in Box 21.3. These values are often called "shadow prices." Thus, inclusive wealth (IW) is defined as

$$IW = P_{MC} \times MC + P_{HC} \times HC + P_{NC} \times NC \qquad (21.1)$$

where P_{MC}, P_{HC}, and P_{NC} represent the shadow prices of manufactured, human, and natural capital, respectively.

The inclusive wealth has been calculated for most countries of the world, and results for three regions with respect to 1992 until 2010 are shown in Figure 21.3.

BOX 21.3 Approaches for Determining Shadow Prices

Economics relies on representing all flows in terms of money. Monetary value is meant to quantify human preference for one resource as compared to an alternative. For goods and service that are in the market, such as manufactured products and services, their market price captures human preference. However, many goods and services that are essential for sustainability, such as those from ecosystems, are outside the market and have no monetary value. Shadow prices are meant to capture the preference that people have for such goods and services. Economists rely on various methods for assigning shadow prices, such as the following.

- *Contingent valuation.* This approach relies on surveys to determine monetary value. It quantifies the monetary value of a resource by asking people about their willingness to pay for it.
- *Hedonic pricing.* This approach is used to determine the value of those goods and services that have an effect on market price. For example, the additional money society is willing to pay for a house with a view or a nice garden indicates the monetary value of these amenities. The value of reducing emissions may be quantified by the savings in health costs due to the emissions.
- *Benefit transfer.* This approach uses valuation results from existing studies of other, similar situations. For example, the value of beaches along Lake Erie may be used to quantify the value of beaches along a similar lake.

Determining shadow prices by methods such as these can also be controversial, for reasons such as the imperfectness of surveys and assumptions underlying these methods.

- For Africa, the IW per capita has decreased owing to a large increase in population along with a large decrease in natural capital per capita, and because there has not been enough increase in human and manufactured capitals per capita to compensate for the lost natural capital. This conveys the unsustainability of development in Africa, which is not conveyed by the GDP per capita. In fact, the GDP conveys a very rosy situation since it has increased by nearly 40 percent.

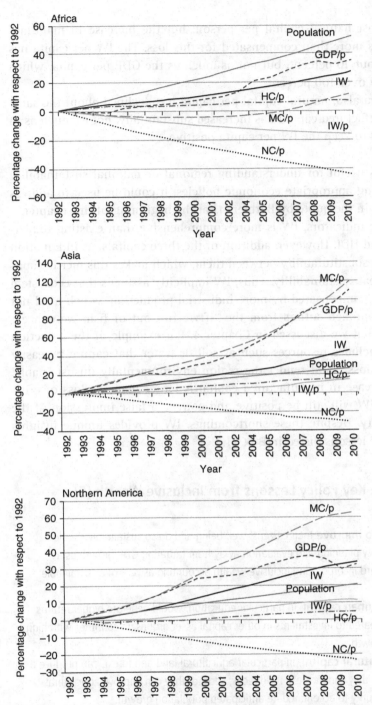

Figure 21.3 Annual mean wet sulfate deposition in 1985, 2003, and 2017. Darker shading indicates less deposition. From http://nadp.slh.wisc.edu.

- Asia has also lost natural capital per person, but the increase in manufactured capital has more than compensated for this loss. The IW per capita has increased by about 10 percent, but not as much as the GDP per capita, which has increased by over 100 percent.
- In North America also, natural capital per capita has declined about the same as in Asia. Manufactured capital has increased substantially, not as much as in Asia but enough to keep the IW per capita positive.

Such information is useful for understanding regional or national sustainability trends and developing appropriate economic policies. It could be used to address the issues identified in the quote by Barbier [1] at the beginning of this chapter.

Among economic indicators, IW is more comprehensive than existing indicators such as GDP and HDI. However, addition of the three capitals, as in Equation 21.1, implies perfect substitutability between them, which makes this metric capable of measuring weak sustainability. Thus, IW implicitly assumes that ecosystem degradation may be compensated by more industrial production, which need not be true. Such an approach suffers from the "if we run out of fish, we can eat boats" assumption that we discussed in Chapter 4. For example, if the reduction in natural capital includes resources such as soil or water that cannot be easily replenished, such a loss could render the region environmentally unsustainable. However, if the increase in manufactured capital is larger than the loss of critical natural capital, IW will still be positive and convey a false impression of the region's sustainability. Despite these shortcomings, IW provides unique insights and some key policy lessons, listed in Box 21.4.

BOX 21.4 Some Key Policy Lessons from Inclusive Wealth [5]

Countries striving to improve their citizens' well-being and do so sustainably should reorient economic policy planning and evaluation away from targeting GDP growth as a primary objective toward incorporating inclusive wealth accounting as part of a sustainable development agenda.

Investments in human capital, in particular education, would generate higher returns for IW growth as compared to investments in other capital asset groups, in countries with high rates of population growth.

Investments in natural capital, in particular agricultural land and forest, can produce a twofold dividend: first, they can increase IW directly; second, they can improve agricultural resiliency and food security to accommodate anticipated population growth.

Investments in renewable energy can produce a triple dividend: first, they can increase IW directly by adding to natural and produced capital stocks; second, they improve energy security and reduce risk due to price fluctuations for oil-importing countries; third, they reduce global carbon emissions and thus carbon-related damage.

Investments in research and development to increase total factor productivity, which decreased in 65 percent of countries, can immediately contribute to growth in inclusive wealth in nearly every country.

Countries should expand the asset boundary of the present System of National Accounts (SNA), which currently captures only 18 percent of a country's productive base, to include human and natural capital, which are now measured only through satellite accounts if at all.

21.3 Summary

Economics plays a critical role in enabling sustainable development by approaches such as regulation, environmental taxes, emissions trading, and holistic indicators. Creating markets for ecosystem goods and services has become one of the most popular approaches for internalizing economic externalities, and it allows markets to find the appropriate balance between the use of ecosystem goods and services and their protection. The US acid rain program is among the most successful examples of such markets, but the global carbon trading program has been less successful. Indicators such as inclusive wealth are more complete than metrics such as GDP and the HDI, and provide useful insights about trends in manufactured, human, and natural capitals. This insight is useful for policies that encourage sustainable development.

Key Ideas and Concepts
- Command and control
- Environmental tax
- Payment for ecosystem services
- Best available technology
- Emissions trading
- Inclusive wealth

21.4 Review Questions

1. Define and provide an example of environmental externalities.
2. State some shortcomings of command and control environmental regulations.
3. Why is the increase in price due to an environmental tax less than the tax itself?
4. How do emissions permits provide a "right to pollute"?
5. Which are the types of capital included in inclusive wealth?

Problems

21.1 It is estimated that, at the present rate, by 2050 there will be more plastic in the ocean than fish. Addressing the issue of plastic pollution is an urgent need and many approaches are being tried. Suggest some economics-based approaches and discuss their pros and cons. Also describe some technological innovation that could help in the success of each economic policy you have suggested.

21.2 Answer the following questions about payment for ecosystem services. Videos available at www.youtube.com/watch?v=gzNWnREZ2xI and http://wqt.epri.com/overview.html may help.

1. In a typical PES scheme, who pays money to whom, and why?
2. Is an entity such as a government or some other institution necessary for PES schemes to be successful?
3. The water quality trading scheme can be effective in reducing nutrient pollution. Does it have any disadvantages such as the shifting of impacts to other domains?

21.3 The emissions trading of sulfur dioxide has been very effective in reducing the total SO_2 emissions in the USA. However, it has resulted in "hotspots." Explain the reasons for this phenomenon.

21.4 A study of ethanol as a transportation fuel reveals that the competitive equilibrium is expected to be at a price of $4 per gallon and a consumption rate of 100 million gallons/day. For a production rate of 10 million gallons/day, the marginal cost is found to be $1 per gallon. Also, at a price of $10 per gallon the demand is 10 million gallons/day. Answer the following questions for this system.

1. Determine the equations for the demand and marginal cost lines.
2. Calculate the consumer and producer surplus for the market equilibrium.
3. It was discovered later that the above information ignored a government subsidy of 50 cents per gallon. How will the demand and marginal cost lines, and the competitive equilibrium, change if this subsidy is removed?

21.5 The demand curve for a lawn fertilizer may be represented by $P = -3Q + 75$, while the marginal cost curve is given by $C = 2Q$. Here, P and C are the price and marginal cost in dollars, respectively, while Q represents the quantity of the product. Calculate the competitive equilibrium. If the external marginal social cost of $25 is also included, calculate the new competitive equilibrium.

21.6 In March 1989, the *Exxon Valdez* spilled 38,000 metric tons of crude oil in Prince William Sound, Alaska. Until the more recent Deepwater Horizon spill, this was the largest oil spill in the USA. Damage from this spill was estimated to include the deaths of 250,000 seabirds, 2800 sea otters, and 300 harbor seals. To estimate the cost of this damage, the shadow price of a sea otter was determined on the basis of the value of its pelt, and that of sea birds from the value of their feathers. Does this type of valuation capture the true value of these species? Do people value these animals and birds for things other than their pelts and feathers? If so, how could the full value be estimated?

21.7 Evaluate the approach of inclusive wealth using the six requirements for sustainability assessment listed in Box 3.3. Can any of the shortcomings of this approach be overcome with the help of other sustainability assessment methods, such as life cycle assessment or techno-ecological synergy?

21.8 Discuss the pros and cons of metrics for quantifying weak and strong sustainability. Can you suggest ways of modifying the inclusive wealth metric to make it stronger?

References

[1] E. B. Barbier. Economics: account for depreciation of natural capital. *Nature*, (515):32–33, 2014.

[2] EPA. EPA acid rain program, 2003 progress report. Environment Protection Agency, 2004.

[3] L. G. Chestnut and D. M. Mills. A fresh look at the benefits and costs of the US acid rain program. *Journal of Environmental Management*, 77(3):252–266, 2005.

[4] Forest Trends, Katoomba Group, and UNEP. Payment for ecosystem services: getting started, a primer, 2008.

[5] UNU-IHDP and UNEP. *Inclusive Wealth Report 2014: Measuring Progress Toward Sustainability*. Cambridge University Press, 2014.

22 Societal Development

In the end we will only conserve what we love; we will love only what we understand; and we will understand only what we are taught.

Baba Dioum

Ultimately, the success of any technology, policy, or other mechanism for enabling sustainable development depends on society's willingness and ability to adopt it. In previous chapters, we have learned about many potential solutions for transforming human activities toward sustainability. These include new technologies, systems inspired by ecosystems, and economic policies. However, a significant obstacle in the success of any solution is that modern society seems to be quite oblivious to the dangers of ecological degradation and the resulting threats such as that of climate change and loss of biodiversity, as we learned in Chapter 7. Societal factors seem to be among the biggest obstacles facing the transformation of human activities toward sustainable development. These factors include:

- the cultural narrative in society that encourages consumption and economic growth at any cost;
- ethical, moral, and religious aspects that are not always aligned with environmental protection;
- the loss of ecological knowledge and literacy with increasing prosperity;
- political systems that seem unable to address factors contributing to unsustainability, such as ecological degradation and rising inequality.

In this chapter we will learn about the role of society in the transformation toward sustainable development. We will focus on the role of individual actions, belief and value systems, worldviews, and global efforts by organizations such as the United Nations.

22.1 Individual Action

The most effective and lasting approach for bringing about change is by individual action at the grassroots level. This would be possible if each individual behaved in a manner that helped transform society toward sustainability. It would follow the principles and approach proposed by thinkers and philosophers such as Gandhi:

Be the change you wish to see in the world.

Another quote from Gandhi in Chapter 3 conveyed the need to consume only what we need since taking more than that is tantamount to stealing from others. Similarly, Jefferson suggested that we should not consume more than can be replenished in a generation.

To many individuals, the need for changing their own actions is not apparent since they may be oblivious to the broader environmental and societal implications of their own activities. People tend to focus on the "here and now" and discount effects that are more removed in space or time. However, many individuals are more aware of the impacts of their activities and behave or wish to behave in ways that enhance sustainable development. Nevertheless, they often feel isolated and consider the number of other such like-minded individuals to be relatively small. Barriers to individual action and change include the feeling that an individual's effort is likely to be insignificant, along with the human desire to fit into societal norms or the dominant cultural narrative. Despite these barriers it is important to keep in mind that small but persistent efforts are needed for societal change, which can happen suddenly and quickly.

Societal norms are passed on to other individuals and groups. The underlying bases of such transfers are called memes. These are analogous to genes, which are essential for the transfer of information about biological traits across generations. Like genes, memes self-replicate and spread across society. Memes that encourage sustainable development include the spread of eco-efficiency efforts in corporations across the world, the construction of "green buildings" by following LEED (Leadership in Energy and Environmental Design) standards, and the rapid adoption of gas–electric hybrid and electric vehicles. Such ideas that catch on in society are very quick in getting over barriers of the cultural narrative by spurring individuals to action.

An approach for bringing about change at the level of households and individuals that is being embraced by many energy companies is to provide a monthly home energy analysis, like the one shown in Figure 22.1. This analysis compares a home's energy use with that of similar homes in the neighborhood. It then provides resources and rebates for reducing energy use, such as energy audits and more efficient bulbs. Studies on the effect of such social comparisons on the energy consumption of households [1] have found that on average a household reduces its energy consumption by 2 percent or 0.65 kWh per day. Thus, such efforts based on social norms and comparisons can encourage individual action toward sustainability.

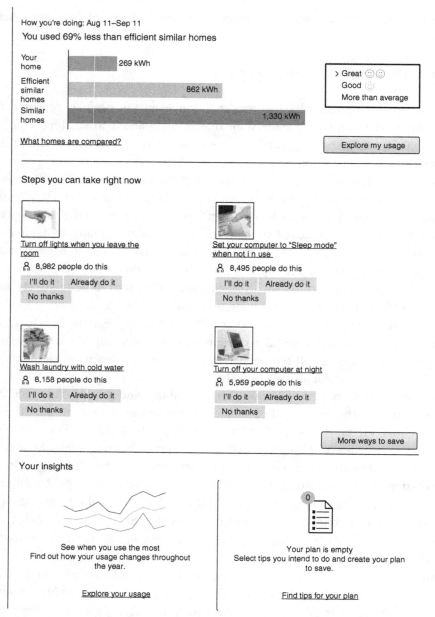

Figure 22.1 Typical home energy report provided by some energy companies to encourage lower energy consumption.

Education is a slow but effective and long-lasting approach for bringing about societal change by affecting the behavior of individuals. An example relevant to sustainable development is the introduction of environmental education in schools across the world, with the resulting changes in efforts toward increasing recycling and reducing pollution. Courses like the one you may be enrolled in while reading this book are also important steps toward introducing sustainability considerations in engineering and other activities.

22.2   Belief and Value Systems

Belief and value systems play an important role in the behavior of individuals and groups. We learned in Section 7.4 how such systems can contribute to unsustainable activities and have also been ignoring nature. In this section we will focus on how these systems are being adapted to encourage sustainable development.

Belief and value systems are often determined by religions. All religions encourage protection of the Earth and ecosystems, but many of these teachings seem to be forgotten or are misinterpreted. We learned about ways in which religion can contribute to unsustainability in Section 7.4. However, this situation is changing since leaders of many religions have actively been focusing on rediscovering aspects of their religion that encourage sustainable development. A prominent example of such an effort in Christianity is Pope Francis' encyclical on the proper interpretation of the Bible and about people's responsibility to care for the Earth. Excerpts in Box 22.1 convey the importance of learning from nature about how to close material loops and the fallacy of anthropocentrism that exploits and depletes natural resources. Many of the approaches covered in this book are needed to address the concerns raised by the Pope. Statements and attitudes of some other religions to encourage sustainable development are summarized in Box 22.2. The more recent movement of "deep ecology" considers the intrinsic value of ecosystems, as opposed to the anthropocentric value considered by shallow ecology. Thus, according to deep ecology, the value of a tree is not just because of the goods and services it provides to people, but because it is a part of nature. This is an eco-centric view. Such an approach does not focus only on pollution prevention and impact minimization, but questions the fundamental reasons behind ecological degradation and unsustainable activities. Deep ecology also embraces the societal and behavioral virtues of humility, reverence, and love. Some ongoing efforts toward sustainability in many religions are summarized in Box 22.3.

In many older and non-anthropocentric religions and indigenous cultures, practices and taboos were developed for sustainable management of the environment and natural resources. In such systems, some of which are still in use today, nature and people are considered together, and it is often difficult to distinguish between cultural and religious practices. Such systems are based on continuous learning about the local environment and human needs over many years. The resulting

BOX 22.1 Excerpts from Pope Francis' Encyclical, "Laudato Si: On Care for Our Common Home" [2]

Need for sustainable engineering. "These [pollution] problems are closely linked to a throwaway culture which affects the excluded just as it quickly reduces things to rubbish. To cite one example, most of the paper we produce is thrown away and not recycled. It is hard for us to accept that the way natural ecosystems work is exemplary: plants synthesize nutrients which feed herbivores; these in turn become food for carnivores, which produce significant quantities of organic waste which give rise to new generations of plants. But our industrial system, at the end of its cycle of production and consumption, has not developed the capacity to absorb and reuse waste and by-products. We have not yet managed to adopt a circular model of production capable of preserving resources for present and future generations, while limiting as much as possible the use of non-renewable resources, moderating their consumption, maximizing their efficient use, reusing and recycling them. A serious consideration of this issue would be one way of counteracting the throwaway culture which affects the entire planet, but it must be said that only limited progress has been made in this regard."

Our responsibility toward the planet. "We are not God. The earth was here before us and it has been given to us. This allows us to respond to the charge that Judaeo-Christian thinking, on the basis of the Genesis account which grants man 'dominion' over the earth (cf. Gen 1:28), has encouraged the unbridled exploitation of nature by painting him as domineering and destructive by nature. This is not a correct interpretation of the Bible as understood by the Church. Although it is true that we Christians have at times incorrectly interpreted the Scriptures, nowadays we must forcefully reject the notion that our being created in God's image and given dominion over the earth justifies absolute domination over other creatures. The biblical texts are to be read in their context, with an appropriate hermeneutic, recognizing that they tell us to 'till and keep' the garden of the world (cf. Gen 2:15). 'Tilling' refers to cultivating, ploughing or working, while 'keeping' means caring, protecting, overseeing and preserving. This implies a relationship of mutual responsibility between human beings and nature."

Emissions trading and excessive consumption. "The strategy of buying and selling 'carbon credits' can lead to a new form of speculation which would not help reduce the emission of polluting gases worldwide. This system seems to provide a quick and easy solution under the guise of a certain commitment to the environment, but in no way does it allow for the radical change which present circumstances require. Rather, it may simply become a ploy which permits maintaining the excessive consumption of some countries and sectors."

BOX 22.2 Hinduism, Islam, and the Environment

The western Abrahamic religions of Judaism, Christianity, and Islam have a strong anthropocentric worldview, owing to which nature is viewed as being less important. However, these religions also consider people to have responsibilities for the protection and stewardship of nature. This is conveyed in Box 22.1 for Christianity. The convenantal traditions of the Hebrew Bible extend to all of creation. In Islam, humans are vice-regents (*khalifa*) of Allah on Earth with the responsibility of caring for all of God's creations.

Eastern religions have more of an ecocentric worldview. Hinduism considers the world to be a manifestation of the divine, and its many traditions consider rivers, mountains, forests, animals, etc. to be sacred. For example, the Rig Veda (6:48:17) says, "Do not cut trees because they remove pollution." Buddhist teachings about right livelihoods requires everyone to consider the impact of their activities on society and the future. They teach the virtues of non-violence, which includes protecting all living things. Confucianism and Daoism focus on cultivating a harmonious connection between other humans and nature. The philosophy of ch'i provides the basis for this connection between matter and spirit.

Indigenous worldviews are closely integrated with nature. Goods and services from nature, including food, clothing, and shelter, are respected, and there is a feeling of gratitude toward the Creator and the underlying spiritual forces.

BOX 22.3 World Religions and Ecology, by Mary Evelyn Tucker, Yale Forum on Religion and Ecology

As we know, solutions to ecological problems need scientific research, policy studies, legal protection, economic strategies, and technological ingenuity. Engineers are making progress in contributing to these solutions. Along with this, religious worldviews and environmental ethics will also be needed for solutions to be fully effective. All the world's religions are coming into a period where they are making contributions to an ecologically flourishing future. Over the last 20 years the world's religions have composed written statements on the need for care for ecosystems and species. They have participated in conferences that have elevated concern for environmental issues such as climate change, biodiversity loss, toxicity, and ecological justice.

Indeed, for three years a series of ten conferences of world religions and ecology were held at Harvard, from 1995–1998. This included the western religions, Asian religions, and indigenous religions. Ten volumes were published by Harvard out of these conferences. Since then many other books and articles have been written on the topic. These statements and annotated bibliographies of these publications can be seen on the Yale Forum on Religion and Ecology website at Fore.yale.edu.

Of special interest are the Engaged Projects of the world's religions that are listed on this website. A few will be described here.

Hinduism

1. *Yamuna River.* In January 2011 a conference was convened at TERI University in Delhi and Radharaman Temple in Brindavan to examine the ongoing pollution of the Yamuna River. This river is sacred to all Krishna worshippers as Brindavan is considered the birthplace of Krishna. The conference brought together religious practitioners and Hindu scholars, such as Shrivatsa Goswami, David Haberman, and Christopher Chapple along with hydrologists and engineers from TERI, such as Bhavik Bakshi, and social scientists and historians of religion from Yale University, such as Harry Blair, John Grim, and Mary Evelyn Tucker. Stimulating papers were delivered at TERI and productive meetings were held with NGO groups in Brindavan. While significant attention has been given to the state of the Ganges River, less has been given to the Yamuna, so this elevated the concerns to clean up the sacred river and slow down some of the development along its banks.

2. *Govardan Ecovillage.* In December 2017 a conference was held at Govardan Ecovillage in the countryside 100 miles outside of Mumbai. This is the site of a community dedicated to Krishna, with Bhakti devotion performed nightly along a river that imitates the Yamuna. The village is an exemplar of sustainable practices, with brick buildings, vegetarian food from produce raised on-site, and a sanctuary for cows, donkeys, goats, and horses. In addition, there is outreach to the local community for empowerment of women and education for young people, especially orphans. A number of groups, including the Bhumi Project and the Centre for Hindu Studies at Oxford, plan to carry forward the collaborative spirit that emerged at the conference. The many stimulating papers may build on the *Hinduism and Ecology* volume published by Harvard.

Confucianism

One of the most promising movements in China is the revival of Confucianism, along with Daoism and Buddhism. Confucianism is a 5000-year-old tradition that has had its classical phase with Confucius and Mencius, its Neo-Confucian phase from tenth to twentieth centuries, and the New Confucian phase from 1949 to the present. While Confucianism was nearly destroyed by Mao in his attempt to eliminate it, Confucian scholars who escaped to Hong Kong and Taiwan kept the Confucian tradition alive and reinvented it in relation to Western philosophy and modernity. This has resulted in a revival of Confucianism in the People's Republic of China on various levels, including education, political philosophy, and popular culture. The Chinese classics are once again being taught in schools and universities. Tu Weiming, now at Beijing University and one of the leading New Confucian scholars, has been instrumental in this revival. One of the results of this revival is an examination of Confucianism as a source of environmental ethics and practices. Indeed, the Harvard conference book on *Confucianism and Ecology* (along with the books on Daoism and

Buddhism) has been translated into Chinese. The former deputy vice minister for the environment, Pan Yue, feels that these Chinese traditions are critical dimensions of environmental protection.

Christianity

1. *Laudato Si.* Pope Francis released an encyclical in 2015 that has had a major impact on the conjunction of Christianity and ecology. Called Laudato Si (meaning Praise Be), it is addressed not only to all Christians (some two billion people), but to all people on the planet. It brings together concern for the environment with concern for justice in a framework called "integral ecology." In this context, the "cry of the Earth and the cry of the poor" are seen as one – the sustainable future of both people and the planet are inseparable. The message of the encyclical is now being taught in Jesuit high schools and colleges around the world. This will ensure that future generations will inherit this message and pass it on.

2. Ecumenical Patriarch Bartholomew. The Head of the Greek Orthodox Church has been speaking out on "ecological sin" and "crimes against creation" for many years. Since 1995 he has organized symposia on environmental issues, especially focused on water. These symposia were held in Europe, in Greenland, in the Amazon, and on the Mississippi River. The organizers gathered high-level UN and EU officials, scientists, and religious leaders to learn about and advocate greater awareness of the pollution of rivers and seas. Father John Chryssavgis and Maria Beckett have been principal organizers of these conferences. Chryssavgis is an author of many books on Christianity and ecology and is the primary editor of the Patriarch's speeches on the environment. The well-known Orthodox theologian, Metropolitan John of Pergamon, has been an inspiration for these conferences with his talks and writings.

3. *Evangelical Environmental Network.* This network based in the USA has worked to bring awareness of environmental issues, including climate change, to evangelical Christian groups. It has encouraged "care for creation" and "stewardship" to protect ecosystems from pollution and assist the poor who are being adversely affected by climate change. Some of its leaders, including Richard Cizik, have been influential in the political arena in the USA.

Interreligious Projects

1. *Friends of the Earth Middle East (FOEME).* This group has been working for several decades to bring together the Abrahamic traditions of Judaism, Christianity, and Islam to work for ecology, justice, and peace in the Middle East, particularly in Israel. They have created a special project to protect the Jordan River against further exploitation and degradation. They have also worked in a Peace Park to encourage cooperation among the Abrahamic traditions for a flourishing future.

2. *Interfaith Power and Light*. This organization, founded by Reverend Sally Bingham in San Francisco, now has chapters in more than 40 states in the USA. These groups work to reduce the carbon footprint of congregations such as churches, synagogues, and temples. They also lobby for legislation and organize political demonstrations and action on climate change.
3. *Earth Ministry*. Based in Seattle, this group has participated in the cleaning up of rivers and the protection of ecosystems in the Pacific Northwest. They have held conferences, talks, and services to educate people on the environment. In recent years, they have been working closely with Native American and First Nation groups to stop pipelines and block the building of oil ports in the Pacific Northwest.

Table 22.1 Rules of thumb used in indigenous societies for sustainable management of natural resources. [3]

1.	The total protection of certain selected habitats.
2.	The total protection of certain species of animals or plants.
3.	Prohibitions concerning vulnerable stages in the life-histories of certain species.
4.	The practice of the monitoring of populations and their habitats.

rules tend to be quite simple and easy to understand and implement, but capable of sustaining socioecological systems over long periods. For example, four rules of thumb used in indigenous societies in India and other countries are given in Table 22.1. We learned about some such systems in Chapter 20 as examples of traditional ecological knowledge. These systems seem capable of overcoming the tragedy of the commons, which we learned about in Chapter 4 as a major reason for unsustainable human activities.

The in-depth study of traditional and modern methods for resource management has resulted in general insight about ways by which a tragedy of the commons may be prevented. Political scientist Elinor Ostrom identified eight principles that need to be satisfied for managing common-pool resources (CPRs) in a sustainable manner [4]. These are summarized in Table 22.2. These principles are based on studies of numerous systems developed all over the world for managing common resources. Some examples of such systems that are in use even today are described briefly in Box 22.4. If we evaluate systems that suffer from a tragedy of the commons in terms of these governing principles, we can quickly identify the underlying challenges and potential solutions.

Table 22.2 Ostrom's design principles for sustainable management of the commons. Reproduced with permission from [4].

1. Clearly defined boundaries

   Individuals or households who have rights to withdraw resource units from the CPR must be clearly defined, as must the boundaries of the CPR itself.

2. Congruence between appropriation and provision rules and local conditions

   Appropriation rules restricting time, place, technology, and/or quantity of resource units are related to local conditions and to provision rules requiring labor, material, and/or money.

3. Collective choice arrangements

   Most individuals affected by the operational rules can participate in modifying the operational rules.

4. Monitoring

   Monitors, who actively audit CPR conditions and appropriator behavior, are accountable to the appropriators or are the appropriators.

5. Graduated sanctions

   Appropriators who violate operational rules are likely to be assessed for graduated sanctions (depending on the seriousness and context of the offense) by other appropriators, by officials accountable to these appropriators, or by both.

6. Conflict-resolution mechanisms

   Appropriators and their officials have rapid access to low-cost local arenas to resolve conflicts among appropriators or between appropriators and officials.

7. Minimal recognition of rights to organize

   The rights of appropriators to devise their own institutions are not challenged by external government authorities.

   For CPRs that are parts of larger systems

8. Nested enterprises

   Appropriation, provision, monitoring, enforcement, conflict resolution, and governance activities are organized in multiple layers of nested enterprises.

BOX 22.4 Principles for Sustainable Management of the Commons and Some Examples

High mountain meadows of Switzerland [4]. Swiss village communities in the Alps have developed ways of managing their common property of meadows that are used for grazing their cows. These meadows are typically at high altitudes, receive little precipitation, and are highly exposed to sunlight. Each village follows strict rules about using common grazing land

and bars outsiders from using it. The villagers themselves are not allowed to send more cows to the meadow than they can feed during the winter. This rule is administered by a local official who can levy fines on violators. The inheritance system allows ownership to pass from parents to children. A period of high population growth in the nineteenth century put pressure on the limited land, but resource use was managed by population control measures such as late marriages, long gaps between children, and emigration.

Bishnois of Rajasthan. This community in the Indian state of Rajasthan has developed a set of 29 rules for managing their community and common lands. These rules are attributed to a religious leader who formalized them over 500 years ago. Rules associated with sustainable management of their communal resources include being compassionate toward all living beings and not cutting green trees. Such rules have allowed the Bishnois to live in harmony with nature by having significant natural areas along with their farming practices. Today, these lands form an ecological oasis where wildlife thrives alongside people, and water is relatively plentiful, in an otherwise semi-arid region. The contrast between Bishnoi lands and other lands managed in "modern" ways is often quite striking.

Example 22.1   Your local lake suffers from harmful algal blooms, which are causing deterioration of the CPR of water quality. The reason for the algal blooms is phosphorus runoff from surrounding farms and lawns that are located in the lake's watershed. Using Ostrom's design principles, suggest a scheme for managing this CPR.

Solution
Using the principles in Table 22.2, the following scheme may be appropriate for addressing this problem of algal blooms.

1. The boundaries are defined by the watershed. This should be known or be determined by a hydrological study.
2. Local conditions need to be studied to determine how the CPR of water quality is being used and how fertilizer use is affecting it. Options that could be considered include limiting fertilizer use, restricting use before rain events if they can be predicted, changing land use to less intensive methods, and/or using some land for buffer strips or wetlands that can intercept and treat runoff, and a taxation or trading scheme.
3. All members of the watershed should be able to participate in the decision making.
4. The water quality and behavior of those who affect it should be monitored. This may require an entity at the watershed scale, such as a government agency.

5. Sanctions for violators should be determined, and a mechanism for applying them should be available.

6. A system will be needed for resolving conflicts. This could be the local judicial system.

7. If the residents of the watershed wish to develop their own institution for implementing these tasks, they should be allowed to do so without intervention by the government or other entities who are outside the watershed.

8. If the region considered does not include the entire watershed, then a nested approach will be needed to cover larger spatial scales.

For systems such as Lake Erie, which we learned about in Box 4.1, it would be interesting to determine which of these actions are not yet being taken.

Given the global nature of today's environmental challenges, there is a need to scale-up such efforts from the local and regional scales, where they have succeeded in the past, to global scales. One such effort is the Paris Accord for addressing climate change. Unlike previous efforts that relied more on markets, the Paris Accord relies more on encouraging social change across the world. More details are provided in Box 22.5.

BOX 22.5 Kyoto to Paris: From Market Mechanism to Social Pressure

Efforts to combat climate change started with the Kyoto Protocol, signed in 1997. It relied primarily on market mechanisms such as emissions trading across countries and the Clean Development Mechanism, which allowed carbon offsets. For a variety of reasons discussed in Section 21.1.2, including difficulties in implementing carbon trading schemes, this effort did not achieve the desired success.

The more recent Paris Accord, signed in 2016, adopts more of a societal approach for addressing the challenges of climate change. It aims to limit warming to 1.5–2 °C. Countries are encouraged to voluntarily propose their own nationally determined contributions toward combating climate change, and to devise appropriate mechanisms within the country. Each country's plans are made publicly known and monitored on a regular basis. This creates societal pressure on each country to develop commitments that match its historical and current greenhouse gas emissions, and then to deliver according to this plan. Each country may devise its own approach, and many are following elements of the market mechanism proposed in the Kyoto protocol combined with ecosystem protection and restoration to enhance carbon sinks. Once a country has signed this agreement they may leave, but the procedure is not easy. How well this strategy, based on social comparison and norms, works is yet to be seen.

22.3   Worldviews and Future Scenarios

We end this chapter and the book by envisioning possible futures for our planet. Of course, no one knows the future, but such a thought experiment can help us think about the big picture and collectively determine the type of future we should seek as a society. First, we relate the ideas developed by ecological economist Robert Costanza [5].

22.3.1   Worldviews: Technological Optimist or Skeptic?

We consider two common worldviews,

- *Technological optimist.* This view assumes that technology will continue to develop and provide solutions to all problems facing humanity. For example, the challenge of clean energy may be met by discovering an unlimited, carbon-free source such as "warm fusion" or renewable technologies. This will allow us to avoid the worst impacts of climate change, will permit continued economic growth, and will result in ways of overcoming all kinds of resource constraints. Thus, this worldview has strong faith in the ability of humanity to innovate our way out of difficulties by developing new technologies. This view is not constrained by resource limits or any environmental challenge since technology will overcome them by finding alternatives. In this view, innovation and progress are encouraged by competition, substitutes exist for all resources, and will continue to be developed, and the market is able to find appropriate solutions and be the guiding principle. The engineering paradigm of dominating and controlling nature and maintaining business as usual is the appropriate one.
- *Technological skeptic.* This is a less rosy view than that of the technological optimist. It assumes that the technological breakthroughs needed for clean, unlimited energy may never happen, or at least not soon enough to avoid many of the negative impacts of climate change. This worldview accepts fundamental limits that cannot always be overcome by human ingenuity. It is skeptical of the ability of technology to overcome all challenges, and prefers to utilize technology along with social and community development to address limits. This view also relies on markets, but not as the primary guiding principle. Instead, it relies on cooperation to devise goals and markets and a possible mechanism to achieve them. It assumes limited substitutability between resources and considers that the engineering paradigm needs to shift toward respecting nature's constraints and working with ecosystems.

For each worldview, we now consider outcomes for the cases when each view is either correct or incorrect. Since we do not know which worldview is correct, such analysis can help in selecting the appropriate worldview.

Table 22.3 Potential scenarios for the future [5].

		Real state of the world in the future	
		Optimists are right	Skeptics are right
Current world view & *policies*	Technological optimism	Star Trek (sustainable)	Mad Max (unsustainable)
	Technological skepticism	Big Brother (sustainable)	Ecotopia (sustainable)

- *Worldview: technological optimist; future reality, technological optimists are right.* If we adopt the technological optimist worldview and this view turns out to be correct, then we will end up with a "Star Trek" future. In this case, technologies will be able to reverse and prevent environmental challenges such as ecological degradation and climate change. This will allow human population to increase and expand beyond Earth. This would be a *sustainable* outcome.

- *Worldview: technological optimist; future reality, technological skeptics are right.* In this view, we are not able to develop technologies to satisfy our resource needs and address environmental degradation. This could result in a "Mad Max" future where the Earth is ravaged, population has collapsed, and everyone is in a fight for their own survival. Clearly, such an outcome is *unsustainable.*

- *Worldview: technological skeptic; future reality, technological skeptics are right.* According to technological skepticism, we do not just rely on technological development but also conserve resources and learn to live with nature. If this view turns out to be correct, we would have an "Ecotopia." Human activities are within Earth's carrying capacity, and consumption and population have stabilized. This is a desirable and *sustainable* outcome.

- *Worldview: technological skeptic; future reality, technological optimists are right.* In this case, we develop a system to regulate environmental impact and resource use, but these turn out to be unnecessary since technology is able to overcome all limits. Then we end up with a highly regulated society that may have unnecessarily constrained economic growth. However, this "Big Brother" situation is also *sustainable.*

These outcomes are summarized in Table 22.3. Note that three outcomes are sustainable, while one outcome, Mad Max, is unsustainable. The question to ponder is, which worldview should we adopt today? Since the decision about the correct worldview could be critical to the sustainability of human activities, and since an incorrect decision could be catastrophic, a risk-averse decision would make the most sense. Among the four outcomes, the Mad Max outcome is the least desirable, so it should be avoided at all costs. This implies that the worldview of technological skepticism makes more sense for society to adopt today. If it turns out to be correct, we will avoid a major catastrophe and a Mad Max outcome and result in a highly desirable Ecotopia. If it turns out to be incorrect, we can always change to a technologically optimistic worldview after we have more information, that is, after the needed technologies are developed. The worst-case scenario is Big Government, which is not ideal, but still sustainable. The expectation that technology will solve all problems and give us a Star Trek future is the dominant worldview today. Costanza compares it with leaping off a tall building and hoping to develop a parachute before hitting the ground. He argues in favor of the skeptic's policies, which would keep the option of this innovation open without completely depending on it.

22.3.2   Toward a Good Anthropocene

Upon realizing the formidable challenges facing humanity in enabling a transition to sustainability, it is common to think about potential dystopian futures such as the Mad Max scenario in the previous subsection. The fact that we are in the Anthropocene means that we are likely to be in an era of low predictability about climate, extreme events, and ecosystem response to such perturbations. The rate of societal transformation may seem to be too slow to address the many challenges of the new geological epoch. However, maintaining a positive attitude and working toward solutions is essential. The worldview of technological skepticism described in Section 22.3.1 needs to avoid thoughts of dystopian future possibilities since they can be paralyzing and hinder societal change. After all, many dire warnings about food shortages and famines that were made before the 1970s did not materialize, owing to various factors, including human ingenuity. With the principles of sustainability incorporated in all disciplines, including engineering, guarded optimism about developing toward a good Anthropocene may be justified. Global efforts based on such thinking are already under way, such as the United Nations Sustainable Development Goals, summarized in Box 22.6. These are being adopted across the world with the aim of achieving them by 2030.

BOX 22.6 United Nations Sustainable Development Goals

In September 2015, many countries adopted the Sustainable Development Agenda to end poverty, protect the planet, and ensure prosperity for all. In each category, only the first goal out of several to be met over the next 15 years is listed below.

1. *No poverty.* By 2030, eradicate extreme poverty for all people everywhere, currently measured as people living on less than $1.25 per day.
2. *Zero hunger.* By 2030, end hunger and ensure access by all people, in particular the poor and people in vulnerable situations, including infants, to safe, nutritious, and sufficient food all year round.
3. *Good health and well-being.* By 2030, reduce the global maternal mortality ratio to less than 70 per 100,000 live births.
4. *Quality education.* By 2030, ensure that all girls and boys complete free, equitable, and quality primary and secondary education leading to relevant and effective learning outcomes.
5. *Gender equality.* End all forms of discrimination against all women and girls everywhere.
6. *Clean water and sanitation.* By 2030, achieve universal and equitable access to safe and affordable drinking water for all.
7. *Affordable and clean energy.* By 2030, ensure universal access to affordable, reliable, and modern energy services.
8. *Decent work and economic growth.* Sustain per capita economic growth in accordance with national circumstances and, in particular, at least 7 percent gross domestic product growth per annum in the least developed countries.
9. *Industry, innovation and infrastructure.* Develop quality, reliable, sustainable, and resilient infrastructure, including regional and transborder infrastructure, to support economic development and human well-being, with a focus on affordable and equitable access for all.
10. *Reduced inequalities.* By 2030, progressively achieve and sustain income growth of the bottom 40 percent of the population at a rate higher than the national average.
11. *Sustainable cities and communities.* By 2030, ensure access for all to adequate, safe, and affordable housing and basic services, and upgrade slums.
12. *Responsible consumption and production.* Implement the ten-year framework of programs on sustainable consumption and production, all countries taking action, with developed countries taking the lead, taking into account the development and capabilities of developing countries.
13. *Climate action.* Strengthen resilience and adaptive capacity to climate-related hazards and natural disasters in all countries.
14. *Life below water.* By 2025, prevent and significantly reduce marine pollution of all kinds, in particular from land-based activities, including marine debris and nutrient pollution.
15. *Life on land.* By 2020, ensure the conservation, restoration, and sustainable use of terrestrial and inland freshwater ecosystems and their services, in particular forests, wetlands, mountains, and drylands, in line with obligations under international agreements.

16. *Peace, justice, and strong institutions.* Significantly reduce all forms of violence and related death rates everywhere.
17. *Partnerships for the goals.* Strengthen domestic resource mobilization, including through international support to developing countries, to improve domestic capacity for tax and other revenue collection.

22.4   Summary

Appropriate societal development is essential for meeting the goals of sustainable development. Society can contribute to sustainability in many ways: individual action, belief and value systems, and worldviews are some of them. Individual efforts are essential for lasting change at the grassroots and for getting over the threshold of change. Beliefs and value systems can play an important role in enabling sustainability. All religions aim to encourage environmental protection and human well-being, and these aspects are being rediscovered and highlighted. A worldview that relies on technological skepticism is more likely to allow humanity to address the challenges of sustainable development. In addition, it is important to avoid pessimistic and dystopian views about the future to prevent paralysis or inaction. Efforts like the United Nations' Sustainable Development Goals are examples of global efforts to address poverty, environmental degradation, and human well-being. The principles and practice of sustainable engineering that we have covered in this book are essential to ensuring a good Anthropocene.

Key Ideas and Concepts

- Individual actions
- Deep ecology
- Technological skeptic
- Ostrom's principles
- Technological optimist
- Sustainable Development Goals

22.5 Review Questions

1. How are some utilities relying on social pressure to encourage individuals to reduce their home energy use?
2. How is deep ecology different from shallow ecology?
3. What is Pope Francis' position on carbon trading?

4. What is the Mad Max scenario for the future?
5. Why is technological skepticism the recommended worldview?

Problems

22.1 Does taking a course on sustainable engineering affect individual behavior toward greater sustainability? Why or why not?

22.2 Describe recent efforts in at least two religions toward sustainable development. Explain the motivation behind these efforts and the barriers to their success.

22.3 Does Pope Francis' encyclical in Box 22.1 align with shallow ecology or deep ecology? Justify your answer in detail.

22.4 Compare the fishing practices of the Cree with the scientific approach of maximizing yield. How do both strategies address the multiscale and complex nature of managing fish as a common resource?

22.5 Despite the insight provided by Ostrom's design principles for the sustainable management of common resources, tragedies of the commons continue to occur. One prominent example is the increasing concentration of greenhouse gases in the Earth's atmosphere. On the basis of the eight principles in Table 22.2, discuss which principles are being violated in managing the global atmospheric commons.

22.6 Unsustainable use of groundwater is causing the depletion of many aquifers across the world. Groundwater is a common pool resource that cannot be seen and is often developed by precipitation over many centuries. Using Ostrom's principles, suggest an approach for managing this resource. Can you find examples anywhere in the world of the successful and sustainable management of aquifers?

22.7 Explain the possible outcomes if we adopt a technologically optimistic world view. Which outcomes are sustainable?

22.8 Which Sustainable Development Goals connect most closely with sustainable engineering? Provide a detailed justification and discuss the methods that can be used. Take a look at the UN SDG websites for more details about each goal.

22.9 Consider an emerging technology that has the potential to greatly benefit people but is also likely to cause ecological degradation. Discuss how society may respond to the adoption of such a technology from the worldviews of technology optimism versus technology skepticism.

22.10 Give examples of practices that were once widely accepted but are now unthinkable. Select one such practice and learn about its history. What role did individual action play in enabling the change?

References

[1] H. Allcott and T. Rogers. The short-run and long-run effects of behavioral interventions: experimental evidence from energy conservation. *American Economic Review*, 104(10):3003–3037, 2014.

[2] Pope Francis. *Laudoto si* – on care for our common home, 2015.

[3] F. Berkes. *Sacred Ecology*, 3rd edition. Routledge, 2012.

[4] E. Ostrom. *Governing the Commons: The Evolution of Institutions for Collective Action*. Cambridge University Press, 1990.

[5] R. Costanza. Visions of alternative (unpredictable) futures and their use in policy analysis. *Ecology and Society*, 4(1):5, 2000.

Index

Printed in the United States
by Baker & Taylor Publisher Services